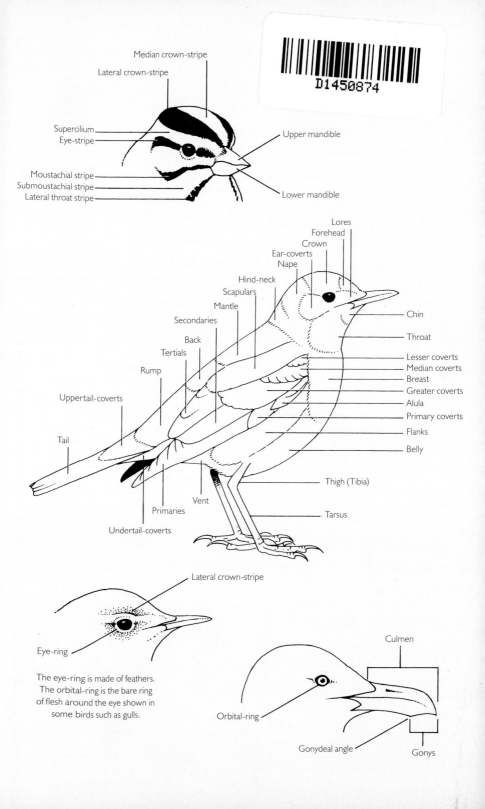

Median crown-stripe

Lateral crown-stripe

Supercilium

Eye-stripe

Upper mandible

Moustachial stripe

Submoustachial stripe

Lateral throat stripe

Lower mandible

Lores
Forehead
Crown
Ear-coverts
Nape
Hind-neck
Scapulars
Mantle
Secondaries
Back
Tertials
Rump

Uppertail-coverts

Tail

Chin

Throat

Lesser coverts
Median coverts
Breast
Greater coverts
Alula
Primary coverts
Flanks
Belly

Thigh (Tibia)

Tarsus

Vent

Primaries

Undertail-coverts

Lateral crown-stripe

Eye-ring

The eye-ring is made of feathers.
The orbital-ring is the bare ring
of flesh around the eye shown in
some birds such as gulls.

Culmen

Orbital-ring

Gonydeal angle

Gonys

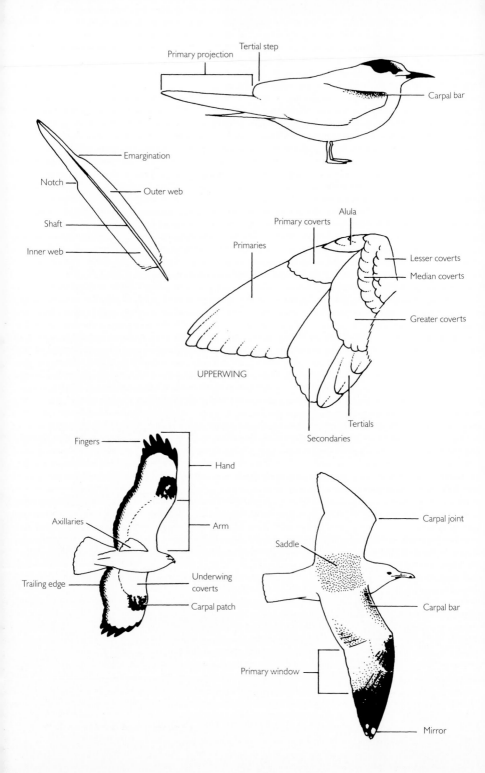

Primary projection

Tertial step

Carpal bar

Emargination

Notch

Outer web

Shaft

Inner web

Primaries

Primary coverts

Alula

Lesser coverts

Median coverts

Greater coverts

UPPERWING

Secondaries

Tertials

Fingers

Hand

Arm

Axillaries

Trailing edge

Underwing coverts

Carpal patch

Carpal joint

Saddle

Carpal bar

Primary window

Mirror

The Helm Guide to Bird Identification

The Helm Guide to Bird Identification

AN IN-DEPTH LOOK AT CONFUSION SPECIES

Keith Vinicombe, Alan Harris
and Laurel Tucker

CHRISTOPHER HELM
LONDON

Laurel Tucker

Laurel died on 20 June 1986, having suffered a cerebral haemorrhage from which she never regained consciousness. She was 35. We had been together for nearly three years and she embarked upon the first version this book with great enthusiasm, excited at the prospect of a joint project together. In the event, she completed only 18 of the plates. Laurel was a remarkable woman, whose vivid and intense personality was somehow reflected in her illustrations. It is hoped that this new edition and its predecessor, together with the *Handbook of Bird Identification* by Mark Beaman and Steve Madge, will serve as a testament to her remarkable ability, both as a bird illustrator and as a birder.

It is, of course, dedicated to her memory.

K.E.V.

First published in 2014

Bloomsbury Publishing Plc, 50 Bedford Square, London WC1B 3DP

www.bloomsbury.com

Bloomsbury Publishing, London, New Delhi, New York and Sydney

A CIP catalogue record for this book is available from the British Library

Publisher: Nigel Redman
Project Editor: Jasmine Parker
Design by Julie Dando, Fluke Art

ISBN (print) 978-1-4081-3035-3
ISBN (ebook) 978-1-4729-0553-6

Printed in China by C&C Offset Printing Co Ltd.

This book is produced using paper that is made from wood grown in responsibly managed forests. It is natural, renewable and recyclable. The logging and manufacturing processes conform to the environmental regulation of the country of origin.

10 9 8 7 6 5 4 3

Contents

* illustrations entirely or mainly by Laurel Tucker; the remainder are by Alan Harris

Acknowledgements

For this new edition – *The Helm Guide to Bird Identification*, Keith Vinicombe would like to thank Dominic Mitchell and all the staff at *Birdwatch* for their help, encouragement and generosity over many years. Some of the texts in this book have been revised and adapted from articles originally written for *Birdwatch* and Keith is grateful to Dominic for allowing him to do this. He is also extremely grateful to Pete Fraser for providing up-to-date and unpublished statistics for scarce migrants. Thanks also to Alan Bone, Robin Chittenden, Lee Evans, Rupert Higgins, Brian Lancastle, Sid Massey, Martin McGill, Nigel Milbourne, Richard Mielcarek, Killian Mullarney, Andy Musgrove, Mark Ponsford, Brian Rabbitts, Chris Trott and Jack Willmott for their help in various ways. Alan Harris would like to offer special thanks to his birding companions of several decades, namely the members of the Rye Meads Ringing Group and the North Thames Gull Group.

We are, of course, grateful to all the people who helped with the original book: Dr Tim Sharrock for all his help and guidance; the *British Birds* Identification Notes Panel who scrutinised the text and plates; Alan Dean, the late Peter Grant, Steve Madge and Dr Malcolm Ogilvie for their invaluable comments, and editor David Christie. At the Wildfowl and Wetlands Trust, we should like to thank Martin Brown, James Godfrey, Nigel Jarrett and Martin McGill for showing us around the breeding pens at Slimbridge and for discussing at length the ageing and sexing of wildfowl. Peter Colston at what is now the Natural History Museum at Tring was as helpful as always during our various visits to check skins. Keith Vinicombe is particularly grateful to Grahame Walbridge, whose unparalleled knowledge was regularly tapped, and to the following, who helped in various ways: Andy Barber, Martin Cade, Tim Cleeves, Andy Clements, Andy Davis, Pete Fraser, Andy Hawkins, Pete Hopkin, John Marchant, Chris Newman, Antony Merritt, Andy and Sally Middleton, Tony Pym, Dick Senior, Chris Stone and Nigel Tucker. He is also eternally indebted to Chris and Theresa Stone and to Geoff and Sarah Upton for attempting to keep him sane. Several people have loaned original artwork for this edition, we are indebted to Tim and Ann Cleeves, Mike Harris, Vera Harris, Ian Kendall, Jane Marshall, Paul Roper, Chris and Theresa Stone, Paul Tout and Malcolm Wilson for this kindness.

The author and artist, however, take full responsibility for any errors that may have crept in from the first version of the book. Last but not least, Keith is eternally grateful to his wife Jane for all her help, support, patience and understanding.

Introduction

This book is based on the *Macmillan Field Guide to Bird Identification*, which was first published in 1989. The original book was born out of a meeting in 1985 between Laurel Tucker and Dr Tim Sharrock, the latter in his capacity as natural history consultant to Macmillan. The idea was to produce a book that would tackle in depth the problems of identifying 'difficult birds'. It was felt that the standard field guides, mainly because of lack of space and, in some cases, poor illustrations, could not do justice to the problem of separating similar species. The result was, in essence, a series of well-illustrated 'mini' identification papers. The original book proved to be an enormous success but, unfortunately, it has been out of print for many years. Choosing the original subjects was problematical, but it was decided to include mainly those regularly occurring British species that present a problem for the 'average birder' and to include only those rarities that are frequently confused with something more common.

In this new version, we have expanded the range of species included. The taxonomy generally follows Mitchell & Vinicombe (2011). Most of the additions simply reflect the remarkable changes in the status of many species over the last 20 years, coupled with changes in taxonomy, partly brought about by advances in the study of DNA. As with the original book, inconsistencies have been inevitable, the most difficult problem being 'species creep'. For example, it is difficult to discuss the identification of the regularly occurring Melodious and Icterine Warblers without dealing with Olivaceous Warbler (now split into Eastern and Western) and, in turn, Booted Warbler, which in itself leads to the unavoidable discussion of the recently split Sykes's Warbler. In other cases, discussion has been confined to the commoner species, as to have gone into detail on the extreme rarities would have been beyond the scope of the book. Although in recent years there has been a marked trend towards the almost forensic examination of minute feather detail, this book offers advice on how to identify birds in the field, when the instinctive evaluation of shape, flight, behaviour and call is just as important – this overall impression that a bird gives is often termed its jizz. It is hoped that, as with the original, the end result will appeal not only to relative beginners but also to those more seasoned observers with gaps in their knowledge.

How to use this book

In order to facilitate the efficient exposition of difficult identification problems, a rigid format has been abandoned. Each 'chapter' has been arranged in the form of a 'mini' identification paper to best suit the topic under discussion. As it is important to be aware of the likelihood of any particular species occurring in a given area at a given season, each article is prefaced with a short section entitled 'Where and when', but it must be stressed that this is only a generalised outline that cannot account for every eventuality. Many accounts end with a short list of references. These point the way to books and articles that contain useful additional information or which go into various aspects of the identification in more detail.

Bird names

Since the first edition, there have been a number of changes to English bird names, some of which have been widely accepted, others almost universally ignored. The names used in this book attempt to reflect current normal usage by 'ordinary birders' but, for the sake of clarity, some epithets (such as 'Barn' Swallow) have occasionally been used.

Statistics

The rare breeding bird, scarce migrant bird and rare bird statistics have been taken from the official record published in *British Birds* (see *Bibliography*) and relate to Britain (but exclude Ireland). Their use attempts to quantify the likelihood of any given species being seen at a specific time and place. All are taken from the most recent reports: rare breeding bird data being accurate to 2010, the rare bird data to 2011 and the scarce migrant data to 2007. 'Current averages' are based on the most recent ten years. Waterbird counts are taken from the annual report *Waterbirds in the UK* and are accurate up to 2010/11.

Short cuts to identification

Knowing the common birds

The only way to become an expert birder is to spend as much time as possible in the field and, as in all things, practice makes perfect. However, time should be spent not just in acquiring a large life list but in getting to know common birds. The secret to identifying a rarity is the ability to eliminate the common confusion species: how, for example, can you identify a Ring-billed Gull if you have never bothered to take a close look at a Common Gull?

Keep an open mind

When faced with something unusual, do not automatically assume that it is a rarity. Consider all of the other possibilities. Could its appearance be due to individual variation? Is it something common in an unfamiliar plumage? Is it aberrant? Is it a hybrid? Is it an escaped cagebird? If you are stuck, take a full description or a photograph, but always try to get someone else to look at the bird. This is especially important with rarities, as substantiated observations will have a much better chance of acceptance. On the other hand, try to make up your own mind about an identification and do not be bamboozled by so-called 'experts' who are not.

Calls

Most good birders identify a large proportion of birds by calls and songs. Unfortunately, there is no shortcut to learning calls, but try to follow up all the unidentified bird calls that you hear. Bird recordings may help, but they are no substitute for hearing the real thing within the context of its natural environment.

Fieldcraft

Standards of fieldcraft have declined over the years, presumably because so many observers now birdwatch from hides or turn up at a 'twitch' to find their quarry already under observation. Remember the old rules of wearing sombre clothing and moving quietly and smoothly. Also, remember the benefits of patient stalking, and do not be afraid to crawl on all fours or to duck down below vegetation to avoid flushing the birds that you are pursuing. It is often possible to get close to birds by using even scanty vegetation as cover. Conversely, on modern bird reserves, many species have become so used to people walking along regular trails that they do not feel threatened by our presence and can be approached very closely.

Note-taking

When faced with an unusual bird, it is imperative that you take exhaustive field notes *on the spot*. If this is not possible, then write some notes on the same day, before you go to bed (once you have slept, the brain seems to file away your 'photographic' mental image into a different intellectual part of the brain, so you will no longer be able to accurately recall what you have seen). It is also useful to take notes on unfamiliar plumage types and on anything else that attracts attention, as note-taking aids the learning process. The easiest way to take notes is to make an annotated sketch: if you cannot draw, just use two circles, one for the head and one for the body. Drawing the bird ensures that all parts are checked. When drawing, it does not matter quite so much if you forget the names of the feather tracts as these can be checked when you get home, but try to familiarise yourself with how the feathers lie, particularly on the wing (it is essential to know which are the median coverts, greater coverts, tertials etc.).

Description-writing

In the modern digital age, description writing is rapidly becoming a forgotten art. But if you fail to obtain a frame-filling digital image, you will need to convince a records committee of your sightings. This is when field notes are essential. Remember to organise your description into a logical sequence (start with the head and work back) and the use of accurate topographical terminology is essential. Remember that records committees have to look at large numbers of descriptions, so make your notes concise, readable and interesting. Start with an introduction, outlining the circumstances of the observation, and write a short summary of the bird's appearance, emphasising the salient features and overall impression, before embarking on a detailed feather-by-feather account. Try to include a drawing if possible, as it is usually easier for a committee to visualise a bird from a drawing than from a description. Furthermore, describe what you actually saw, not what you think you should have seen, and do not trot out standard text book clichés: it is those descriptions that include a minor feature not mentioned in the books that are often the most convincing. Finally, do not forget the bird's behaviour, habitat and calls, as well as weather details, distance, optics, names of other observers and details of your previous experience. All of these points are important if a record is to be accepted.

Photographs

Nowadays of course, description writing has been largely sidelined by digital photography and digiscoping. There is no doubt that this has revolutionised birding, it being possible to take excellent record shots with an ordinary digital camera and a telescope. However, beware of falling into the trap of spending so much time taking photographs that you fail to take a proper look at the bird itself. Also remember that digital images have their limitations and can be misleading, particularly in the evaluation of subtle plumage tones.

Making mistakes

Remember that *all* birders make mistakes, even experts. Rather than dwelling on misidentifications, remember that mistakes are all part of the learning process and are nothing to be ashamed of. Inevitably, the more experienced you become, the fewer mistakes you will make, but do not lose sight of the fact that the identification of many of the species covered in this book really is difficult, particularly in testing field conditions.

Glossary

Arm Inner wing between body and carpal joint (often used for raptors).

Brownhead Term applied to encompass female, juvenile and immature plumages of Smew, Red-breasted Merganser and Goosander.

Chum Revolting mixture of dead fish, fish oil and, sometimes, popcorn poured onto the sea to attract seabirds on pelagic trips.

Culmen The ridge along the top of the upper mandible of the bill.

Eclipse Non-breeding plumage of male ducks in late summer.

Filoplume A hair-like feather.

Gonys The ridge of the lower mandible between the gonydeal angle and the bill tip (often used in reference to large gulls).

Gular Pertaining to the throat.

Hand Outer wing between carpal joint and tip (often used for raptors).

Leucism All-white plumage or all-white feathers mixed with normal-coloured ones. Always normal-coloured eyes. Other colour aberrations (such as black plumage that is normally pale brown) have a variety of terms; see van Grouw (2013).

Overshooting When a migrant bird flies further than it should (e.g. a Subalpine Warbler heading from Africa to Spain, but 'overshooting' to Britain.)

Pishing 'Shushing' noise made by birdwatchers to attract warblers and other small passerines.

P10, P9 etc. The primaries on some species, particularly gulls, are referred to by a letter and number, P10 being the outermost primary, P9 the second outermost and so on to P1, which is the innermost primary immediately adjacent to the secondaries. Confusingly, this system is usually reversed in passerines, where the outermost primary is referred to as P1.

Rectrices Main tail feathers.

Remiges Main flight feathers (primaries and secondaries).

Tertial crescent Large white crescent-shaped fringe to the closed tertials, seen on adult gulls at rest.

Tertial step Refers to a distinct 'step' on the closed wing between the tertials and the primaries when viewed side-on (commonest on large gulls).

Wreck Occasional disaster affecting seabirds that are swept ashore or inland by gales and/ or food shortages.

Topographical tips

See topography drawings on pp. 1–2 at front of book.

Learning the topography of a bird is fundamental to understanding and describing the features that you see. It is important to stress that feathers on a bird are not haphazard, but are grown in organised rows or groups, overlapping like roof tiles. These feather tracts have specific standardised names illustrated on the front endpapers. Many birders confuse some of the terminology and the following lists some areas that most frequently cause problems.

Culmen, gonys and gonydeal angle The ridge along the top of the upper mandible of the bill is the 'culmen'; hence, a bird may show a 'strongly decurved culmen'. On the lower mandible, some birds, particularly gulls, have a distinct angle, about three-quarters of the way along towards the tip; this is the 'gonydeal angle'. The ridge of the lower mandible between the gonydeal angle and the bill tip is called the 'gonys'. Thus, the heavier-billed Great Black-backed Gull has a much stronger gonydeal angle than a Lesser Black-backed Gull.

Eye-stripe and supercilium The eye-stripe is exactly that: a stripe through the eye. When present, this is a dark line that usually extends from the bill back through the eye. On some birds, it extends just from the eye back, leaving the lores (the area between the eye and the bill) pale and unmarked. This may seem an insignificant difference, but unmarked lores produce a different facial expression – a rather pleasant, gentle, 'open-faced' appearance – whereas those species that have a line across the lores tend to look more 'severe'. Species for which this difference is significant include Reed and Little Buntings, and Tawny and Richard's Pipits. The supercilium is the *pale* line extending back from the bill *above* the eye and *above* the eye-stripe; as it is a Latin word, the plural is supercilia. The dictionary definition of the word is 'eyebrow' which is what, quite sensibly, it is called in North America. Some species have the supercilium extending only from the eye back, while others, such as Spotted Redshank, basically show only a fore-supercilum from the bill to the rear of the eye.

Crown-stripe Some species possess a pale stripe on the centre of the crown known as the 'crown-stripe'; occasionally it may be more precisely referred to as the 'median crown-stripe'. Some birds show a *dark* line on the side of the crown, immediately above the supercilium, and this is called, quite logically, the 'lateral crown-stripe'. Common Snipe, Aquatic Warbler and Firecrest show both.

Eye-ring and orbital-ring Many birds show a ring pale of feathers around the eye known as the 'eye-ring'. Whereas an eye-ring comprises feathering, the 'orbital-ring' is a narrow ring of *bare skin* immediately surrounding the eye. In some species, this may be swollen, colourful and very obvious, classic examples being the yellow orbital-rings shown by Lesser White-fronted Goose and Little Ringed Plover.

Chin and throat The 'chin' is the small area of feathering immediately below the base of the lower mandible of the bill. As such, it is not usually necessary to differentiate it from other areas of a bird's plumage. The 'throat' is the large, triangular area of feathers between the chin and the upper breast.

Moustachial stripe, submoustachial stripe and lateral throat-stripe These three areas are often confused, but a recent development has clarified the position quite considerably. The 'moustachial stripe' is a dark line shown by many birds, most obviously pipits and buntings, which runs obliquely from the gape down and along the lower border of the ear-coverts. Many birds have another dark line below this, bordering the throat, properly referred to as the 'lateral throat-stripe'. Between the dark moustachial stripe and the lateral throat-stripe is usually a pale area known as the 'submoustachial stripe'. So, from top to bottom there is: (1) the dark moustachial stripe, (2) the pale submoustachial stripe and (3) the dark lateral throat-stripe. Note that the confusing 'malar stripe' is no longer in use.

Nape and hindneck The rear of the head, between the crown and the mantle, is called the nape (upper) and the hindneck (lower).

Mantle and back These two terms are often confused. The 'mantle' is the forward part of the upper side of a bird, immediately below the nape, whereas the 'back' is the lower part, sandwiched between the mantle and the rump. One complication, however, is that with adult gulls and terns, the whole of the grey area of the upperparts, which involves the mantle, back, scapulars, wing-coverts and tertials, is often referred to in its entirety as 'the mantle'.

Tertials On the closed wing, between the wing-coverts and the primaries, are the 'tertials'. They are in fact the three innermost secondaries that are elongated to cloak the bases of the primaries at rest; they often show contrasting pale edges or fringes (hence: 'tertial fringes'). On some birds, such as gulls, the secondaries are often invisible at rest, but on others, such as warblers, the secondaries are visible immediately below the tertials. On some species, such as Icterine Warbler, the tertials and secondaries have contrasting pale fringes that form a rectangular 'wing-panel'.

Speculum The iridescent speculum on dabbling ducks is on the secondaries and this area too is often readily visible at rest.

The open wing This can be divided into the flight feathers and the wing-coverts. The flight feathers comprise the primaries, secondaries and tertials. The feathers that overlay the primaries, towards the bend of the wing, are the 'primary coverts' (in fact comprised of lesser, median and greater primary coverts but, for identification purposes, these are not usually differentiated). The line of larger coverts overlying and immediately in front of the secondaries are the 'greater coverts', while the line of smaller coverts overlying the greater coverts are the 'median coverts' (not *medium* coverts). In front of these is a larger, more random looking area of small feathers: the 'lesser coverts'. The large oval-shaped area of 'shoulder' feathers between the open wing and the mantle/back is the 'scapulars'.

The closed wing What is less straightforward is how these feathers lie when the wing is closed. The first point to remember is that, at rest, the lesser coverts are largely concealed by the overlying scapulars and sometimes by the fluffed-up flank feathers, so that all that can usually be seen at rest are the median and greater coverts; on many passerines, these feathers have pale tips, which form two wing-bars. On many non-passerines, the scapulars are often the largest area of 'non-flight feathers' visible on the upper side of a bird at rest, although this depends to some extent on the bird's posture. On swimming ducks, the wing-coverts are usually completely hidden by the scapulars and flank feathers.

Primary projection At rest, the visible primaries beyond the tertials are referred to as the 'primary projection'. As a general rule, the longer the primary projection, the longer distance a bird migrates. In some species pairs, for example Willow Warbler and Common Chiffchaff and Icterine and Melodious Warblers, accurate assessment of the primary projection is one of the best ways to separate them. What you have to do is to gauge the length of the primary projection *compared* to the length of the overlying tertials. Thus, on a Common Chiffchaff, the primary projection is about half the overlying tertial length, whereas on the longer-winged Willow Warbler it is about three-quarters to equal. Some birds show no primary projection at all, their primaries being completely cloaked by the overlying tertials. The most obvious examples are larks, pipits and wagtails.

Remiges and rectrices 'Remiges' is a collective word given to the primaries, secondaries and tertials combined; as such it is a useful abbreviation when referring to those three feather groups, which comprise the 'flight feathers'. 'Rectrices' is the equivalent term for the tail feathers, but is less frequently used.

Moult and ageing

1. The basics

The correct ageing of a bird is often fundamental to its identification. To facilitate this, an understanding is required of the precise terminology that describes the various ages, which are related to a bird's moult cycle.

Juvenile plumage Newly hatched birds fall into two categories: (1) naked and blind, remains in the nest and is fed by its parents (altricial), or (2) covered in down, has its eyes open and is soon able to leave the nest and feed itself (precocial). All young birds must grow as quickly as possible and acquire the power of flight. In some small passerines this can take as little as ten days, which is an extraordinary physiological feat. In that time, they must grow their first covering of feathers, which is known as *juvenile plumage*. In order for it to grow so quickly, something has to be sacrificed: the strength and durability of that plumage. Consequently, the juvenile feathers of many species, particularly passerines, are weak and this is why young birds have an endearing fluffy appearance. The parents continue to feed them once they have left the nest but, gradually, the young become independent. The feeding is good in the long, warm days of late summer, making this the ideal time for young birds to quickly lose their inferior juvenile plumage and grow a stronger set of feathers that will see them through the winter. Thus, in late summer and early autumn, most young birds undergo *post-juvenile moult*. This usually involves just a body moult in which all the head and body feathers are replaced. The juvenile flight feathers (primaries and secondaries) and the tail feathers are retained. A young Robin, for example, loses its spotted brown juvenile body plumage – which made it difficult to detect during its inexperienced days in the shady undergrowth – and acquires its red breast and plain brown upperparts. By late autumn, it resembles its parents and, to all intents and purposes, is in adult plumage.

Immature plumages While a Robin may be adult-like by late autumn, other birds take much longer to acquire adult plumage, and larger birds may pass through a number of immature plumages before reaching adulthood. As a rule, the larger the bird, the longer it takes to mature and the more immature plumages it has. A group that demonstrates this particularly well is the gulls. Larger gulls, such as Herring Gull, usually take about four years to reach maturity, the medium-sized Common Gull takes three and the small gulls, such as Black-headed Gull, take two. To take Common Gull as a straightforward example, its juvenile plumage is essentially brown, but it gradually moults much of its body plumage during autumn, acquiring a whiter head and underparts, and grey back, mantle and scapulars. This new plumage is given a specific name, which is, quite logically, *first-winter plumage*. In spring, it has another body moult into *first-summer plumage*. It is not until late summer, when the bird is one year old, that it has its first complete moult, replacing all of its wing and tail feathers for the first time and its body feathers for a third. Its *second-winter plumage* is much more adult-like but still shows traces of immaturity, such as brown in the primary coverts. This is followed by another spring body moult into *second-summer plumage*, which is then followed by another complete moult into *adult plumage*, when it is just over two years old. Each plumage is more adult-like than the previous but remember that the different plumages may vary individually, particularly in larger birds (such as Herring Gulls) so do not expect all birds to conform exactly to those illustrated in field guides. Bare parts too – eyes, bill and legs – also change as a bird gets older and, although their progression is roughly in sync with their plumages, there is great individual variation in the acquisition of bare-part colours.

Juveniles and immatures At this point, it is essential to clarify one particular area of confusion. Any bird to that is not mature (i.e. not an adult) is an 'immature'. Literally, 'immature' simply means 'not mature'. Thus, a juvenile bird is an immature, as is a first-winter, a second-summer and so on. The term 'juvenile' should be confined to a bird's very first plumage worn for a short time after leaving the nest. Although many larger birds migrate whilst still in juvenile plumage, most passerines do not. Therefore, you will not see juvenile Redstarts or Yellow-browed Warblers in September falls on the east coast. If these birds are not adults, then they are 'first-winters'. Similarly, *subadult* is also to be avoided if possible, as again more accurate ageing can usually be determined.

Spring body moult This has an important and obvious function both for adults and many first-years: it enables them to change from their drab, functional winter plumage into brightly coloured breeding or summer plumage.

Calendar-year ageing Some large birds, such as cormorants and large birds of prey (e.g. eagles) have a fairly continuous moult in which several generations of feathers are present simultaneously. Precise ageing terms, such as 'first-winter' and 'second-winter' then become redundant so, to circumvent this problem, larger birds are often aged in relation to the year in which they hatched. A bird hatched in May 2013 will be called 'first calendar year' up to 31 December 2013, 'second calendar year' from 1 January to 31 December 2014, and so on.

Feather wear When identifying birds and assessing ageing, bear in mind the effects of wear, abrasion and bleaching. Old feathers wear and fade, and significant plumage features (such as wing-bars) can disappear through wear, while some species alter plumage tone (for example, fresh autumn Meadow Pipits possess a green tint, whereas worn breeders are much browner). Immature gulls are notoriously prone to wear and bleaching, and may look particularly 'tatty' and faded in summer, when their wing and tail feathers are nearly a year old. Some species acquire summer plumage not by moult, but by the wearing away their duller feather fringes (see below).

2. Waders and other non-passerines

Juvenile waders As noted in part 1, most passerine migrants have a body moult before they head south in autumn, but what happens if there is not time to complete it? This is the problem that faces all birds that nest in the Arctic. Although food there is abundant, summer is short, so young birds do not have time to complete a post-juvenile body moult. They therefore have no option but to head south while still in juvenile plumage. Consequently, nearly all southbound autumn waders are in full juvenile plumage. In many ways, this plumage is a halfway house between adults' summer and winter plumages. Like summer adults, juveniles tend to be brown and patterned on the upperparts, providing good camouflage when viewed from above, but the underparts are mainly white, like the adults' winter plumage. Many species really do look intermediate. For example, juvenile Black-tailed Godwits, particularly those of the Icelandic race *islandica*, are usually quite bright orange on the underparts, while juvenile Spotted Redshanks appear rather dusky grey-brown, somewhere between the smart black of summer plumage and the grey and white of winter. It was no doubt this juvenile plumage that gave rise to its now redundant alternative name of 'Dusky Redshank'.

Adult waders in autumn Because young waders' feathers are all grown at the same time and because they are all very fresh in late summer and autumn, juvenile waders always look neat. In contrast, adults in late summer often undertake body moult while on migration, so they look tatty and 'moth-eaten' compared to their immaculate offspring. Most young waders do not moult until they arrive on their wintering grounds, after which they look like winter adults. There are, however, a number of notable exceptions, the most obvious being juvenile Dunlin, which often acquire grey scapulars on migration. Grey Phalaropes are also transitional by the time they reach Britain, with a patchy grey-and-black appearance to their upperparts (Red-necked Phalaropes moult somewhat later). For two rarities, Semipalmated and Western Sandpipers, moult timing can be fundamental to their separation.

Other Arctic migrants Birds such as divers, geese, ducks, skuas and terns also migrate in juvenile plumage and gradually start to moult into first-winter plumage once they have reached their winter quarters. Interestingly, whereas most mid-latitude gulls soon start to acquire fresh first-winter feathering, Arctic gulls such as Glaucous and Iceland retain their juvenile body plumage well into the winter. Another useful example is provided by Common and Arctic Terns: any 'Commic Tern' showing primary or secondary moult in autumn will be a Common Tern; the more northerly breeding adult Arctics do not normally commence wing moult until they reach Antarctica.

Tail and primary moult With a few exceptions, juveniles do not moult their primaries and secondaries or tail feathers in late summer. Therefore, almost any bird showing wing and tail moult in autumn is an adult. This means that, for example, distant Curlews, Common Buzzards and Ravens with symmetrical gaps in their flight feathers can be aged at considerable distances, even in flight.

3. Passerine exceptions

When Lars Svensson first published his iconic *Identification Guide to European Passerines* in 1970, the ageing of passerines was purely the domain of ringers. In recent years, however, improved optics and the advent of digital photography has meant that field birders are in a much better position to try their hand at the subtleties of ageing. This may seem like a step too far, but the accurate ageing and sexing of a bird is as much a part of the identification process as its specific or racial identity. The following provides some tips on the moult and ageing of passerines, but it must be stressed that this is a difficult and complicated subject, bedevilled by significant individual variation.

Single moulters (acquisition of summer plumage by feather abrasion) Several passerines do not moult twice a year, only once. The most obvious examples are some of the chats and finches. In late summer, they lose their colourful summer plumage and acquire a plumage that is much drabber. This is because the colourful bases of the new feathers are obscured by paler and browner fringes. As winter progresses, these dull tips gradually wear off so that, by spring, their bright and colourful 'summer plumage' is revealed. Most of these species are resident or relatively short-range migrants, e.g. European Stonechat, Common Starling, Linnet, Chaffinch and Brambling.

Species that migrate in juvenile plumage A small number of passerines migrate while still in full juvenile plumage, their post-juvenile moult not commencing until arrival in the winter quarters. Some warblers fall into this category, including Sedge and Aquatic (but not Reed) and *Hippolais* warblers (Melodious and Icterine). Another bird that migrates in juvenile plumage is Common Rosefinch. Peculiarly, Red-rumped Swallow migrates in juvenile plumage, whereas 'Barn' Swallow has a body moult before it heads south.

Species that migrate in juvenile *or* first-winter plumage Some species vary in the acquisition of first-winter plumage. For example, young Richard's and Tawny Pipits from early broods have time to moult into first-winter plumage prior to migration, whereas later broods may not. This means that many later-hatched individuals migrate while still in juvenile plumage. Others may suspend moult and migrate in a mixture of the two plumages.

Adults that moult in their winter quarters Some long-distance adult passerine migrants do not moult until they arrive in their winter quarters, the most obvious being hirundines (although some start their moult in Europe). Other species that follow this strategy include Reed Warbler and Spotted Flycatcher. Adult Spotted Flycatchers are in fact surprisingly rare in Britain in autumn, when they look decidedly worn and abraded compared to the more pristine first-winters.

Juveniles that have a complete post-juvenile moult A final group of exceptions includes a disparate bunch of unrelated species in which juveniles undertake a complete post-juvenile moult. One thing that they have in common is that most are sedentary, and because they do not migrate, they have time to renew all their feathers. These include the larks, Bearded and Long-tailed Tits, Common Starling, House and Tree Sparrows, and Corn Bunting.

Ageing first-winters Following post-juvenile moult, most young passerines are basically adult-like. Indeed, many, such as *Phylloscopus* warblers, are virtually impossible to age even in the hand. Some families of birds are easier to age than others and there are some useful pointers to remember. **1 BUFF TIPS TO THE GREATER COVERTS** Many first-winter thrushes (and related species such as nightingales) retain buff tips to their juvenile greater coverts, forming a narrow wing-bar. It should be noted, however, that whilst the presence of such tips indicates a first-winter, their absence may not prove the reverse since first-winters may lose them through abrasion. Remember though, that Song Thrushes and Robins can show buff tips even in adult plumage. **2 DIFFERENT LENGTH GREATER COVERTS** Some young birds moult their greater coverts in autumn and winter, the consequence being that their new adult inner greater coverts are longer than their old juvenile outer feathers, so there is a distinct 'step' between the two ages (often referred to as 'moult limit'). This is particularly useful with thrushes and pipits. On thrushes, there is an additional difference in that the adult inners are fairly plain, whereas the juvenile outer feathers show distinct pale fringes. On larger pipits, such as Richard's and Tawny, the adult inners are diffusely brown-fringed, whereas the juvenile outers are sharply fringed with white. **3 TAIL FEATHER SHAPE** In some passerines adults and first-winters have different-shaped tail feathers: adults have broad tips with rounded corners, whereas first-winters retain more pointed juvenile feathers. This can be particularly useful for ageing well-photographed vagrants. **4 FAULT BARS** Young birds in the nest sometimes undergo periods of food shortage, particularly during bad weather. This may manifest itself in 'fault bars' across the tail, which are inconspicuous but distinctive narrow bars across each feather, representing structural deficiencies brought about by poor diet. As adults re-grow their tail feathers in sequence, they do not show fault bars across the entire width of the tail (unless they re-grow a new tail after losing it through an accident).

Whooper and Bewick's Swans

Where and when Whooper Swan *Cygnus cygnus* has a more northerly winter distribution than Bewick's *C. columbianus*, occurring mainly from September to April in Scotland and Ireland. Paradoxically, however, the largest concentrations are in England: Lancashire and the Ouse Washes in Norfolk/Cambridgeshire. A few pairs breed in Scotland, and occasionally elsewhere. Bewick's Swan has a more southerly distribution, occurring mainly from October to March in England, with the Ouse and Nene Washes being the main sites. Mute Swan *C. olor* is a widespread and common resident.

Structure Bewick's is the smallest swan, with rather goose-like proportions. Compared to Whooper, the head is more rounded, the bill smaller, the neck shorter and proportionately thicker. Whooper has a long neck, a large wedge-shaped bill and it tends to show a more bulging breast. Most of these differences are also apparent in flight, but caution is then required.

Bill patterns Bewick's has a smaller bill, at least half of which is black; the amount and shape of the yellow at the base varies individually, but the yellow is rounder or squarer than on Whooper. On the latter, the yellow extends in a sharp point towards the tip so that, from a distance, most of the bill appears yellow. Young juveniles of both species have ivory or cream at the base, the distal part being pink (and the tip itself black). At a distance, therefore, young Bewick's can appear to have a large, pointed pale area, recalling Whooper, and lone individuals can be surprisingly tricky. However, the point to remember is that, in both species, the ivory or cream area at the base is similar in shape to the yellow on the adults' bills. Thus, on Bewick's it is cut off squarely from the pink, behind the nostril. On Whooper it extends in a sharp point past the nostril, a smaller patch of pink being restricted to the area behind the black tip and immediately in front of the nostril. In both species, the ivory or cream basal patch may gradually become yellower as the winter progresses. In addition, both species gradually lose the pink, which turns black and is subsumed into the black bill tip, the extent of which mirrors that of the respective adults. However, unlike Whooper, many juvenile Bewick's retain the deep pink into their first-summer. Juveniles and first-winters are otherwise best separated by size and structure, although their identification is usually facilitated by the fact that the young normally stay with their parents throughout their first winter (many Mute Swan families split up in autumn).

Plumage *Adults* As a result of feeding in iron-rich pools, Whoopers often show orange head staining. *Juveniles* Uniformly pale ashy-grey (Whoopers tend to be slightly darker) and even when the white first-winter feathering appears, they still look rather uniform as a result of their overall paleness; indeed, distant young are not always obvious amongst the adults. Compared with juvenile and first-winter Mute, they are much paler, more uniform and smoother looking. Juvenile Mute is darker and browner, and the gradual appearance of white first-winter body feathering creates a uniquely patchy appearance. However, the amount of white varies (some are still uniformly dull brown in late winter) but it is usually similar within a brood. In flight, such birds still show a patchy contrast between the whitish flight feathers and the brown wing-coverts and body. Note also young Mute's dark grey, black-based bill. Rarely, all-white juvenile Mute Swans occur (so-called 'Polish Swans').

Calls Bewick's has variety of calls, most being softer and more muted than Whooper; perhaps

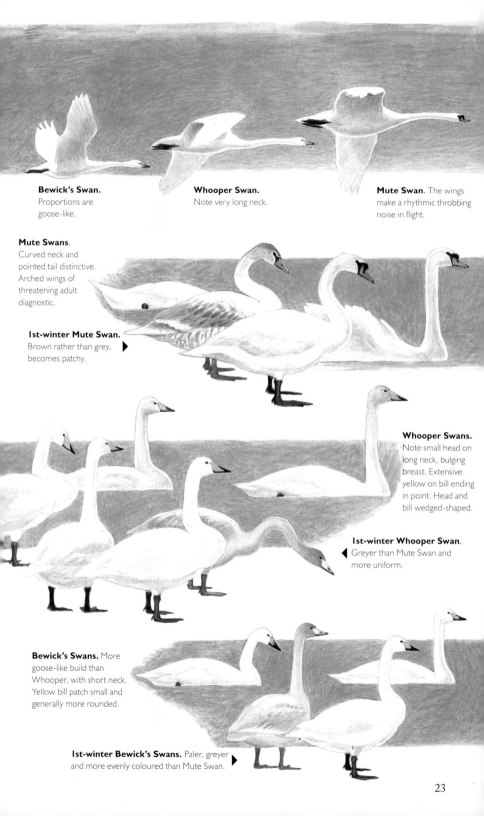

Bewick's Swan. Proportions are goose-like.

Whooper Swan. Note very long neck.

Mute Swan. The wings make a rhythmic throbbing noise in flight.

Mute Swans. Curved neck and pointed tail distinctive. Arched wings of threatening adult diagnostic.

1st-winter Mute Swan. Brown rather than grey, becomes patchy. ▶

Whooper Swans. Note small head on long neck, bulging breast. Extensive yellow on bill ending in point. Head and bill wedge-shaped.

◀ **1st-winter Whooper Swan.** Greyer than Mute Swan and more uniform.

Bewick's Swans. More goose-like build than Whooper, with short neck. Yellow bill patch small and generally more rounded.

1st-winter Bewick's Swans. Paler, greyer and more evenly coloured than Mute Swan. ▶

23

most characteristic is a soft, pleasing, rather hollow *oop oop*. Whooper's calls are typically louder, deeper and more trumpeting. Mute occasionally gives a deep but quiet, wheezy grunt, a goose-like honk and a loud hiss; in flight, its wings make a distinctive rhythmic musical throb.

Distant separation from Mute Swan Note Mute's generally more curved neck and, particularly, its longer, pointed tail; the orange bill with a black knob is visible at some distance. Bewick's and Whoopers are sleeker, do not arch their wings in threat like Mute, but engage in trumpeting displays with outstretched necks and open wings.

'Whistling Swan' There are two British records (1986 and 1998) of Bewick's showing characters of the nominate North American race, known as 'Whistling Swan'. It can be distinguished by its almost all-black bill, often with tiny amounts of yellow at the base. When faced with a 'black-billed Bewick's', very detailed attention should be given to the bill in order to eliminate the possibility of discoloration. The exact pattern of any yellow should be noted and the bird photographed.

Grey geese

Where and when When identifying grey geese, it is important to be aware of each species' main distribution patterns and traditional wintering areas. Most wild geese arrive in Britain in September and October and remain until March (although some Pink-footed Geese remain into May or even June). Being the only grey goose that breeds here, Greylag Geese may be encountered throughout the year. As distribution is a vital clue to their identification, this is covered in the individual species accounts.

General approach Geese often occur in non-traditional areas and the careful scrutiny of flocks sometimes reveals the presence of rarer species. Frustratingly, however, geese also escape from captivity, so always check out-of-context individuals for rings, wing-clipping and unusual tameness. When identifying geese, concentrate on size, structure, head and neck colour, and body patterning, but most important are the bare-part colours. Note that males of all species average about 5% larger than females and about 10% heavier; this is particularly relevant when judging the size of lone geese. In flight, geese can be identified by a combination of structure, forewing colour and calls, although the latter require practice (and there is much variation).

Ageing of grey geese Juveniles have weak and rather fluffy head, neck and body feathering, duller than the adults with duller and buffer feather fringes on their upperparts; these extend right around the edges of the feathers to produce a *scalloped* pattern. Adults have paler, more contrasting and more noticeable pale fringes, but these are strongest across the *tips* of the feathers. Consequently, adults' upperparts are strongly but neatly *barred*. Early in autumn, fresh-plumaged juveniles tend to look paler and plainer than the adults, with their heads similar in colour to their bodies. During the autumn, however, juveniles commence a body moult, starting with the head, neck, lower scapulars and flanks. The moult varies individually, with some birds renewing all of their body feathers by late winter, at least some tail feathers and some median coverts, but others retain some juvenile body feathering into the following summer (Cramp & Simmons 1977). The moult means that, in the field, young birds often appear less immaculate than their parents; this is because they gradually show a mixture of old worn, scalloped juvenile feathers and fresh barred adult ones.

Greylag Goose *Anser anser*

Distribution Greylag Goose is the only species of grey goose breeding in Britain. Wild birds breed in n. Scotland, particularly the Outer Hebrides, Orkney and Shetland, but re-established birds are now common over much of Britain and in a wide variety of freshwater habitats; in fact the two populations are no longer distinguishable. Winter immigrants from Iceland occur in large numbers, mainly from October to April in Scotland (with smaller numbers in NE England); they are surprisingly rare further south.

Structure The largest, heaviest and most thickset of the grey geese. It is worth remembering that Greylag is the ancestor of the domestic 'Farmyard Goose' and it shares many of its characteristics, most notably its loud, honking calls.

Plumage A rather pale goose with a distinctly grey tone to its plumage, particularly to the head and underparts. It shows a heavy, thick-based, mainly orange bill and pink legs. Note too the very strong crenellations that run down the sides of the neck, standing out against the pale grey plumage. Most distinctive is its extensive pale grey forewing in flight (see below).

Eastern race *rubrirostris* The Greylags breeding and wintering here are of the nominate race but birds showing characters of the eastern race *rubrirostris* may also occur (e.g. a small influx in November 2011). Pure *rubrirostris* has a pink bill, which is slightly longer and slimmer than *anser*. Consequently, the bill shape may be reminiscent of Taiga Bean Goose. Its plumage is also a paler, cleaner grey with more obvious white barring, both above and below, producing a strongly scalloped impression. There is a problem, however, in that *anser* and *rubrirostris* intergrade across Europe, so birds with intermediate characters occur, such as a pink bill with an orange base. It is, therefore, difficult or impossible to be certain of the racial integrity of any individual seen in Britain although, when all the birds in a flock show *rubrirostris*-type characters, it is possible to be more confident about their Continental origin, especially if they exhibit 'wild behaviour'.

Confusion with White-fronted Goose Many Greylags show traces of black on the belly and white feathering around the base of the bill (vaguely recalling White-front). Indeed, lone Greylags can be confused with juvenile White-front: note that Eurasian White-front has a pink bill and orange legs, the opposite to Greylag (Greenland White-front has an orange bill and legs). White-front is smaller, slighter, smaller-headed, smaller-billed and generally browner and plainer than Greylag, lacking the latter's strong, crescent-shaped whitish fringes to its body plumage.

Flight identification Greylags are heavy and thickset with a heavy bill and a large head, sometimes emphasised by their 'pinched-in' neck; most obvious are the forewings, which are strikingly pale grey. They also show pale grey underwing-coverts that contrast with dark under-remiges (all other grey geese have uniformly dark underwings).

Calls Their loud honking calls, which recall domestic geese, allow for instant and confident identification.

Pink-footed Goose *Anser brachyrhynchus*

Distribution The most abundant wintering goose, found mainly in southern and central Scotland, NW England, the Wash and n. Norfolk, where they form huge and impressive flocks. Present from late September to April, with stragglers into May or even June. At least two pairs bred in Scotland in 2010.

Greylag Goose. Large with pinched neck, large head and broad-based wings. Note pale grey forewing, pale underwing diagnostic.

Pink-footed Goose. Short neck, small dark head and pale grey wings.

Bean Goose. Dull wings, long dark neck, dark rump.

White-fronts with single Greylag Goose (second from right) and Taiga Bean Goose (third from right).

Greylag Goose

All three species may show white at base of bill.

Greylag Goose. Note size and bulk, large bill and colour of bare parts.

Eastern race *rubrirostris*.

Western race *anser*.

Taiga Bean Goose. Long-billed with wedge-shaped head and long dark neck. Strong white tips to upperpart feathers. Bill dark with variable orange band, bright orange legs.

Juvenile Taiga Bean Goose. Dappled underparts, upperpart fringing less distinct than in adult. Note leg and bill colour.

Taiga Bean Goose. Wedge-shaped bill. ▼

Tundra Bean Goose. Compact, shape closer to Pink-footed Goose.

Juvenile Tundra Bean Goose

Pink-footed Goose

Two Tundra Bean Geese showing variability of orange bill markings.

Adult Pink-footed Goose. Note dumpy body, short neck and small round dark head, greyish mantle. Small dark bill with pink band, pink legs.

Juvenile Pink-footed Goose. Note dark mantle and mottled underparts. In early autumn legs maybe ochre.

26

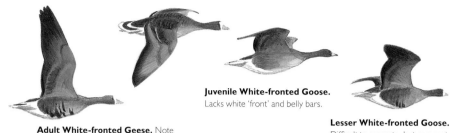

Juvenile White-fronted Goose. Lacks white 'front' and belly bars.

Adult White-fronted Geese. Note long wings, white 'front' and belly bars.

Lesser White-fronted Goose. Difficult to separate, but compact and short-necked.

White-fronts with a Pink-foot (fifth from right), a Lesser White-front (second from the right), and a Taiga Bean Goose (extreme right).

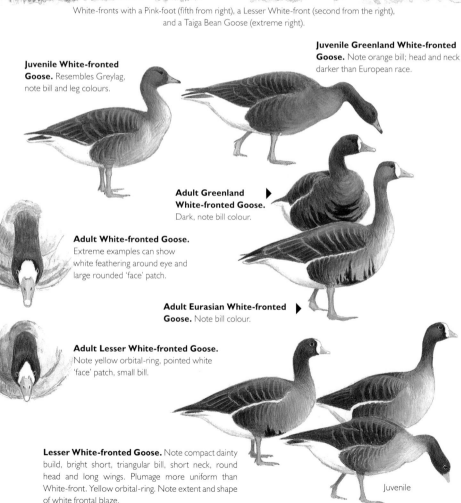

Juvenile White-fronted Goose. Resembles Greylag, note bill and leg colours.

Juvenile Greenland White-fronted Goose. Note orange bill; head and neck darker than European race.

Adult Greenland White-fronted Goose. Dark, note bill colour. ▶

Adult White-fronted Goose. Extreme examples can show white feathering around eye and large rounded 'face' patch.

Adult Eurasian White-fronted Goose. Note bill colour. ▶

Adult Lesser White-fronted Goose. Note yellow orbital-ring, pointed white 'face' patch, small bill.

Lesser White-fronted Goose. Note compact dainty build, bright short, triangular bill, short neck, round head and long wings. Plumage more uniform than White-front. Yellow orbital-ring. Note extent and shape of white frontal blaze.

Juvenile

Structure Much smaller and more compact than Greylag, being more similar in size to White-front. It has a distinctly short, thick neck, a rounded head and a short, triangular bill which is crossed with a pink band (sometimes extending back along the cutting edge); its legs, of course, are also pink.

Plumage The very dark brown head and neck contrast strongly with the pale body, which is strongly suffused with grey over the upperparts. This contrast facilitates even distant recognition. Some adults show traces of white feathering around the base of the bill. Juveniles are duller and browner on the upperparts and less strongly patterned. Particularly dull and 'messy' first-winters can be confusing, but they should be identifiable by structure and bare-parts coloration, although the pink on the bill and legs may be duller than in adults.

Flight identification Pink-feet look small, stocky, short-necked and pointed-winged in flight. The dark head contrasts strongly with the pale body and the forewings are noticeably grey (but not as pale as Greylag).

Calls High-pitched, musical and usually double-noted: *wink-ink*, sometimes very squeaky (or trebled *wink-ink-ink*); also a distinctive, deeper *ang-ank*. Large flocks make a tremendous noise.

Tundra Bean Goose *Anser serrirostris*

Distribution Tundra Bean has no regular wintering populations in Britain but Musgrove *et al.* (2011) estimated a wintering population of about 320 birds. It occurs erratically in small parties (rarely more than ten) at a wide variety of disparate sites, mainly in e. England. Consequently, it is far more likely to be encountered than Taiga Bean, which is largely confined to two discrete areas (see below). Small influxes may occur during severe freezing weather.

Structure and plumage It is helpful to think of Tundra Bean as being similar to Pink-foot, the two species having similar ecological requirements (the Arctic tundra: Pink-feet in Iceland and Greenland, Tundra Bean in N. Europe and Asia). A bulky, heavy goose with a rather thick neck (and crenellations on the side), a rounded head and a deep, rather triangular bill which, at any distance, may appear quite stubby. It averages slightly larger and bulkier than both Pink-foot and White-front, with more of a side-to-side swagger when walking. Like Pink-foot, it has a dark brown head and neck but it lacks the abrupt contrast between the head and body, the latter being much darker (lacking Pink-foot's grey suffusion). In addition, adults are more heavily and more contrastingly patterned, the body feathers being coarsely tipped or fringed whitish, creating an obvious scaly effect; strong white fringes to the tertials can be particularly useful in picking out back-on individuals from other geese. Adults can show traces of white feathering around the base of the bill. Juveniles appear much more uniform, their narrow dark buff feather fringes failing to contrast significantly with the rest of the plumage. Concentrate on the bill and legs: adult has bright orange legs and a discrete orange band across the bill. On some, the orange extends back along the cutting edge (suggesting Taiga Bean). The orange is slightly duller on juveniles. Bear in mind that orange and pink can be surprisingly difficult to differentiate in the field.

Flight identification Tundra Bean has dull grey forewings, lacking obvious contrast between the wing-coverts and the flight feathers; thus the forewings are far less contrasting than those of Pink-feet.

Calls Resemble Pink-feet, but are lower-pitched, harder and more strident.

Taiga Bean Goose *Anser fabalis*

Distribution Taiga Bean Goose is largely confined to two areas: the Yare Valley in Norfolk (currently around 120 individuals from November to February) and on the Slamannan Plateau in central Scotland (currently around 270 from October to February); it is very rare elsewhere and should be identified with great care. As its name suggests, it breeds in bogs, marshes and pools in the taiga forest belt of Europe and Asia.

Structure and plumage It is larger, longer-billed and longer-necked than Tundra Bean, their ecological separation being similar to that of Whooper and Bewick's Swans. This is eerily reflected in their structural and bill colour differences (the Bewick's/Whooper analogy is worth bearing in mind when separating the two species). Plumage is extremely similar to Tundra Bean, and, like Greylag, it shows fairly obvious crenellations down the sides of the neck. To separate the two, concentrate on structure. Taiga Bean is a large goose, similar in size to Greylag; it averages 6% larger than Tundra Bean and about 15% heavier. It shows a long, swan-like neck (alert Taiga Beans tower above accompanying Pink-feet). Taiga also has a longer, less triangular-shaped bill than Tundra, often creating a flatter-headed impression (conversely, when at rest the head shape can often look rather peaked). It also has a noticeably bulging breast, as well as a long, bulky body with a heavy bulge in the ventral area. It is slow and ponderous in its movements, having a lumbering gait with a side-to-side swagger. All these features combine to create a distinctly swan-like impression, especially when the bird is alert. At closer range, another feature is that Taiga's bill lacks a 'grinning patch'. On Tundra Bean, the lower mandible bulges, creating a distinct oval gap between the two mandibles. On Taiga Bean, the mandibles fit together more closely so that it lacks this prominent gap. Most Taiga Beans have extensive orange on the bill, either from the tip back along the cutting edge to the base, or sometimes covering almost the entire bill, such individuals being very distinctive. It is important to emphasise, however, that a minority (about 5% in one flock studied) have a bill pattern very similar to Tundra Bean, with a discrete orange band near the tip; conversely, some Tundra Beans have orange extending back in a point to the base of the bill.

Habitat This may be a useful clue. Whereas Tundra Beans tend to feed in similar environments to Pink-footed and White-fronted Geese (agricultural land), Taiga Beans are more likely to be found in longer, coarser vegetation in wetter, marshier environments, where they feed on the roots and seeds of aquatic plants, often burrowing their head into the mud to do so. Consequently, whereas Tundra Beans tend to associate with Pink-feet or Eurasian White-fronts, lone Taiga Beans often associate with Greylags.

Flight identification Whereas Tundra Bean suggests Pink-foot (being rather stocky with a short neck) Taiga is heavier-bodied, noticeably longer-necked and has slower wingbeats. However, flight identification in Britain should be attempted with great care.

Calls Distinctive, deep and guttural, perhaps more reminiscent of Greylag, and obviously deeper than Tundra Bean: *an-angk, ar-an-angk* or *ur-ur-ur*.

Eurasian White-fronted Goose *Anser albifrons albifrons*

Distribution The nominate Eurasian race occurs in s. Britain, mainly in Norfolk, north Kent and Gloucestershire. It has declined in recent decades, with more now wintering in the Netherlands. However, numbers may increase considerably in severe winters. It arrives in October, departing mainly in February and March.

Structure and plumage A rather evenly proportioned, pale, grey-brown goose with a slightly gingery-brown head. Note in particular the pink bill and orange legs. Adults are easily identified by their white facial blaze and black belly barring (there is individual variation in the amount of black, with some showing a huge and solid area). Juveniles lack both the white facial blaze and the belly barring, and are noticeably paler than the adults, being pale grey-brown. Consequently, lone juveniles can appear bland and rather featureless, prompting confusion with Greylag Goose (concentrate on bare-part colours and structure). They gradually start to acquire a white 'front' as the winter progresses. In flight, White-fronts appear rather evenly proportioned with dull grey forewings; the adults' black belly barring and white facial blaze should be obvious given a decent view.

Calls High-pitched, yelping and rather musical, especially in flocks, when they may recall a pack of distant dogs: *yi-yip, yi-yi-yip, ar-yip-yip-yip* and so on.

Greenland White-fronted Goose *Anser albifrons flavirostris*

Distribution The Greenland race *flavirostris* winters chiefly in SW Scotland (mainly on Islay and the Mull of Kintyre) with small numbers in W. Wales. The main wintering site, however, is the Wexford Slobs in Ireland. They arrive in October and depart in April, later than Eurasian birds.

Bare parts The most obvious difference from Eurasian is the bill colour, which is orange (on young juvenile, strongest at the base and somewhat pinker towards the tip). Note, however, that bill colour can be surprisingly difficult to distinguish in the field, so caution is required. Greenland's bill is also about 10% longer than Eurasian's, so it can appear quite long, wedge-shaped and not particularly deep at the base.

Structure and plumage If seen well, Greenland can be surprisingly different from Eurasian. It averages about 5% larger and is distinctly darker, more chocolate-brown, with narrower and darker buff feather fringes that produce a more uniform appearance. Grey wing-coverts are also rather dark. The darkness of their plumage renders the white rump and undertail-coverts more contrasting in flight. On average, Greenland has more extensive black barring on the belly, but this is individually variable in both races. Note that juvenile Greenland with a mud-stained bill can suggest one of the bean geese.

Lesser White-fronted Goose *Anser erythropus*

Distribution Formerly regular, Lesser White-front is now a major rarity with just three records between 2000 and 2012. It has traditionally occurred in flocks of White-fronts (mainly in Gloucestershire) but recently also with Taiga Beans in Norfolk. Introduced birds from Scandinavia now winter in the Netherlands and these too could conceivably occur; frustratingly, escapes are not infrequent.

Structure and plumage Not only the rarest grey goose, but potentially the most problematical. Before identifying Lesser White-front, it is essential to note *all* the differences, particularly structural. Finding a Lesser usually requires prolonged and patient scrutiny of White-front flocks and, without previous experience, the species can be surprisingly difficult to pick out. The adult has a more extensive white blaze and also a yellow orbital-ring, but neither may be obvious, particularly at any distance, so other clues are needed. The white blaze extends well up onto the forehead, but the shape of the white may be useful: viewed front-on, it is narrow and usually forms a point on the forehead (more rounded on most White-fronts, although some also show a more pointed blaze). It is a distinctly smaller bird in direct comparison (about four-fifths the size) with a shorter and proportionately thicker neck (perhaps suggesting Pink-foot), a steeper forehead and rather square head. The bill is small, delicate, rather triangular and generally brighter pink than White-front's (like White-front, sometimes with a grey area on the upper mandible). The head and neck are slightly darker and, with practice, this is a useful character, particularly coupled with its plainer body plumage and smaller area of black on the belly. More subtle characters include slightly longer wings and a quicker feeding action. Juveniles are particularly difficult but, as with adults, the combination of shape, structure, head and neck colour, and yellow orbital-ring is most useful. Like White-fronts, juveniles lack black belly barring and are plainer above, but most first-winters gradually acquire a white forehead blaze, similar in shape to the adults', but slightly smaller.

Pitfalls Some White-fronts show a particularly large white blaze, some even show yellow orbital-rings and occasional individuals show white feathering around the eyes which, at a distance, may suggest the yellow orbital-ring of Lesser. Sleeping White-fronts reveal pale eyelids which, momentarily, can suggest the rarer species.

Call A three or four-noted *ee-wee-wee-weeut*, shrill and penetrating compared to White-front.

Other Confusion Species

When identifying out-of-context grey geese, be aware of escaped birds (such as blue-phase Snow *A. caerulescens*, Bar-headed *A. indicus* and Swan Geese *A. cygnoides*) as well as hybrids (such as Canada *Branta canadensis* × Greylag and Snow × Barnacle *B. leucopsis*) which may show peculiar combinations of plumage and bare-part colours.

References Cramp & Simmons (1977), Holt *et al.* (2011), Musgrove *et al.* (2011), Ogilvie & Wallace (1975).

Snow and Ross's Geese

Where and when Small numbers of Snow Geese are annual in Britain and Ireland, usually accompanying Icelandic Greylag Geese *Anser anser* or Greenland White-fronts *A. albifrons flavirostris*. Their status is, however, muddied by the presence of escaped and feral birds, with populations on the islands of Mull and Coll in the Inner Hebrides and in Lancashire, Yorkshire, Oxfordshire and Sussex (about 180 birds; Musgrove *et al.* 2011). Since 2001, Ross's Geese have also been seen with some regularity, usually with flocks of Pink-footed Geese *A. brachyrhynchus* that gradually move south from Scotland to traditional wintering areas in Lancashire and, particularly, Norfolk.

Background There are two races of Snow Goose, both of which have undergone a population explosion in recent decades. The most abundant is the nominate Lesser Snow Goose *caerulescens*, which breeds in n. Canada from Baffin Island west to NE Siberia. They use the Pacific and Mississippi Flyways and do not normally occur on the Atlantic Flyway. Unlike Lesser, there is only one population of Greater Snow Goose *atlanticus*, which breeds on the islands of ne. Canada and in n. Greenland, having expanded its breeding range eastwards in recent decades. In winter, it is restricted almost entirely to the Atlantic Flyway. Although traditionally wintering in California, Ross's Goose appeared in the Mississippi Flyway in the mid 1970s and in the Atlantic Flyway in the mid 1990s. It too has recently undergone a population explosion (2,000–3,000 in the early 1950s rising to perhaps two million at present) and it has spread eastwards to breed on Baffin Island. This coincides with their recent regular appearances in Britain, the original vagrancy presumably taking place in the Arctic. The birds subsequently arrive with 'carrier' species that originate from the same general area. Snow and Ross's Geese may rarely occur among goose populations from n. Europe and Siberia.

Snow Goose *Anser caerulescens*

Dimorphism There are two morphs (or phases) of Lesser Snow Goose: white and 'blue'. Western birds are predominantly white, the eastern ones predominantly 'blue'. Unlike Lesser Snow Goose, 'blue' morph Greaters are quite rare, comprising less than 4% of the population.

Identification The plumages of the Snow Geese are similar. Both morphs may show orange staining on their faces, which they acquire on their tundra breeding grounds. ***White-morph adult*** All white, except for black primaries. ***White-morph juvenile*** Variable, but basically pale grey on mantle and scapulars; the wing-coverts and secondaries are dark grey (feathers fringed white), the crown and nape greyish and the underparts greyish-white. The plumage becomes whiter during the winter post-juvenile moult. ***Blue-morph adult*** White head and upper neck but exact pattern varies. Most have white extending down the neck-sides in a point, and a thin grey line of dark extending up the nape to the rear head. The body is dark blackish-grey, with elongated, curved and pointed inner greater coverts (black, prominently fringed white) which droop downwards at rest. The belly is white (extent variable). In flight, they have grey upperwing-coverts (white along the leading edge) contrasting with the black remiges. The rump, upper- and undertail-coverts are white and the tail is grey. The underwings are also bicoloured: white lesser and median underwing-coverts contrasting with dark greater coverts and remiges. ***Blue-morph juvenile*** Also variable but most are uniformly dark smoky grey-brown, the only pale feathering often confined to the chin and throat; like the adults, they too show paler grey wing-coverts and black tertials and greater coverts, fringed whitish. There is a variable post-juvenile moult during the winter.

Bare parts Adults have a pink bill and legs. Juveniles start life with yellowish legs and a dark grey bill, which becomes pinker by spring. The bills of Snow Geese (but not Ross's Geese) have a serrated edge and a distinctive elongated oval gap between the mandibles, known as the 'grinning patch'. They can be separated as follows:

Greater Snow Goose *A. c. atlanticus*

Large size, being only slightly smaller than Greylag Goose. It is about 20–30% heavier than Lesser with a 10–15% longer bill. Consequently, it looks longer-faced with a more strongly wedge-shaped head and a longer, deeper bill. Like Lesser (but unlike Ross's) it has a curved line of demarcation between the bill and the face, as well as an obvious black 'grinning patch'. The blue morph is rare.

Adult 'blue morph' Ross's Goose. This rare morph has restricted white on head, and white wing-coverts.

Adult Ross's Goose has small triangular bill with vertical line of demarcation at bill base and no 'grinning patch'. Bill base bluish. ▼

Ross's Goose

Juvenile Ross's Goose. Whiter than juvenile Snow Goose. ▼

Lesser Snow Goose (race *caerulescens*).

Adult Ross's Goose. Small and stocky.

Most British records of Snow Geese are of the larger race *atlanticus*, the Greater Snow Goose.

Greater Snow Goose (race *atlanticus*). Note 'grinning patch', common to both races of Snow Goose.

Adult 'blue morph' Snow Goose. ▶ Note extent of white on head and neck, grey wing-coverts. The grey underparts are variable in extent. ▼

Juvenile 'blue morph' Snow Goose

Juvenile 'white morph' Snow Goose. Dingy and greyer than adult.

Adult 'white morph' Snow Goose. Note bill structure.

Before claiming a Ross's or Snow Goose take care to eliminate albino, white or partially white geese of domestic ancestry and also hybrids.

Lesser Snow Goose *A. c. caerulescens*

Medium size, similar to Pink-footed Goose. Larger than Ross's Goose with a more wedge-shaped head and bill profile. Unlike Ross's, note that the line of demarcation between the bill base and the head feathering is *strongly curved* and that the bill shows an obvious black 'grinning patch'.

Ross's Goose *Anser rossii*

Resembles a miniature Snow Goose, being about 80% the size of Lesser Snow and about half the weight. It is in fact similar in size to Lesser White-fronted Goose *A. erythropus*. It has a short neck with a rounded head and a small, triangular bill (pink with a bluish base in adults). In flight, it looks very stocky and short-necked. Note in particular that, unlike Snow Geese, it has a vertical line of demarcation between the head feathering and the bill base (not rounded) and there is *no 'grinning patch'*. **White-morph juvenile** Much whiter than juvenile white-morph Snow Goose, with pale grey shading confined to a band through the eye and variably on the hindcrown, neck, nape and mantle, sometimes with grey shaft streaks on the inner greater coverts. It may also show grey on the bill base. **Blue-morph adult** Rare. Adults differ from 'blue' morph Lesser by a combination of blacker neck, back and scapulars, whiter wing-coverts and tertials, and a restricted white face.

Hybrids

Note that hybrids sometimes occur between Lesser and Ross's and these are intermediate in size, bill shape and 'grinning patch'. A hybrid seen in Norfolk was generally considered to have been a cross between Ross's and Pink-footed Geese. It suggested a blue-morph Lesser Snow Goose but it was *completely* white on the head and underparts (apart from scattered grey feathers on the head and grey upper-flank feathers, fringed with white). The upperparts, however, resembled Pink-foot, being grey-brown with white feather fringes, but the greater coverts and tertials were more like blue-morph Ross's, being broadly fringed with white. Its bill was black with a pink tip, more like Pink-foot.

Farmyard geese, aberrant geese and Canada Goose hybrids Also bear in mind the occasional occurrences of escaped or hybrid geese that resemble Snow Goose. White 'farmyard geese' are often encountered but they normally show an orange bill and white primaries (they lack the contrasting black primaries of white-morph Snow Goose). White leucistic geese may also occur but, again, such birds fail to show the smart black-and-white contrast of Snow Geese. Other hybrids may resemble 'blue' Snow Goose. Canada × Snow Goose hybrids may show a similar combination of dark body and white head and neck, but check the wing pattern (mainly dark) and, in particular, bill colour (usually dark on adults, not pink). Harrison & Harrison (1966) described a hybrid Greylag × Canada Goose that, remarkably, showed a white head and neck and a fleshy-grey bill; such a bird could easily have been mistaken for a 'blue' Snow Goose.

References van den Berg (2004), Cramp & Simmons (1977), Harrison & Harrison (1966), Madge & Burn (1987), Musgrove *et al.* (2011), Sibley (2003).

Cackling and Canada Geese

Background Introduced into Britain in the 17th century, Canada Goose *Branta canadensis* is a common and familiar species in a wide variety of freshwater habitats. Introduced birds are of the large nominate Atlantic race *canadensis*. In North America, Canada Goose occurs in a variety of forms, traditionally separated into 11 subspecies. In 2004, however, the American Ornithologists' Union split the species into two: the large ones remaining as Canada Goose *B. canadensis*, the small tundra forms becoming Cackling Goose *B. hutchinsii*. Here in Britain, wild Canada or Cackling Geese occur most winters, accompanying 'carrier species' that breed in Greenland: principally Barnacle Geese *B. leucopsis* and occasionally Pink-footed Geese *Anser brachyrhynchus*.

A new approach In addressing the problem of their identification, Garner (2008) challenged the official view of Canada Goose classification, referring to a monumental work on the subject by Hanson (2006, 2007). Hanson concluded that the complex could be split into six species (not two) with 162 subspecies. Regardless of the perspicacity of Hanson's conclusions, Garner makes some interesting points. **1** The original splitting of Canada Goose into 11 races was based on limited data, a review of the taxonomy in 1946 being based on just 359 specimens, of which only 8% were from the breeding grounds (Hanson's conclusions were based on 1,800 specimens, 60% from the breeding grounds). **2** Hanson concluded that each local population is, effectively, reproductively isolated and has evolved in response to its own particular environment. This isolation is reinforced by the fact that the geese breed alone or in small localised colonies, they mate for life, return to the natal area to breed and remain faithful to the same staging and wintering locations. This complexity is now compounded by the fact that, in some areas of North America, escaped and feral birds have mixed with wild populations and clouded the position still further.

The need for caution Sibley (2007) also stressed the problems of identifying the Canada/Cackling Goose complex and he too warns that the currently described subspecies should be considered as only one interpretation of their taxonomy, adding that the information on these forms remains patchy and incomplete. A number of specific points also need to be kept in mind. **1** SIZE When judging size, bear in mind that males of all forms are larger than females. **2** NECK LENGTH This depends on posture and on what the bird is doing. **3** BODY PLUMAGE Each form shows a range of colour variations. **4** NECK COLLAR Some forms (e.g. *hutchinsii*) tend to show a white neck collar, but a significant minority may lack it. **5** BLACK CHIN-STRIPES Similarly, a black dividing line down the centre of the chin/throat (dividing the white face patch into two halves) is often shown in some forms, but this too is variable.

Vagrancy potential From a European point of view, it is important to concentrate on those forms most likely to occur on this side of the Atlantic. Broadly speaking, the larger subspecies are found in the interior of North America, in the south of the breeding range. As these birds have the shortest migrations, they are the least likely to reach Britain. 'Canada Geese' become smaller towards the Arctic and it is the small tundra-breeding forms that are the longest migrants and those most likely to cross the Atlantic. Between these extremes is a range of intermediate-sized birds, including ones that breed in the subarctic zones of Baffin Island, n. Hudson Bay and Greenland. As these are also strongly migratory, they too must be regarded as likely vagrants. The forms outlined below are considered to be the most likely candidates for vagrancy, but remember that escaped birds may also confuse the situation. Notwithstanding Hanson's conclusions, for the sake of simplicity the summaries below adhere to the AOU's two-way split and the names given to the officially recognised races.

Ageing As with grey geese, juvenile Canada and Cackling Geese have a weak and rather fluffy body plumage which is usually quickly moulted in late summer. However, the more northerly tundra-breeding Cackling Geese may not have time to moult prior to migration so some young birds arrive on their wintering grounds still in full juvenile plumage. Those immature vagrants largely in adult-like first-winter plumage are not easy to age, but it may be possible to detect a few retained buff-fringed juvenile scapular and wing-covert feathers. These are rounded in shape (not square like the adults) and, although tipped paler, the feathers are plainer, lacking both the adults' conspicuous *thick* pale buff tip and their darker subterminal band (close views and/or detailed photographs are necessary to confirm their ageing).

Cackling Goose *Branta hutchinsii*

Where and when Richardson's Cackling Goose *B. h. hutchinsii* is by far the most likely Cackling Goose race to reach Britain. This is because it breeds in n. and ne. Canada, including parts of Baffin Island. Our vagrants usually arrive with Greenland Barnacle Geese, which winter in w. Scotland (principally on Islay) as well as in w. Ireland. There have also been records of *hutchinsii* occurring independently, but such birds are usually dismissed as escapes. However, a cluster of three around the Severn Estuary in the 'American autumn' of 2011 may suggest that such records should be taken more seriously.

Identification Cackling Geese are easily separated from *canadensis* Canada Geese by their very small size: about half the size of Canada and more similar to Barnacle Goose (in direct comparison, about 10% smaller than the latter). They are also structurally distinctive, with a short neck, a very square head and a tiny, triangular bill. The conventional view of *hutchinsii* is of a rather pale greyish-brown goose with a pale whitish-buff breast and little if any neck-ring. This form is, however, variable, both in its breast colour and in the presence or absence of a white neck collar. Sibley (2004) pointed out that one study revealed that 50% of adults and 25% of first-winters show a neck-ring. Some have a very dark chocolate-brown or slaty-brown breast with a prominent collar, some have a pale grey-brown breast and others are somewhere in between. The white face patch is triangular (rather pointed at the top) and many have distinctive blackish 'trousers' (thigh feathering at the top of the legs). They usually, but not always, lack a black line down the centre of the chin and throat. In flight, they appear small, slim-winged and thin-necked. The calls are high-pitched and yelping.

Canada Goose *Branta canadensis*

Various other types occur, some approaching nominate *canadensis* Canada Goose in size, others Cackling Goose. Such birds may occur in Britain with Barnacle and Pink-footed Geese and are undoubtedly wild. The racial identification of these birds is extremely difficult and it is often unwise to be too dogmatic about their identity.

Todd's Canada Goose *B. c. interior*

Breeds around Hudson and James Bays and its range has recently expanded into Greenland, rendering it a particularly likely vagrant. It is a large bird, similar in size to nominate

Canada Goose typical of the British feral population. Large and pale-bellied.

Canada Goose of the race *interior*, Todd's Canada Goose. Slightly smaller than British feral birds, browner breast, long thin neck and long bill.

Canada x Greylag Goose hybrid. This frequently encountered hybrid shows mixed features with diffuse head pattern and is often darker than either parent. Note pale bare parts. ▼

Canada Goose of the race *parvipes*, Lesser Canada Goose. Intermediate in size, tends to show narrow white 'face' patch. ▼

Canada x Barnacle Goose hybrid is typically dark breasted and a dingier grey than a pure Barnacle Goose.

Three **Cackling Geese** of the race *hutchinsii* showing plumage variation. All are very small, short-necked, square-headed with a tiny bill.

canadensis, but it is usually slightly smaller with a browner breast concolorous with the rest of the body. It thus appears darker and drabber overall than *canadensis* (although some are more similar). It may show a proportionately long thin neck and a long shallow bill (Batty & Lowe 2001) as well as slightly longer primaries, falling just short of the tail tip.

Lesser Canada Goose *B. c. parvipes*

Birds resembling Lesser Canada Goose *parvipes* have also occurred (this form breeds from central Alaska east to Hudson Bay). Both plumage and structure are similar to *canadensis*, but it is smaller. A bird in Somerset that resembled this form was about 80–90% the size of *canadensis* but only 75% its weight, with a thinner neck and a smaller, slimmer bill (but larger and longer than that of Cackling Goose). Most notably, it showed a much narrower white face patch, which was kinked backwards and distinctly pointed in the upper rear corner. Some *parvipes*, however, are darker-breasted. Particularly small individuals may look similar to *hutchinsii* Cackling Goose, with which it often intergrades (Sibley 2004). Such intermediates would provide a significant identification problem on this side of the Atlantic.

References Batty & Lowe (2001), Garner (2008), Hanson (2006, 2007), Sibley (2004).

Brent Geese

Where and when Brent Goose occurs in four forms. Dark-bellied *bernicla* breeds in Siberia and is locally abundant in winter in e. and s. coastal areas, from the Humber Estuary to Devon (rarer elsewhere). Pale-bellied *hrota* breeds in Arctic Canada and Greenland. Arriving from the west, most winter in Ireland with small and variable numbers appearing in w. Britain, sometimes in small influxes (as early as September). Another population occurs on Svalbard and Franz Joseph Land, wintering mainly in Northumberland. Pale-bellied Brents can, however, occur almost anywhere on the British coast, usually in flocks of Dark-bellied. Black Brant *nigricans* breeds in n. Canada, Alaska and ne. Siberia. Most arrive here from the east with Dark-bellied Brents, and they occur regularly in very small numbers in e. and s. England; there have also been records from the west with Pale-bellied Brents. Grey-bellied Brent has not been not assigned subspecific status. It breeds on Melville and Prince Patrick Islands in the w. Canadian Arctic, wintering on the Pacific coast of the USA. Examples of this form have been identified in Northern Ireland.

Identification A small, compact, thick-necked goose, similar in size to a Shelduck *Tadorna tadorna*. Easily identified by its black head, neck and breast (with a variable white neck collar), dark grey upperparts and a variable grey to black belly; the adults have a white 'flash' along the flanks. In flight, it is short-necked and compact, with pointed wings and a short rear end that is extensively white (a large white V-shaped rump/uppertail-covert patch, with limited black visible in the closed tail). They regularly occur in large, dense and noisy flocks. The calls are distinctive: a deep *ur* or a rolling *k-r-r* (sometimes almost bugling).

Ageing Adults have plain upperparts, but juveniles show strong pale cross-barring, produced by prominent pale buff or white feather tips that are gradually moulted during the winter

Adult Dark-bellied Brent Goose race *bernicla*. Plain upperparts and white neck collar. Juvenile (middle) lacks collar and has pale fringes to wing-coverts and upperparts.

Black Brant (third from right) with Dark-bellied Brent Geese.

Juvenile Dark-bellied Brent Goose ▶

1st-winter Dark-bellied Brent Goose. Note plain upperparts and collar are moulted in as the winter progresses.

Adult Black Brant race *nigricans*. Note large collar, black breast merges into belly, upper flanks contrastingly white. Dark chocolate-brown tone to plumage.

Adult Black Brant with **Dark-bellied Brent Goose** (behind). Note extensive collar and white flanks.

Juvenile Black Brant with adult-type collar moulting through.

Adult Pale-bellied Brent Goose race *hrota*. Very pale underparts contrast with dark breast. Pale between legs.

Two **Grey-bellied Brent Geese.** Intermediate between Pale-bellied Brent Goose and Black Brant, with brownish-grey underparts extending back between the legs.

(moulting juveniles in midwinter often look 'moth-eaten'). There is, however, much variation in their moult. Juveniles also lack white on the neck (sometimes just a hint) and young juvenile Dark-bellied and Black Brants also lack white on the flanks.

Dark-bellied Brent Goose *Branta bernicla bernicla*

The 'default' race in Britain. Dark smoky grey-brown upperparts show little contrast with similarly coloured underparts. Also, there is relatively little contrast between the black head/breast and the dark grey-brown body. Both the white flank patch and the white neck patch are relatively weak.

Pale-bellied Brent Goose *B. b. hrota*

Most distinctive is the strong contrast between the black head/breast and the pale body. The upperparts are a rather pale grey-brown and the belly is obviously pale (greyish-white or brownish-white), the flanks showing variable brown barring, mainly at the front (usually light but sometimes heavy) and at the rear. Unlike Dark-bellied, the pale body shows strong contrast with the black tertials, secondaries and primaries. Note that, unlike the other forms, the dark of the belly does not extend back between the legs. On early winter juveniles, the belly is buffier, delicately and regularly mottled with brown. In flight, whereas Dark-bellied shows a striking contrast between the dark body and the white rear-end, Pale-bellied lacks this contrast; instead, the contrast is always towards the front of the bird, between the black head, neck and breast and the pale body (Stoddart 2008).

Black Brant *B. b. nigricans*

Most similar to Dark-bellied Brent, but a strikingly 'black-and white' bird, strong plumage contrasts permitting easy location, even at a distance. The following are the main features. **1** BODY Appears very black (very dark chocolate-brown at close range, females slightly browner than males) showing very little contrast with the black head and breast. The black on the belly is clearly visible between the legs when up-ending, contrasting strongly with the pure white undertail-coverts. A key feature is the huge, pure white flank patch (prominently barred with black) that contrasts strongly with the black plumage. Note, however, that the flank patch is variable, being most striking on males but less obvious on some females, which may have a slightly smaller, less well-defined patch, sometimes more heavily barred and/or sullied with buff. **2** NECK-RING This is deeper and far more prominent than on Dark-bellied: usually with two parallel white lines with obvious white vertical crescents or 'spokes' connecting the two. Unlike Dark-bellied, the neck-patches join around the front of the neck to form a deep collar (but not at the rear). Males have a much stronger and deeper neck collar than females; in most females the collar is weak, sometimes joined at the front by only a thin white lower line or hardly joined at all. Bear in mind, though, that the neck-ring is less obvious when the bird is at rest with its neck retracted. **3** SIZE AND STRUCTURE Black Brant averages slightly larger, bulkier and thicker-necked than Dark-bellied Brent but, in direct comparison, males are larger than females. **4** INTERGRADES Beware of intergrades between Dark-bellied Brent and Black Brant, which should show 'unconvincing' intermediate characters (separating such birds from poorly-marked female Black Brants may be extremely difficult). **5 JUVENILE** Matt black head and breast plumage, with only a very faint suggestion of a dark buff neck-ring. The body is very dark chocolate-brown, contrasting only slightly with the black breast. The feather fringes on the upperparts are dark buff, showing little contrast with the feather bases. There is no white flank patch. As winter progresses, however, they acquire varying amounts of first-winter plumage and, by the New Year, they may start to show the deep neck-ring and contrasting white flank patch.

Grey-bellied Brent Goose

The taxonomic position of this form is currently unclear. It would appear to represent a stable intermediate form between Black Brant and Pale-bellied Brent, but the position may not be that simple (Garner 2008). Very variable: some extremes resemble Pale-bellied, others Black Brant. Typically, however, Grey-bellied is intermediate between the two, showing a more heavily pigmented brown or grey-brown belly. This tends to extend backwards between the legs or just beyond (unlike Pale-bellied, but like Dark-bellied and Black Brant). The upperparts vary from brown to brownish-grey. A narrow neck collar may or may not join at the front.

References Garner (2008), Stoddart (2008).

Mandarin and Wood Ducks

Where and when Mandarins breed on pools, streams and rivers in wooded areas, but they have recently shown signs of spreading into the wider countryside. Breeding records have occurred throughout Britain, and in Ireland, but it is commonest in SE England from London to Hampshire, and in the Forest of Dean, Gloucestershire. The British population is estimated at 7,000 birds (Musgrove *et al.* 2011). Wood Ducks sometimes escape from captivity and such birds occasionally breed (currently less than five pairs; Holling *et al.* 2011); it is also a potential vagrant (several have occurred in the Azores).

General appearance Males of the two species are among the most attractive birds in the world, but females, eclipse males and juveniles are much duller, less distinctive and potentially confusing. Both species have a relatively long, full tail, a bulging 'mane' of feathering on the rear of the head and a small, triangular bill. Females and juveniles are brown above, streaked on the breast and spotted with buff on the flanks. Most distinctive is a prominent white eye-ring and narrow white fringes to the primaries. In flight, the wings are plain with a white trailing edge to the secondaries. Both species nest in holes in trees and are often found sitting on low branches overhanging the water. They often forage on the woodland floor and are usually confiding, frequently joining tame Mallards *Anas platyrhynchos* being fed by humans.

Mandarin Duck *Aix galericulata*

Plumage *Female* Soft grey head with a prominent but narrow white eye-ring, and a white 'tear-line' extending back behind (occasionally has a very broad eye-ring; see below). There is also a narrow white vertical line on the lores (adjacent to the bill) and a white chin and

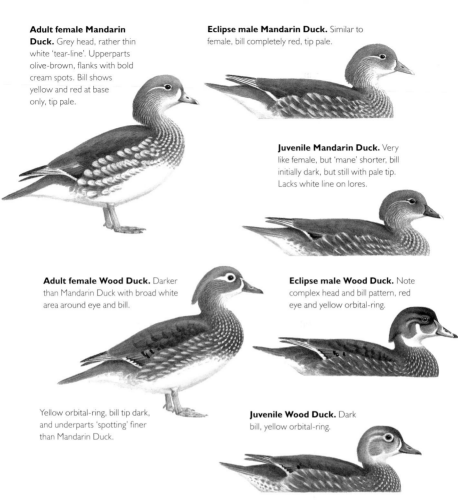

Adult female Mandarin Duck. Grey head, rather thin white 'tear-line'. Upperparts olive-brown, flanks with bold cream spots. Bill shows yellow and red at base only, tip pale.

Eclipse male Mandarin Duck. Similar to female, bill completely red, tip pale.

Juvenile Mandarin Duck. Very like female, but 'mane' shorter, bill initially dark, but still with pale tip. Lacks white line on lores.

Adult female Wood Duck. Darker than Mandarin Duck with broad white area around eye and bill.

Eclipse male Wood Duck. Note complex head and bill pattern, red eye and yellow orbital-ring.

Yellow orbital-ring, bill tip dark, and underparts 'spotting' finer than Mandarin Duck.

Juvenile Wood Duck. Dark bill, yellow orbital-ring.

Head pattern more subdued than female.

throat. In fresh plumage, there are subdued whitish lines running diagonally across the face (like whiskers). A thick 'mane' of feathering hangs down the nape. The bill is black with limited and variable dark red and dull yellow at the base (above and behind the nostril) plus a whitish nail at the tip (darker in late summer). The upperparts are quite a rich, greenish-brown and the breast is immaculately lined with pale buff, merging into prominent buffish-white spots on the flanks. The belly and undertail-coverts are prominently white and the legs are a fairly bright yellow. The pattern of the closed wing is complicated: a green speculum is thickly bordered with white; the outer tertial is glossed with blue and the green-glossed primaries have prominent white fringes to the outer webs. In flight, the wings appear largely plain with a noticeable white trailing edge to the secondaries. They are quite vocal in flight, females giving a soft *wack* (recalling Coot *Fulica atra*) and a quiet, abrupt *ick*, but courting males utter a distinctive upslurred, husky whistle *wueep*, as well as a strange buzzy sneeze. *Eclipse male* Very similar to female but retains its obviously red bill; however, this is usually a duller pinky-red, variably suffused with black (pinky-red most prominent around the base and sides). The head tends to be greyer than the female's, the eye-ring and tear-line duller and weaker and the white around the bill base is lacking; it also tends to show a stronger, more bulging 'mane'. *Juvenile* Similar to adult female but the head is a subtly softer powdery grey and the 'mane' is much shorter (so the head appears squarer). The eye-ring and line behind the eye are much weaker and it lacks the vertical white line adjacent to the bill. The background colour to the underparts is buffer and the flanks are streaked buff, rather than spotted. Juvenile males soon start to acquire traces of orange or pinkish-red on the bill base. *First-winter male* Acquires red bill and moults into female-like first-winter plumage prior to the gradual acquisition of adult-like plumage from autumn onwards.

Wood Duck *Aix sponsa*

Plumage *Female* Similar to Mandarin but the most distinctive difference is the eye-ring, which is larger and thicker, appearing not so much as a ring but as a prominent white tear-shaped patch surrounding the eye. Note, however, that some female Mandarins show a thick eye-ring (often extending back as a 'split' double tear-line behind the eye), so do not rely on this feature in isolation. Like Mandarin, female Wood Duck shows a narrow white vertical line on the lores, bordering the sides of the bill but, unlike Mandarin, this also extends around the top of the bill base. It also shows a narrow yellow orbital-ring, which Mandarin lacks. The bill is longer and less triangular than Mandarin's and it always has a black nail and bill tip (former white on Mandarin, but duller in late summer). There are variable amounts of yellow on the bill, most frequently along the cutting edge, but also diffusely over the tip (equally, the yellow can be largely absent). Female Wood Duck is darker and more chocolate looking than Mandarin. It is rather blackish-grey on the crown; the mantle and scapulars show a bronze or purple sheen and the wing-coverts a blue iridescence (all these feathers also have a black border or terminal band). The primaries are glossed green. The breast and flank spotting is smaller than Mandarin's, consisting of lines of buff streaks running down the centre of each feather. *Eclipse male* Very distinctive. Retains bright red eye and yellow orbital-ring, as well as the colourful bill (red and yellow, with a black patch in the centre and a black tip). Face pattern complicated: the forecrown is black and the rest of the face mostly grey but, most distinctively, the large white throat extends back as two spurs, the lower forming a partial collar around the neck and the upper projecting vertically towards the rear of the eye. It retains a broad white broken eye-ring, with a diffuse white tear-line behind, but there is also a crescent of black curving from the lores back below the eye-ring. Like the female, the lower wing-coverts are strongly glossed blue (with thick black feather tips) while the back and scapulars show a blue, green or bronze iridescence. Like the female, the chocolate-brown breast and flanks show only narrow buff streaking. *Juvenile* Resembles female but immaculate. The breast is neatly and finely streaked buff and the belly is greyish-white, subtly but more strongly mottled dark. The face pattern is subdued and diffuse compared to adult female, lacking the clear-cut thick white spectacle around the eye as well as the thick white line around the bill base. The bill is dark grey and there is a narrow yellow orbital-ring. Juvenile males may show a hint of the facial spurs that develop in first-winter. *First-winter male* Moults into an intermediate first-winter plumage similar to eclipse male (including a red eye, red and yellow on the bill and white on the chin and throat, extending back as two spurs towards the nape and the eye). Adult-like plumage is gradually acquired from autumn onwards.

References Holling *et al.* (2011), Musgrove *et al.* (2011).

Eurasian and American Wigeons

Where and when Eurasian Wigeon breeds mainly in Scotland and n. England, with small numbers in e. England. Returning migrants start to appear in late summer and, by midwinter, it is our second most numerous duck, congregating both on the coast and on lakes and flood-waters inland. American Wigeon is a rare vagrant, mainly in autumn and winter. It currently averages 19 records a year.

Eurasian Wigeon *Anas penelope*

General features A stocky, short-necked duck with a steep forehead, a small grey bill (tipped and edged black) and a pointed tail. In winter, it typically grazes on land, often in large dense flocks. Newly arrived late-summer migrants, however, tend to feed on water weed with Coots *Fulica atra* and Gadwall *A. strepera*, when their very white belly is obvious while up-ending. Males utter a very distinctive *wee-oo* whistle, females a grating growl.

Flight identification Stocky and short-necked, with a pointed tail. Often flies at some height, frequently in large flocks. Adult male easily identified by its large area of white on the wing-coverts. Both female and first-year male lack the white forewing patch, but they may show a subdued paler patch. All have an obscure speculum, faintly bordered with white in females. The white belly patch is obvious.

Plumage *Adult male* Very distinctive and easily identified by its reddish head, with a yellow forehead, pale grey body, black-and-white rear-end and large white patch on the wing-coverts. Some show a shiny iridescent green area around the eye and, on some extreme individuals, this extends back as a broad band down the head-sides (being iridescent, more obvious at some angles than others). Note that this is not indicative of hybridisation with American Wigeon. In all ages and sexes, Eurasian shows *greyish*-white axillaries and median underwing-coverts (see American Wigeon p. 46). *Adult female* Plumage tone varies individually but most are dark reddish-brown with little or no contrast between the head and body; however, some are less saturated, being slightly paler or greyer on the head with a paler, more orangey-brown body. At close range, the head is peppered blackish (difficult to see at any distance) and some show dark around the eye; on a minority, this may extend back in a thick, but diffuse, curved band. It lacks the male's large white wing-covert patch; instead, the wing-covert feathers are dark brown, narrowly but obviously fringed with white (thus forming a subdued patch). The underwings appear rather dusky. The tertials are blackish with narrow buff or whitish fringes, but most distinctive is that the lower (outermost) tertial is largely white, forming a distinctive white line along the rear of the wing at rest. *Adult male eclipse* Easily sexed as it retains its large white wing-covert patch (lacking in first-summer males). Distinctive in its own right, the head and body being a rich, deep, mahogany-red, peppered with black. On

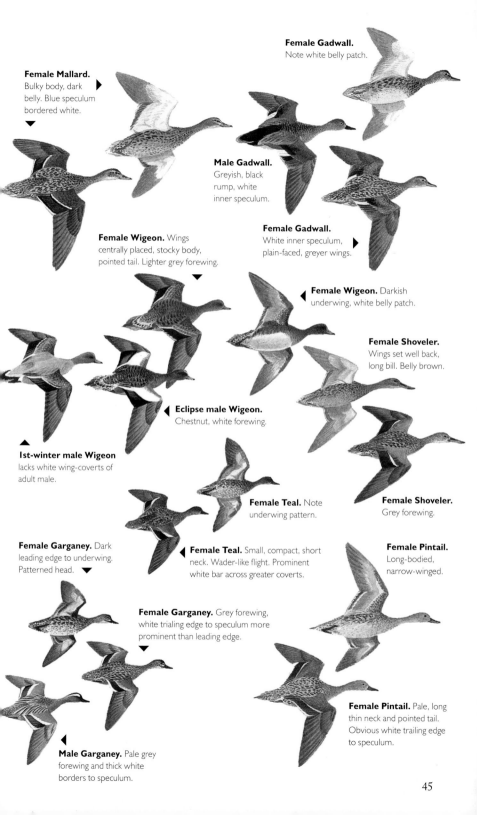

Female Gadwall. Note white belly patch.

Female Mallard. Bulky body, dark belly. Blue speculum bordered white.

Male Gadwall. Greyish, black rump, white inner speculum.

Female Wigeon. Wings centrally placed, stocky body, pointed tail. Lighter grey forewing.

Female Gadwall. White inner speculum, plain-faced, greyer wings.

Female Wigeon. Darkish underwing, white belly patch.

Female Shoveler. Wings set well back, long bill. Belly brown.

1st-winter male Wigeon lacks white wing-coverts of adult male.

Eclipse male Wigeon. Chestnut, white forewing.

Female Teal. Note underwing pattern.

Female Shoveler. Grey forewing.

Female Garganey. Dark leading edge to underwing. Patterned head.

Female Teal. Small, compact, short neck. Wader-like flight. Prominent white bar across greater coverts.

Female Pintail. Long-bodied, narrow-winged.

Female Garganey. Grey forewing, white trialing edge to speculum more prominent than leading edge.

Female Pintail. Pale, long thin neck and pointed tail. Obvious white trailing edge to speculum.

Male Garganey. Pale grey forewing and thick white borders to speculum.

the back and scapulars, some show fine grey vermiculations, with mahogany feather fringes, but others are more like females, having black back and scapular feathers with broad chestnut-orange fringes, as well as a paler chestnut-orange body. They may show a hint of a pale forehead. The tertials are black, noticeably fringed white (browner-edged on females and juveniles). *Juvenile* Very similar to typical adult female, but plumage obviously neat and immaculate. The best ageing feature is a smooth, rich orangey-white belly, clearly not as white or as contrasting as the adult's. It often has a slightly paler body, with a more fulvous-toned breast and flanks and a slightly paler head. The upperparts appear darker and more uniform, with narrower dark orangey feather fringes. The breast is also more uniform and only faintly spotted. Differences are, however, rather subtle. *First-year male* Adult body plumage is gradually acquired by early winter but it appears duller and less immaculate, less advanced individuals retaining obvious traces of brown juvenile plumage. Advanced birds are largely in full body plumage by the New Year but all are easily aged by their lack of a white wing-covert patch, which is not acquired until their first complete late-summer moult, when one year old.

American Wigeon *Anas americana*

General features Similar to Eurasian, but it tends to look slightly larger-headed, with a steeper forehead, and a slightly longer tail.

Plumage *Adult male* Easily separated by its dark pinkish-brown body, which contrasts with its distinctive head, which displays the following features. **1** FACE Buff, peppered black; **2** FOREHEAD Pale creamy-white (sometimes tinged yellow) usually extending over the crown towards the nape and **3** HEAD-BAND A broad iridescent green band extending from the eye and down the neck-sides; the prominence of this varies according to the angle (bronze tones may even be detectable). The male's call is slightly huskier than Eurasian's. *Adult female* Superficially similar to Eurasian and difficult to pick out. In an initial view, it has a rather pale greyish-white head, liberally and coarsely peppered black. It shows a dark area around the eye, sometimes extending back as a dark shadow towards the nape. The forehead, lores and throat are whiter and, on some, look particularly white when viewed head-on. The pale head contrasts quite noticeably with the body, which is orangey-brown, rather paler than most Eurasians. Because the head and neck of Eurasian are darker, (a) they contrast less with the body and (b) the dark peppering on the head does not stand out (even the palest, most washed-out Eurasians show a distinct brown component to the head colour). Conversely, some Americans are slightly browner-headed than others and thus less easy to identify. Faced with a potential American, check the axillaries and median underwing-coverts: on American they are *gleaming white* (vermiculated grey on Eurasian, giving an overall greyish effect in the field). Prolonged, close-range observation should also reveal subtle *on-average* differences, including some or all of the following. **1** BREAST More coarsely mottled on American. **2** BACK and SCAPULARS American tends to have thicker feather fringes which, being slightly paler, make the upperparts look more coarsely patterned. **3** TERTIALS American tends to have more prominent white tertial fringes. **4** WING-COVERTS Generally whiter on American, often forming a subdued whitish patch on the open wing, mirroring that of adult male. **5** RUMP Tends to look contrasting and colder on American, there being more contrast between the dark brown feather bases and the whitish fringes (on Eurasian, the rump looks more uniform as the feathers are usually dark brown, edged brownish, although

Eclipse male Wigeon. Deep chestnut, retains white forewing.

Wigeon shows small grey bill, steep forehead, rounded head, stocky body and pointed tail. Note white belly patch in all plumages.

Female Wigeon. Variable. Breast and flanks dull chestnut to orange. Head varies from brownish to greyish.

Female Wigeon

Juvenile Wigeons are darker above, plainer, more orange on flanks. Breast almost unmarked. ▶

Adult male American Wigeon in eclipse. Some very similar to female, but most retain green stripe behind eye. Flanks paler than male Wigeon.

In flight, note diagnostic white axillaries of American Wigeon (left). Eurasian Wigeon (right) has greyish axillaries.

Adult female American Wigeon. Very similar to some Eurasian Wigeons, but note pale greyish-white head peppered darker, dark shadow around eye, orangey breast and flanks. Always check colour of axillaries.

some do show pale fringes). These differences between the two species are due largely to the fact that American's entire plumage is much less saturated than Eurasian's; however, beware of a minority of female Eurasians that are also less saturated, such individuals looking more similar to American. *Eclipse male* Most are rather dark orange on the body with a brownish-grey head, coarsely and profusely peppered with black; they show a subdued and diffuse broad blackish-green band extending back from the eye, mirroring the pattern of full plumage. The tertials are very black with contrasting narrow white fringes. Some individuals look practically identical to adult females (although the forehead is often whiter); if in doubt, they are easily sexed by the white wing-covert patch. *Juvenile and first-year male* Similar to adult female. Juvenile/first-winter female has its wing-coverts similar to female Eurasian, whereas young males show more white. First-year males usually start to show adult-like plumage by late autumn/early winter, but they too fail to acquire the white forewing patch until their second winter.

Chiloé Wigeon and hybrids

When identifying American Wigeon, it is advisable to eliminate South American Chiloé Wigeon *A. sibilatrix*, which is frequently kept in wildfowl collections. Both sexes of Chiloé have a black head, glossed green, a white face, orange flanks and black upperparts and breast, with thick white feather fringes. Hybrids are a further possibility, so always check for anomalous characters. Eurasian × American Wigeon is a possibility and such birds should show intermediate characters. There have also been records of Eurasian × Chiloé Wigeon (Harrison & Harrison 1968). The latter have involved males, which showed a superficial resemblance to adult male American, but had black-centred feathering on the mantle and scapulars, a yellower (square-cut) forehead and green head-bands that coalesced on the rear crown (on American, they usually meet on the lower nape, the flecked cream crown extending as a point down the back of the head and nape).

Reference Harrison & Harrison (1968).

Large dabbling ducks in late summer and autumn: Mallard, Gadwall, Pintail and Shoveler

Moult and ageing The identification of male dabbling ducks in full plumage is straightforward, but females, juveniles and eclipse males are more difficult. Furthermore, their immature plumages have not been studied in great detail. This is undoubtedly due to both the subtlety and complexity of the subject. From late spring onwards, adult males lose their distinctive finery and acquire 'eclipse plumage', which is retained for about four or five months. In fact, 'eclipse' is essentially the birds' winter plumage that is 'shunted forwards' to help camouflage them during their late summer moult, when a period of flightlessness renders them vulnerable. It must be stressed that eclipse males, although similar to females, are not the same. They are easily separable in most species, but their identification is complicated by individual variation and because different males are often in different stages of moult. Juveniles are generally similar to adult females, and their separation is not as easy as it is, for example, with most waders. However, like juvenile waders, the entire plumage of juvenile dabbling ducks is neat and pristine. Most significant is that many species show a very neatly and uniformly *lined* breast (as opposed to mottled) as well as immaculate mottling on the belly. This is most obvious and most relevant in those species that show a white belly patch in adult female and eclipse male plumages (Gadwall, Common Teal, Garganey and closely related species); an exception to this rule is Eurasian Wigeon (see p. 44). The sequence of moult thereafter is similar to that of waders. Juvenile waders moult into first-winter plumage and in the following spring most then moult into summer plumage. With ducks, however, this whole sequence is again shunted forwards and compressed. They remain in juvenile plumage for a relatively short period before starting a partial and variable body moult into first-winter plumage. In young males, this plumage resembles that of the eclipse male. Consequently, in species where eclipse males resemble females (such as Common Teal) this first-winter plumage will come and go unnoticed. However, in species where the males possess a distinctive eclipse plumage (such as Shoveler) the first-winter plumage will be distinctive in its own right. This plumage is, however, also short-lived as, usually in late autumn/early winter, they continue to moult into their adult-like 'full' or 'breeding' plumage. A useful characteristic of young males is that they start to acquire full adult plumage later than the adults; also, their moult takes much longer to complete. This means that, whereas adult males are in full plumage by late autumn or early winter, first-winter males usually retain obvious traces of immaturity well into the winter. Consequently, first-winter males are usually less immaculate and many individuals retain limited brown feathering as late as the following spring. It must be stressed, however, that there is considerable individual variation.

Mallard *Anas platyrhynchos*

Where and when A familiar and abundant resident throughout Britain and Ireland, numbers augmented in winter by visitors from n. Europe.

General features The 'standard' duck of ponds, lakes, streams and rivers, it should act as the yardstick when identifying other members of the genus. The largest and bulkiest dabbling duck, often appearing rather round-backed with a longish bill that forms a continuous curve with the rather rounded head. The most familiar call is the female's quack; the male has a soft, nasal *raab* as well as a short, flat whistle and a grunt in courtship.

Flight identification A bulky duck, with broad wings that produce a soft whistling; the speculum is bright blue, narrowly bordered with black and white. The underparts are *uniform brown*, lacking a white belly patch (cf. Gadwall); white underwing-coverts contrast with darker remiges and the brown body. The tail feathers are white or whitish.

Plumage *Adult female* A brown duck, the pale creamy-brown background colour being boldly mottled and patterned with darker brown. The facial pattern is variable, most having a pale supercilium and a narrow dark eye-stripe. Domesticated varieties may be plainer or, conversely, may show a striking white supercilium and sometimes a dark bar across the lower lores. Bill colour is also variable: most show orange across the tip and often at the base, but some show extensive orange on the sides, recalling Gadwall (but rarely as clear-cut); some domesticated types may lack orange altogether. The ventral region is predominantly brown when up-ending (cf. the whitish belly of Gadwall). *Eclipse male* Easily separated from adult female as it retains a wholly yellow bill (although some have a blue-green bill). Unlike female,

Female Mallard. Note bill pattern, rich brown coarsely marked plumage. Orange legs.

Eclipse male Mallard. Resembles female, but yellowish bill, dark crown and rustier breast.

Juvenile male Mallard. Bill quickly begins to turn yellow.

Differences in tail position of up-ending Mallard (left) and Gadwall (right) sometimes useful.

Gadwall shows whiter vent when up-ending.

Juvenile Mallard. Darker than female, lined on breast.

Juvenile Gadwall. Smaller than Mallard, plainer, greyer head. Note bill pattern. Breast neatly lined, belly speckled.

Eclipse male Gadwall. Richer brown than female, greyer head and tertials. Note bill pattern.

Adult female Gadwall. Smaller, greyer than Mallard, coarse breast markings, white belly patch. Note bill pattern, white wing flash, yellow legs.

Juvenile Pintails are very neat with thick whitish scalloping on flanks and upperparts; finely streaked belly.

Adult male Pintail in eclipse. Note finely patterned grey-toned plumage, rusty head. Belly almost plain. Note bill pattern and grey legs.

Adult female Pintail. Plain, buffy head, long slender neck, coarsely marked body. Bill blackish.

the crown and eye-stripe are blackish, glossed green. The underparts are more cinnamon in tone, and the breast is rather rusty, strongly mottled with black. The rump and uppertail-coverts remain blackish (patterned brown and black in the female). *Juvenile* Tends to look smaller and scrawnier than the adults, and its shape can suggest Gadwall. Plumage is similar to the adult female's, but fresher and neater, while the overall tone is a richer, deeper orangey-brown. Easily separated from adult female as (1) the crown and eye-stripe are black (like eclipse male) and (2) the underparts, particularly the breast, are neatly streaked (more spotted or mottled on female, coarsely mottled on eclipse male). Juveniles have more extensive diffuse orange on the sides of the bill, sometimes including most of the upper surface. Juvenile males soon develop the adult-like bill colour, turning dull green, greenish-yellow and then yellow.

Gadwall *Anas strepera*

Where and when Has increased considerably in recent years and is now a common resident over much of lowland Britain, usually associated with reservoirs and gravel pits.

General features Superficially similar to Mallard, but smaller, less bulky, shorter-necked, flatter-backed and squarer-headed. The bill is shorter and narrower, projecting from a steeper forehead. Mallard usually bends its tail away from the wing-tips when up-ending, whereas Gadwall usually holds its wings and tail together in a vertical plane, producing a different profile when side-on. Male call a peculiar nasal *angh* and also a high-pitched wheezing whistle; female gives a quack, quieter and slightly higher-pitched than Mallard.

Flight identification The *white speculum* is obvious on the male, but it is often reduced to one or two inner secondaries in the female (occasionally almost lacking). On the male, the speculum is virtually surrounded by black and it also has a broad chestnut panel across the median coverts (the latter also shown by some females). Unlike Mallard, it shows an obvious clear-cut white belly, except on juvenile (always brown on female Mallard). Males display frequently, even in eclipse, and small groups often engage in high aerial pursuits when courting or defending territory, the males often giving their distinctive whistling calls.

Plumage *Adult female* Similar to Mallard, but colder looking. The body feathers are black, prominently fringed orangey-buff, the body contrasting somewhat with the greyer head. Conspicuous *clear-cut orange sides to the bill* (unusual on Mallard). The white speculum may be visible at rest (but note that moulting Mallards in late summer may also show white feathering in this area). When up-ending, the belly is noticeably white, contrasting with the browner undertail-coverts (entire underparts brown on Mallard). The legs are yellower than Mallard's. *Eclipse male* Starts to enter eclipse by late May and moults out by mid September. Similar to adult female, but orange-brown feather fringes produce a dark orange tone to the body, contrasting with the greyish head; it also gains orange sides to the bill. The tertials are pale grey (blackish-grey on female). Easily sexed by reference to the open wing, which retains the chestnut and black of full plumage. *Juvenile* Quite easy to age by its rather orangey breast, neatly and evenly lined with black (mottled on female). The crown and eye-stripe are clear-cut and black (usually less well defined and greyer on female); the greyish face contrasts markedly with the neatly lined breast. Overall, it looks darker and more uniform than adult female and has neater, fresher plumage (female messy and mottled in comparison). It lacks female's clear-cut white belly, which instead is heavily spotted. Sexing of juveniles is not easy, juvenile male's open wing being similar to female's, having less black and chestnut than adult male.

Pintail *Anas acuta*

Where and when A rare breeding bird, mainly in n. and w. Scotland and e. England. Returning breeders start to reappear from July onwards, with numbers increasing considerably from late August into autumn. Numerous in winter, but rather localised compared to other common ducks. Largest concentrations occur on certain estuaries, particularly in NW England.

General features Rather slim, with a long, pointed tail projecting well beyond the wing-tips. Also significant is the long and slender neck (obvious compared to Mallard and Gadwall). The head is rather rounded and the bill longish and slim. Not particularly vocal, but the male has a distinctive disyllabic nasal whistle in display, the female a high-pitched quack.

Flight identification A rather slim-bodied, long-necked duck with a long, pointed tail; the wings are set towards the rear of the body. It usually flies at some height. The speculum is obscure (green on male, brownish on female) narrowly bordered with dark buff or white in front and thickly bordered with white behind. The latter forms a *conspicuous and distinctive white trailing edge to the secondaries.*

Plumage *Adult female* At a distance, it appears a pale buffish duck (but a tiny minority of all age/sexes can be quite orange toned). It has a bland, featureless head, often strongly tinged with warm buff or ginger. The upperparts are thickly spangled with whitish or golden and the flanks show strong dark chevrons and white scalloping; the belly is whitish, delicately mottled darker. The bill is blackish-grey, with diffuse grey on the sides. *Eclipse male* Similar to female but easily sexed by its *clear-cut* pale grey sides to the bill and predominantly plain pale grey tertials. The wing-coverts are plain mid grey (mottled buff on female). Plumage variable, paler and greyer overall than female, with finer, more delicate markings, especially on the breast and flanks (but some are more coarsely patterned); like female, the head is obviously tinged dark buff or brown. Males starting to acquire full plumage may look plainer and greyer at a distance. *Juvenile* Less easy to age than other dabbling ducks as females and eclipse males also show mottled underparts. However, as with all juvenile ducks, it is neater and more regularly patterned, with thick whitish scalloping on flanks and upperparts and profuse fine streaking on belly. The bill is paler than adult female's, dull greyish, but juvenile male soon starts to acquire the adult's pattern, albeit less clear-cut. It can also be sexed by reference to the open wing (corresponds to the adults'). First-winter males start to acquire adult plumage by early winter but they remain less immaculate until late winter, retaining traces of immaturity and lacking the long central tail feathers.

Shoveler *Anas clypeata*

Where and when A widespread but localised breeder, commonest in e. England. Much more numerous outside the breeding season, with the first returning breeders arriving in midsummer; numbers peak in September to November, rather than midwinter.

General features A rather dumpy, short-necked duck, about three-quarters the size of Mallard. Easily identified by its huge spatulate bill. Bright orange legs are always conspicuous on land. Feeds by dabbling on the surface, often in large packs, or by up-ending, when combination of orangey-brown underparts and distinctly pointed tail are noticeable differences from Mallard, Gadwall and Pintail. Also dabbles on mud and not infrequently dives for food in shallow water, with an open-wing action. It is rather silent, but males give a quiet *took*, the females quacking noises.

Adult female Shoveler. Plain-'faced' coarsely marked plumage, often with orangey underparts.

Eclipse male Shoveler. Blue forewing, wide white wing-stripe.

Juvenile Shoveler. Similar to female, but note dark smoky appearance. Upperpart and flank feathers fringed dark orange-brown. Belly neatly peppered brown.

Eclipse male Shoveler. Note greyish head, gingery flanks. Yellow eye indicates adult.

1st-winter male Shoveler may appear mottled during the first year. Some develop a whitish half-moon pattern between eye and bill.

Flight identification Very distinctive deep, throbbing wingbeats on take-off. In flight it appears scrawny, with a long bill and a narrow neck; the narrow wings are set far back. It usually flies at some height, like *Aythya* ducks. The male has a bright blue forewing, offset by a thick white bar across the greater coverts, but it lacks a prominent white trailing edge to the secondaries. First-year male has slightly duller forewing. Female's forewing is dull grey, and the white greater covert bar may be virtually absent. All show a dark belly.

Plumage *Adult female* A pale duck, with dark brown feathers thickly patterned with pale buff (especially on the breast and flanks). In fresh plumage this creates a very scaly impression (with a plain orangey-buff belly). The buffish head is rather plain, the eye-stripe petering out behind the eye. The eye is dark (dull orange at close range) with a faint eye-ring. The bill is orange at the base, sometimes extending right across the upper surface. Tail predominantly white. *Eclipse male* Obviously different from female, with a dark orange tone to the flanks and belly (the feathers show dark orange fringes and internal markings). The upperparts are plain blackish-brown (narrowly fringed orange-buff) and the tertials are black with two or three prominent narrow white horizontal lines (sometimes one or even none if the feathers are in moult). In contrast, the female is browner and buffier overall and has brown tertials with only narrow and inconspicuous buff fringes. The male's head is dark greyish, peppered black, with the manic yellow eye conspicuous (female's eye dark). The bill develops a green tint in summer, becoming yellower or dull orange at the base, or even completely orange. Later in autumn, many males develop 'supplementary' feathering, acquiring variable amounts of white on the breast and rear flanks, as well as a large buff or whitish crescent before the eye. *Juvenile* Immaculate. Noticeably darker than female with blackish-brown crown and eye-stripe and distinctly dark chocolate-brown body plumage, fringed with dark orange-brown. Belly buff, delicately peppered brown. Subdued eye-ring and dark eye (but male's soon starts to whiten). Base and sides of bill orange. *First-winter male* Moults from juvenile into first-winter plumage that resembles eclipse male, having black body feathers fringed quite dark orangey-brown (but more cinnamon in tone than adult male). The fringing is thick and scaly on the flanks, but narrower on the upper-parts. The dark orangey body contrasts with a dull buff head. The tertials are black, fringed whitish, but it lacks eclipse male's prominent white horizontal lines. Unlike adult male, the eye is dull and the bill may be extensively orangey-green. The blue forewing is also duller. It gradually acquires full plumage towards spring but the moult is variable, some remaining predominantly brown and female-like into late winter (but developing the adult male's whitish eye).

Small dabbling ducks: Common, Green-winged, Cinnamon and Blue-winged Teals, and Garganey

Common Teal *Anas crecca*

Where and when Abundant in winter (our third-commonest duck) with large concentrations on the coast and inland. As a breeding species it is uncommon, mainly in northern and upland areas; much scarcer in lowland areas.

General features A very small dabbling duck with a short neck and a small, narrow bill. Females, eclipse males and juveniles are rather featureless, but note the short, narrow white streak on the outermost undertail-coverts, immediately adjacent to the tail. They show a bright green speculum, although at certain angles this can occasionally appear bright 'Mallard blue'. A quick, lively, nervous duck, usually seen dabbling on mud or in shallow water, frequently amongst partially submerged vegetation. On occasion, they may persistently dive for food with an open-wing action.

Call The male has a diagnostic high-pitched *crink crink*, the female a quiet quack.

Flight identification Easily identified by its small size, short neck, compact body, narrow pointed wings and quick actions. It often forms tight flocks that twist and turn like waders. The *open wing has a thick white bar across the greater coverts*, immediately in front of the predominantly bright green speculum; a white trailing edge to the wing is narrower and less obvious than the greater covert bar. The male's forewing is dull grey, the female's brown.

Plumage *Adult female* Plumage tone varies considerably, from greyish to brown, coarsely mottled. It appears rather featureless at any distance, the plain head relieved only by a darker eye-stripe behind the eye and a darker crown. There is, however, some variation, a small minority having a distinct pale spot on the lores and/or a darker line across the cheeks. The bill is grey, but it may show a small amount of orange or yellow at the base. *Eclipse male* Not readily separable from female except by the open wing (see above) although the tertials are

Adult female Teal. Small, dark grey-brown duck. Featureless 'face'. Usually shows slight orange at base of bill. Note white stripe down sides of undertail-coverts.

Juvenile Teal. Similar to female but darker with neater streaking below.

Adult female Garganey. Note pale supercilium, pale loral spot and dark line across 'cheeks'. ▼

Dull Garganey (left) and Teal (right). Note subtle differences in shape.

Eclipse male Garganey. Note chalky-grey forewing and white speculum border. ▼

Juvenile Garganey. Similar to female but darker and neater streaked below, often with cinnamon tint to underparts. Some virtually lack dark 'cheek' bar.

Blue-winged Teal

Cinnamon Teal

Female Blue-winged Teal. Plain 'face', white eye-ring and pale loral spot. Thin dark eyestripe.

Juvenile female Cinnamon Teal. May show head pattern similar to female Blue-winged Teal. Note warmer plumage tones, finer breast streaks, long spatulate bill.

Adult female Cinnamon Teal. Very plain face. Note bill shape and plumage tone.

usually longer and greyer, with dark centres. A minority of birds show a strong reddish tint to the head (others have a slight cinnamon tone to the head and underparts). They can also be spotted with black on the lower breast. Like female, it may also show orange at the base of the bill. *Juvenile* Similar to female, but immaculate, and plumage is usually richer and buffier (a minority can be quite orange in tone) neatly streaked or spotted on the breast and neatly patterned on the flanks; the belly is smooth and creamy, not as well defined as the adult's white belly. In comparison, adults look messy and rather coarsely patterned. The sides of the bill usually show orange at the base.

Green-winged Teal *Anas carolinensis*

Where and when A regular vagrant in late autumn, winter and early spring (not realistically separable in late summer and early autumn) with a broad geographical spread. Currently averages 42 records a year. A distinct spring peak is thought to be caused by vagrants heading north from s. Europe.

Plumage *Adult male* Easily identified by a prominent vertical white stripe down the side of the breast, thick at the top but tapering towards the bottom; this replaces the prominent horizontal white scapular stripe of Common Teal. It also shows less of a buff border to the green head patch. The breast is usually a deeper pink than Common, almost burgundy coloured. Two points to remember: (1) Green-winged may show a *rather vague hint* of a buffish-white horizontal line above the black scapular line; however, if an apparent Green-winged shows both a strong white vertical line and a *strong* white horizontal line, then it is probably a hybrid, examples of which have occurred; (2) on some, the buff border to the green eye-stripe may be more obvious than anticipated. Adult males may start to become identifiable from late September/early October, once the white breast stripe gradually starts to reappear. *First-winter males* As with other ducks, first-winter males are much slower in their moult out of juvenile/first-winter plumage, retaining brown feathering into the New Year. On the upper flanks any retained juvenile feathers are pointed, whereas retained adult male eclipse feathers are shorter and rounded (Duivendjik 2011). The breast may also appear paler and less colourful than the adult male's. *Female and juvenile* Females tend to show a stronger facial pattern than most Commons, with a pale spot at the base of the bill, a pale supercilium and a darker cheek-bar, creating a pattern reminiscent of Garganey; others are plainer. However, given the sheer abundance of Common Teal and the fact that some individuals are better patterned than others, the location of a female Green-winged in this country is largely unrealistic. If, by any chance, a candidate is discovered (e.g. one paired to a male), Green-winged apparently shows a narrower, more parallel-sided and more extensively orange-toned greater covert bar (as does the adult male); this creates a plainer-looking wing in flight.

Garganey *Anas querquedula*

Where and when Unlike other members of the genus, a summer visitor, mainly to e. England; a rare breeder. More numerous and widespread on migration (although still uncommon) in March to May and July to September, with stragglers into November (and rarely in winter).

General features Slightly but distinctly larger, bulkier and longer-bodied than Common Teal and, on the water, its rear-end is held higher. The bill is also distinctly longer. Feeds mainly by head-immersion, with little up-ending (it has an odd habit of repeatedly scratching its throat after head-immersion, although other ducks may also do this). On occasion, it may persistently dive for food.

Call In the breeding season, the male has a very distinctive rattling burp, the female a short, sharp quack.

Flight identification Similar to Teal, but looks longer-bodied. Eclipse male easily identified by combination of its pale *chalky-grey* forewing and green speculum, thickly and evenly bordered with white, *forming two parallel white lines.* Juvenile male similar, but forewing slightly duller and slightly darker grey; also, the white lines bordering speculum are marginally narrower. Female less distinctive, with a dull grey forewing; it also has white borders to the speculum, but the front bar is usually (but not always) faint, the hind bar wider, *forming a distinctive white trailing edge to the secondaries*, recalling Pintail *A. acuta*. Note that, on both sexes of Common Teal, the front greater covert bar is obviously the wider, being more conspicuous than the white trailing edge. Garganey also has a darker leading edge to the underwing-coverts. Its distinctive head pattern may be obvious in flight, while juveniles may show distinctly orangey underparts.

Plumage *Adult female* Similar to Teal, but generally paler, buffier and more coarsely and more contrastingly patterned. Easily identified by its head pattern. Unlike the plain-faced Teal, it has a *prominent, clear-cut thick whitish supercilium*, highlighted by a dark crown and eye-stripe. It also has a noticeable *whitish loral spot and throat patch*, separated by a dark line across the lower lores, which often extends back to form *dark bar across the cheeks*: on some, however, the bar is inconspicuous or even lacking, thereby prompting confusion with Blue-winged Teal (see below). ***Adult male eclipse*** Very similar to adult female, but easily sexed by the open wing. Unlike most other dabbling ducks, it remains in eclipse for a long time, with the first signs of new feathering appearing from late October to December, and full plumage is not acquired until midwinter. ***Juvenile*** Similar to adult female, but fresh birds neat and immaculate, darker and less coarsely patterned. Most are easily aged by the less well-defined cream belly patch (very faintly and delicately mottled); the entire underparts are usually tinged orangey-brown, neatly streaked down the breast and neatly patterned on the flanks (bolder than on Teal). The tertials are blackish, thinly but noticeably fringed pure white; they thus appear quite contrasting and can be a surprisingly useful feature. White borders to the speculum are narrower than on adult female. Juvenile plumage is retained well into winter and some first-winters may be distinctly orange-toned on the body.

Blue-winged Teal *Anas discors*

Where and when An American vagrant, mainly in April to May and August to October (extremely rare in winter). Currently averages about three records per year.

Eclipse male, female and juvenile In many ways intermediate in appearance between Common Teal and Garganey (longer-billed than Common Teal). Overall plumage tone similar to Common Teal, usually appearing greyer and colder looking than Garganey. Strongly patterned with buff, particularly on flanks, sometimes forming a contrasting 'crazy paving' pattern, apparent at some distance. Tertails browner and less contrastingly fringed than Garganey. Head pattern distinctive. **1 LORAL SPOT** Prominent whitish loral spot, generally larger and more obvious than on Garganey (but in some juveniles the spot can be quite faint and diffuse). **2 EYE-RING** *Noticeable broken white eye-ring* (which Garganey lacks). **3** FACE Weaker supercilium than Garganey and lacks the latter's dark bar across the lower lores and cheeks (the bar is, however, a variable feature on Garganey and is lacking in some). Before identifying a Blue-winged, check the open-wing pattern and leg colour. **4** OPEN WING The wing is very similar to that of Shoveler. The male has a bright sky-blue forewing (thus brighter and bluer than the chalky-grey forewing of male Garganey) with a broad white greater covert bar; unlike Garganey, it has *no white trailing edge to the secondaries* (this should always be clearly established). The female's wing is similar to the male's, but the blue is slightly duller, the greater covert bar is obscure (feathers blackish, fringed white, but the outer greater coverts are whiter) and the speculum is blacker. Juvenile/first-year male has a white greater covert bar, like adult male, but with small and inconspicuous blackish spots on the feather tips and also within the white. The underwing-coverts and axillaries of both sexes are gleaming white, contrasting with the underparts. **5** LEG COLOUR Yellowish, brightest on adult male but duller greenish-yellow or horn-yellow on female and juvenile male (can be particularly dull on young birds); the legs are grey on Common Teal and Garganey. Juvenile's breast neatly lined black, the scapulars and flanks neatly scalloped pale buff and the belly cream with neat dark flecking. Like Garganey, juvenile Blue-winged retains its juvenile/first-winter plumage well into the winter.

Cinnamon Teal *Anas cyanoptera*

Before identifying a Blue-winged Teal, this important pitfall species must be eliminated. A W North American and South American duck that has escaped from waterfowl collections on numerous occasions and caused confusion (they can be very difficult to separate). It is worth remembering, however, that female and juvenile Cinnamon Teal have a slightly 'Shoveler-like' jizz, a product of the slightly longer, broader and more spatulate bill, a rather rounded head and plain head plumage. The latter is the most obvious difference from Blue-winged, imparting a bland, pleasant and gentle facial expression. Close inspection reveals the following. **1** CROWN Lightly streaked (not strongly contrasting with the face). **2** EYE-STRIPE A variable but rather *faint* dark eye-stripe (sometimes more obvious before the eye, sometimes behind, or sometimes lacking). **3** EYE-RING Only a *faint* pale eye-ring. **4** SUPERCILIUM Lacks an obvious supercilium. **5** LORAL SPOT A pale spot that is smaller, duller, ill-defined and less obvious than on Blue-winged, but pale coloration sometimes extends diffusely to the upper lores, before the eye (but the prominence of the loral spot varies in both species). **6** PLUMAGE TONE The whole plumage of female and juvenile Cinnamon is plainer, less boldly patterned than Blue-winged, and the general plumage tone is a warmer, cinnamon-brown (very intense on some adult females and eclipse males, becoming almost ruddy on the breast); the belly, too, has a cinnamon tint and is less clear-cut than Blue-winged. **7** EYE COLOUR Also distinctive is that the adult male has orange or reddish eyes, even in eclipse (Blue-winged is always dark-eyed). First-winter males start to look very reddish by midwinter as they gradually acquire adult-like plumage, and the eye colour also starts to turn orange. Apparent hybrids between Cinnamon Teal and Shoveler have occurred on several occasions (intermediate in appearance).

References Duivendijk (2011), Garner (2008).

Aythya ducks: Greater and Lesser Scaup, Ring-necked Duck, Ferruginous Duck and hybrids

The problem of *Aythya* hybrids Identification of the rarer *Aythya* ducks is complicated by the frequent occurrence of hybrids. These vary greatly, but tend to fall into distinct categories. Those illustrated represent the four most frequent types: (1) Tufted Duck × Pochard, (2) Tufted Duck × Scaup, (3) Pochard × Ferruginous Duck and (4) Ring-necked Duck × Tufted Duck. It must be stressed that hybrids show great individual variation although, in most cases, an educated guess at the parentage can be made. Many are obviously 'half-and-half' (showing distinct characters of both parents), some show characters of one parent but not the other, but a few may resemble a different species altogether. There is no 'silver bullet' to solve their identification: what is needed is a knowledge of the main features of the parent species, combined with common sense and logic. When the parentage is uncertain, identification should be considered 'possible', 'probable' or 'presumed'. Female hybrids are much more difficult to identify than males, so it is necessary to be even more circumspect about their identification. Such birds are usually best logged as 'Pochard-like hybrid', 'Scaup-like hybrid' and so on, more as an acknowledgement of their general appearance than an attempt to make a definitive identification. It is worth remembering that the most likely hybrid combinations occur between those species that come into close contact during the breeding season: Tufted Duck × Pochard, Tufted Duck × Greater Scaup and Pochard × Ferruginous Duck. Note that, being geographically and ecologically isolated, Pochard × Scaup is particularly rare. Remember also that some hybrids are escapes from captivity, so a degree of lateral thinking may be required: one returning hybrid seen in Somerset was thought to have been a Tufted Duck × New Zealand Scaup *A. novaeseelandiae*. Before identifying an unusual *Aythya*, thorough familiarity with all plumages of the three commonest species − Tufted Duck, Pochard and Greater Scaup − is essential, as is an awareness of individual variability. Never claim a rare *Aythya*, no matter how distinctive it may appear, without considering the possibility of a hybrid. If in doubt, check your notes against detailed texts and photographs or, better still, visit a good wildfowl collection and familiarise yourself with the more obscure confusion species. When faced with a difficult *Aythya*, concentrate on bill-tip pattern, bill and head shape, overall size and structure, eye colour, back and flank colour, and wing-stripe. The less common *Aythya* species and their potentially confusable hybrids are outlined below. As female hybrids are less well known and tend to be overlooked, details refer to adult males unless otherwise stated.

Greater Scaup *Aythya marila*

Where and when Essentially a winter visitor, commonest in coastal bays and estuaries in Scotland (with a large population in Northern Ireland), becoming increasingly scarce further south. Most originate in Iceland, arriving mainly from October to November and departing between March and May. Small numbers are frequent on inland waters throughout the country, even in southern areas, mostly juveniles/first-winters, the peak time being November. The odd bird may, however, be encountered in any month, with tiny numbers of adult males occasionally appearing as early as July.

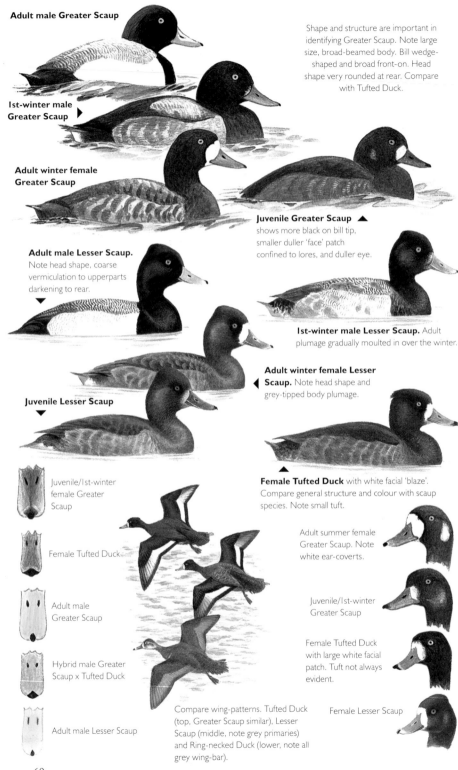

Adult male Greater Scaup

Shape and structure are important in identifying Greater Scaup. Note large size, broad-beamed body. Bill wedge-shaped and broad front-on. Head shape very rounded at rear. Compare with Tufted Duck.

1st-winter male Greater Scaup ▶

Adult winter female Greater Scaup

Juvenile Greater Scaup ▲ shows more black on bill tip, smaller duller 'face' patch confined to lores, and duller eye.

Adult male Lesser Scaup. Note head shape, coarse vermiculation to upperparts darkening to rear.

1st-winter male Lesser Scaup. Adult plumage gradually moulted in over the winter.

Adult winter female Lesser Scaup. Note head shape and grey-tipped body plumage.

Juvenile Lesser Scaup

Juvenile/1st-winter female Greater Scaup

Female Tufted Duck

Adult male Greater Scaup

Hybrid male Greater Scaup x Tufted Duck

Adult male Lesser Scaup

Female Tufted Duck with white facial 'blaze'. Compare general structure and colour with scaup species. Note small tuft.

Adult summer female Greater Scaup. Note white ear-coverts.

Juvenile/1st-winter Greater Scaup

Female Tufted Duck with large white facial patch. Tuft not always evident.

Compare wing-patterns. Tufted Duck (top, Greater Scaup similar), Lesser Scaup (middle, note grey primaries) and Ring-necked Duck (lower, note all grey wing-bar).

Female Lesser Scaup

Adult male Tufted Duck x Greater Scaup hybrid, resembling male Lesser Scaup.

Female Greater Scaup-like hybrid. This female has orange eye and whitish band near bill tip.

Female Tufted Duck x Greater Scaup hybrid (February). ▼

Adult male Tufted Duck x Pochard hybrid. Note bill pattern and finely vermiculated grey upperparts. Head very faintly ▼ tinged brown.

Three Scaup-type hybrids. ▲ Note head and bill shapes. Most show vestigial tufts.

Adult male Ring-necked Duck x Tufted Duck hybrid. Tuft, bill and flanks not quite right for Ring-necked Duck. ▼

▲
Two **adult male Ring-necked Ducks**. Note bill shape and pattern, head shape and shape of flank panel.

All sexes and ages of Ring-necked Ducks have 'boxy' body shape, rather flat back finely tapering bill, longer tail than other *Aythya*.

Juvenile/1st-winter male Ring-necked Duck (late October). Eye dull, wispy crest due to moult. ▼

Adult winter female Ring-necked Duck

Adult male Ring-necked Duck in eclipse (August).

Adult female Ring-necked Duck in summer (August).

Adult male Tufted Ducks moulting out of eclipse.

Juvenile male Ring-necked Duck. Strong buff tones to pale areas.

Adult male Ring-necked ▼ Duck, moulting out of eclipse. Note tail.

61

General features Structure is very important at all ages. A large, bulky, broad-beamed *Aythya*, obviously larger than Tufted Duck *A. fuligula* and about the size of Pochard *A. ferina*. The bill is long and broad, and the head has a rather steep forehead but, most importantly, an evenly rounded rear crown, with no hint of a tuft. The white wing-stripe fades to grey on the outer primaries, similar to Tufted. When diving, it often leaps more strongly than Tufted, sometimes clearing the surface.

Plumage *Adult male* White flanks and the very pale whitish-grey mantle create a very white impression at a distance (delicate vermiculations on the mantle are visible only at close range). The pale blue bill has a small black tip, restricted to the oval nail and is inconspicuous in the field. However, this can occasionally expand slightly around the rim or very rarely show as a small triangular black tip. The head has a green gloss, often obvious in good light. *Eclipse male* Very like a dull full-plumaged male, but black areas are duller and the back and flanks darker grey and less even. A large pale crescent may develop on the lower ear-coverts. It enters eclipse in July and moults out mainly in late October/November. *Adult female winter* A large white facial blaze around the base of the bill is striking (some female Tufted show a prominent white blaze, but never as large as on adult female Scaup). The body plumage is brown with grey vermiculations on the mantle and flanks, creating an obvious grey cast to the body. Some show a trace of a white ear-covert patch. *Adult female summer* Lacks grey vermiculations, becoming brown, rather paler and richer than Tufted, especially on the flanks. The facial blaze is often dark buff, but may be whitish. A large whitish patch on the lower ear-coverts is a conspicuous difference from Tufted. The bill becomes dark grey making the small black nail difficult to see. *Juvenile* In southern areas, most autumn migrants are juveniles (plumage retained into October/November, when body moult begins). The smooth and immaculate body plumage is brown, distinctly paler than female Tufted Duck and similar in tone to female Greater Scaup's summer plumage. However, instead of the adult female's large white facial blaze, juveniles show a large buff patch on the lores, extending over the bill on some; this can be quite a rich dark buff, but on all it gradually turns buffish-white then dull white as the winter progresses, increasing in size and whiteness. Juveniles also show a dull buff ear-covert patch or crescent, prominent on some (sometimes much whiter). The belly is creamy-white faintly and delicately streaked darker at close range (white on adult female). Unlike adult female winter, the upperparts and flanks lack grey vermiculations (on females they gradually appear during the winter, albeit duller than on adults). The bill is blacker than adult female winter, but gradually becomes greyer. Because of the dark base, the black nail is difficult to see. Many show extensions of black around the extreme tip, most prominent when front-on, when it may appear as an extensive black diamond or D-shaped bill tip (sometimes still present even as late as April). Such birds may also show a subtle pale grey subterminal band (may be curved). The eye is much duller than in adults. *First-winter male* Note that adult males show grey body plumage all year and are largely in full plumage by early winter. Juvenile males gradually acquire adult-like plumage from October/November (but rate of moult varies individually). First to appear is the black head feathering and small patches of whitish-grey on the scapulars, which gradually increase in size and then dominate. Early in winter, transitional birds show a messy combination of adult male and juvenile characters (and also some darker, vermiculated, intermediate first-winter feathering). By New Year, most are adult-like, although limited dark feathering may persist within the grey, even as late as March.

As the bill changes from blackish to pale blue with a small oval nail, intermediate patterns may occur; until late winter, some retain dark either side of the nail and some may continue to show a large D-shaped patch at the tip.

Confusable hybrids *Tufted Duck × Pochard* Male is generally smaller than Scaup; it has a peaked head, often with a slight tuft, and the bill is smaller and less broad. The Pochard influence renders the mantle and flanks darker, greyer and more solid-looking, while the head may show a brown tint at close range. The wing-stripe is often obviously grey (a Pochard character). Unlike Scaup, the bill has a large black tip and the eye tends to be darker (orange or reddish). *Tufted Duck × Greater Scaup* A greater pitfall than Tufted × Pochard. Males are often very similar to Lesser Scaup (see below). Most have a squarer head than Scaup, often with a slight tuft. The Tufted influence darkens the grey-vermiculated mantle, which usually looks too solidly dark for either species of scaup and too finely vermiculated for Lesser. The flanks are white. Tufted's influence produces a larger black bill tip than Greater or Lesser Scaup, slightly broader and less oval in shape, covering more than just the nail. The wing-stripe is extensively white, similar to Greater Scaup and Tufted Duck (Lesser Scaup has grey primaries). Lesser Scaup is smaller and flatter-backed than both Scaup and Tufted, with a longer, thinner neck (see below). *Female hybrids* With female or immature Greater Scaup, check the overall shape, bill pattern, eye colour, any hint of a tuft or grey in the wing-stripe: any discrepancies may indicate a hybrid. The female Tufted × Greater Scaup hybrid illustrated on p. 61 is based on two intermediate birds seen in Somerset which, although resembling Tufted in size and structure, showed obvious scaup characters (grey vermiculations across the whole body and a large white facial blaze). Head shape, extensive black on the bill tip and a completely white wing-stripe eliminated Lesser Scaup.

Lesser Scaup *Aythya affinis*

Where and when First recorded in Britain in 1987, it is now a regular vagrant, with as many as 27 in 2007. Most occur in winter and spring, and there has been a wide geographical scatter. Nearly 80% of records have related to males.

General features Lesser Scaup is obviously smaller than Greater Scaup and is slightly smaller and slighter than Tufted Duck. The most significant structural feature is head shape, but this varies significantly, depending on what the bird is doing (remember that the shape of the crown is not related to the skull shape, but is produced by elongated feathering). Compared to both Tufted Duck and Greater Scaup, it shows a high crown, often appearing distinctly domed, but at other times evenly rounded. It often shows a sloping forehead with a rather rounded peak to the forecrown, levelling out into a flat crown before a subtle peak, 'bump' or a *very slight tuft* at the rear (the latter often readily apparent when sleeping). In side-on views, there is often also a slight indentation at the back of the head, immediately below the peak, before falling away to a relatively straight or gently convex nape. When relaxed, however, the head shape can appear surprisingly similar to that of Tufted Duck, with a fairly steep forehead rising to a distinct peak on the forecrown, with an evenly rounded rear crown (this head shape is more normal in eclipse, especially before the rear crown feathers are fully grown). Its head is also narrower from front to back than Greater's (which appears obviously rounded and bulbous in comparison). Lesser, however, often shows noticeably bulbous cheeks when viewed head-on. It often looks flatter-backed than Tufted and, when the neck is extended, it shows a longer, thinner neck and a relatively small head and bill. The latter is smaller, less deep, narrower-tipped and less spatulate than that of Greater Scaup (and indeed Tufted Duck). Another key difference is the wing-stripe: whereas both Tufted Duck and Greater Scaup have an all-white stripe, Lesser has the white confined to the secondaries, the primaries appearing pale grey with an abrupt line of demarcation between the two (this feature should always be carefully checked). In fact, only the inner webs are pale grey, the outer webs being darker (and there is a thick black trailing edge to the entire wing).

Plumage *Adult male* Both species of scaup are basically black-and-white ducks with pale grey upperparts (a product of dark grey vermiculations on a white background). The following are the main differences. **1** UPPERPARTS COLOUR Lesser is slightly but distinctly darker grey above than Greater. This is because the vermiculations are coarser than Greater's, a key difference that needs to be carefully checked (but note that the coarser vermiculations are apparent only at relatively close range; at distances of more than *c.* 100m, the upperparts of the two species can look almost identical, even in direct comparison). **2** BILL Sky blue with a subtly paler area behind the black nail. Although inconspicuous, the latter should be carefully checked in order to eliminate similar-looking hybrids (particularly Tufted × Greater Scaup). Lesser shows a small, discrete, black nail and, most importantly, it is oval in shape (hybrids show more extensive black that tends to be more triangular). **3** HEAD GLOSS An often-quoted difference is that Greater has a green gloss to its head, whereas Lesser has blue. Although basically true, this 'colour' is an iridescence that varies according to the angle of viewing: while Lesser's head gloss usually looks blue, at certain angles it can look strikingly green. *Eclipse male* Similar to adult male but the vermiculated grey upperparts and flanks are much darker and more subdued, with limited brown feathering on the flanks. The head and breast are also duller and browner; some acquire a whitish spot on the lores and a large swathe of whitish across the ear-coverts. The bill becomes greyer, rendering the black nail difficult to see, and the head shape becomes more rounded. *Adult female winter* As with males, concentrate on structure and the wing-stripe. Plumage-wise, winter adult female resembles winter female Greater Scaup, being brown, lightly vermiculated with dull grey on the upperparts and flanks. Like Greater, it has a conspicuous white blaze on the face, but this varies individually. Some show a large blaze that completely surrounds the bill, but on others it is confined to two large patches on the lores. Winter females may have a bright yellow eye, but many show a distinctly duller, more orange tone. Like the male, the bill is pale blue with a small black nail. *Summer female* Like Greater Scaup, female Lesser has a distinct summer plumage that is essentially rich chocolate-brown. They retain a variable area of white or dull buff on the lores (sometimes barely visible). Like summer female Greater Scaup, it has a diffuse buff crescent at the rear of a slightly paler face. The eye turns quite dark orange in summer and the bill becomes much greyer. Structurally, summer females often look quite round-headed and this, combined with their lack of obvious plumage features, renders them very difficult to pick out from Tufted Ducks. *Juvenile* In keeping with other *Aythya*, juvenile plumage resembles adult female's summer plumage, being predominantly brown with a small buff or whitish patch on the lores and a diffuse buff patch on the ear-coverts. Juveniles appear slightly but distinctly paler brown than summer female and note in particular that the breast is also rather a pale brown (distinctly darker on Tufted). The bill is dark grey with the black on the nail inconspicuous, but it may extend to either side of the nail and around the tip of the bill (this may produce an obvious black bill tip on some transitional juveniles/first-winters). They also show very dull orangey eyes. Like other *Aythya*, juveniles are likely to be finely and regularly mottled on the belly (adult has a white belly patch). *First-winter male* Whereas adult males moult quickly out of eclipse in late autumn and are in full plumage by early winter, juvenile males have a much more leisurely and protracted moult, retaining variable amounts of brown juvenile feathering well into the New Year (some until late March). Remember that some transitional individuals may show quite extensive black on the bill tip. By late winter, males may be difficult to age, but some adults (not all) show subtly vermiculated black-and-white inner tertials. *First-winter female* Delicate grey vermiculations appear on the upperparts and flanks during the winter, and an area of white also develops at the base of the bill (varying from a dull off-white patch confined to the lores to a large white blaze). **Confusable hybrids** See also Greater Scaup (see p. 59). The biggest pitfall is presented by hybrids, particularly between Greater Scaup and Tufted Duck, which are relatively frequent. One such bird in Somerset differed from Lesser as follows. **1** Upperparts *delicately* vermiculated and thus appeared more 'solidly' grey than Lesser. **2** It had a slight but distinct tuft on the rear crown. **3** It always showed a strong green gloss to the head, never blue. **4** Although small, the black on the nail clearly extended either side, producing a small triangular-shaped area of black, rather than the discrete oval of Lesser Scaup (but transitional juvenile/first-winter Lesser can show a more extensive black tip). **5** It had a complete white wing-stripe, rather than the grey and white of Lesser.

Ring-necked Duck *Aythya collaris*

Where and when First recorded in 1955, but now a regular vagrant, currently averaging 31 records a year, with a peak of 52 in 2001. Juveniles and first-winters often arrive in October/November but there may be another peak in spring as vagrants move north from s. Europe; however, records occur throughout the year and some individuals remain for many years. It has occurred in small flocks, the record being eight (but up to 15 in Ireland).

General features Similar in size to Tufted, but with a squarer body, a longer bill and a longer tail (often half-cocked). The most obvious shape difference is a pronounced peak at the back of the head, but juveniles, first-winters and even eclipse males can appear flatter-crowned (see below). Unlike Tufted Duck, the wing-stripe is always grey.

Plumage *Adult male* Easily identified by its black head, breast and upperparts, and rounded grey flanks with a striking white 'spur' at the front. The grey bill has a broad white subterminal band and a narrow white basal band. Beware of moulting male Tufted Ducks with greyish flanks. *Eclipse male* By mid June, males begin to moult into eclipse and, by August, are in full eclipse. This resembles a much duller version of full plumage but is nevertheless distinctive. The black on the head and upperparts becomes dull, the flanks dull grey, and the white 'spur' on the fore-flanks is replaced by a diffuse whitish triangular area. The black breast feathers show broad grey-brown fringes, giving an overall blackish-brown or even pale brown impression. The eye remains yellow (no eye-ring) and the bill band becomes duller (at the base, it usually retains a narrow white line above the bill, but not at the sides). A pale loral patch may develop and on some a large swathe of whitish on the ear-coverts, curving down the rear of the face from the eye. Full plumage reappears in September/October. Note that, in eclipse, the crown is not as strongly peaked as in full plumage and October birds that have not regrown their new crown feathers can show a distinctly rounded and often quite a ragged head shape, surprisingly similar to Tufted Duck. **Adult female winter** Much browner than female Tufted Duck, with rather rufous flanks that are noticeably paler at the front, mirroring the male's white 'spur'. The facial pattern is distinctive: a dark crown (sometimes quite blackish), a greyish face, conspicuously pale buffish lores and throat, a dark yellow eye (appears brown at any distance) and a prominent pale eye-ring, usually with a pale extension (or 'tear-line') running back towards the ear-coverts. The facial pattern is thus more suggestive of Pochard than Tufted Duck. Some, perhaps older females, show white flecking on the face and nape. There is an obvious white subterminal bill band. **Adult female summer** The white bill band becomes duller and less obvious, and the face pattern more subdued; it shows a white eye-ring but lacks the obvious white tear-line behind the eye. It may show white flecking on the rear crown. **Juvenile** Similar to adult female but neat, fresh and immaculate with strong buff tones to the pale areas on the head and a neatly mottled breast and belly. **First-winter male** By early October, juvenile males start to acquire pale eyes (but duller than adult's) and blackish head feathering (plus white at the base of the bill). Despite this, they may be surprisingly inconspicuous among Tufted Ducks, particularly since the crown is less peaked than the adult's. By early November, the mantle and breast turn black and grey flank feathering gradually starts to predominate, contrasting with the whiter 'spur' developing on the fore-flanks (although this can be slow to appear). By mid to late winter, it is similar to the adult, but the bill band is often narrower and less well-defined.

Confusable hybrids Not a great problem with Ring-necked as most vaguely similar male hybrids (Tufted × Pochard and Tufted × Greater Scaup) are grey-backed. However, Ring-necked Duck × Tufted Duck have occurred. Males are black-backed, but they fail to adequately show all the obvious Ring-necked features (such as a prominent bill band and a well-defined flank 'spur'). Obvious Tufted characters, particularly a hint of a tuft and white in the wing-stripe, are instant giveaways. Female Tufted × Pochard may cause problems, but attention to structure and the wing-stripe should reveal any discrepancies.

Ferruginous Duck *Aythya nyroca*

Where and when A rare and declining species in its e. European, Asian and N. African breeding range, but it remains an annual winter visitor here, currently averaging 12 a year with as many as 21 in 1987. Most occur from August to April, with a distinct peak in November, but records have occurred in all months. A pair attempted to breed in Somerset in 2003–06 and possibly also in Norfolk in 1992–93.

General features A compact duck, slightly but noticeably smaller than Tufted, with a rather steep forehead, a pronounced central peak to the crown and a rather slender bill. When active, the crown may appear flatter, with a peak towards the rear; it may also look round-headed at times, particularly in eclipse. The tail appears broader, fuller and more rounded than Tufted's. It shows beautiful reddish-chestnut tones in all plumages but most distinctive is a large and striking white undertail-covert patch (in males, this is framed and emphasised by black on the rump and rear flanks). Note, however, that the white is obvious only when the bird is resting with its tail raised: when the tail is held flat on the water (e.g. when diving) the white usually appears as two round spots either side of the rear end (and may disappear altogether). Also note that many female Tufted Ducks also show obvious white undertail-coverts (particularly in autumn) but they are never as extensive or as conspicuous as those of Ferruginous. The white wing-stripe is striking, recalling that of Red-crested Pochard *Netta rufina*. It is more extensive than Tufted's, extending right across the outer primaries (only the outer two are significantly darker) and it contrasts strongly with the blackish wing-coverts and narrow black trailing edge. Note also that the axillaries and underwings are also strikingly white, with blackish along the leading coverts and trailing edge. Adults (but not juveniles) show a contrasting white belly patch (obvious in flight) but some summer males and females may show heavy dark mottling or barring on the belly.

Plumage *Adult male* Bright reddish-chestnut head and breast, smooth and silky in appearance, almost as if highly polished (but can look browner in dull light). A diffuse black ring extends from the nape, forwards around the front of the neck (best seen when neck extended). The upperparts are black. The flanks are often slightly browner and, in fresh plumage, males may be distinctly cinnamon on the flanks (owing to pale feather tips that gradually wear off, revealing darker chestnut-red bases). Its striking white eye is conspicuous (but less so in dull light). The bill is generally blue-grey (blacker on some) fading to a pale grey subterminal area, before a small black tip (restricted mainly to the nail). *Eclipse male* Moults into eclipse in June/July and regains full plumage by late August/September. This is similar to full plumage but duller, lacking strong chestnut tones to the head, breast and flanks, and has a slightly browner mantle. Some show noticeable dark barring or mottling on the belly. The bill may appear uniformly blackish. It retains its white eye and white undertail-coverts. *Adult female winter* Chestnut-brown, but duller and browner than the male, with a dark brown eye. The undertail-coverts are slightly less striking as they are bordered with dark brown, not black. A large but subtle dark orange loral patch is often present, but is not conspicuous. The bill is blackish, with a faint whitish subterminal band before a diffuse dark tip (often more extensive than the male's, extending either side of the nail). *Adult female summer* Varies from rather dark chocolate-brown to chestnut-brown, sometimes with a slightly capped appearance and subtly paler areas on the lores and ear-coverts. The breast and flanks may appear mottled, a result of brown feathers fringed reddish-brown. The white belly may show dark barring. The bill is darker, blackish-grey, with the pale band less distinct or even lacking. *Juvenile* Duller and browner than the female, including the upperparts. It lacks the pure white undertail-coverts, brown flecking creating a rather dirty and messy appearance. The best ageing feature is the belly, which is buff, neatly and regularly speckled with brown (white on adult, but barred/mottled on some in summer). The bill is entirely dark blackish-grey, the dark nail not standing out from the rest of the bill. The eye is dark but males acquire whitish eyes at least as early as September. *First-winter male* Following late autumn/early winter post-juvenile body moult, difficult to separate from adults by midwinter, although at least some first-years retain traces of dark mottling on the belly. Transitional juvenile/first-winters may show an intermediate bill tip pattern, with more extensive black on the tip.

Confusable hybrids *Ferruginous Duck × Pochard* Surprisingly frequent, being more numerous in Britain than pure Ferruginous (such birds are thought to originate in E. Europe and inherit Pochard's westerly migratory instinct). There are two distinct types: 'Redhead type' and 'Ferruginous type'. The former is dealt with below (see Pochard and Pochard-like hybrids). The latter are very similar to pure Ferruginous, so much so that they present significant

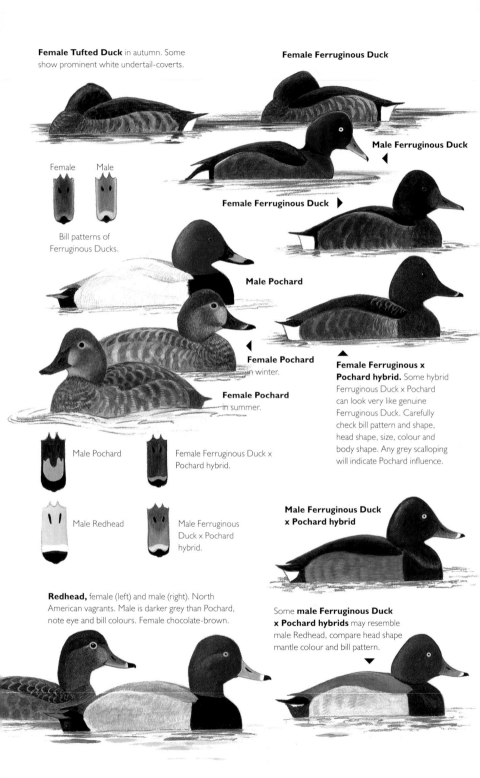

Female Tufted Duck in autumn. Some show prominent white undertail-coverts.

Female Ferruginous Duck

Male Ferruginous Duck ◀

Female Ferruginous Duck ▶

Female   Male

Bill patterns of Ferruginous Ducks.

Male Pochard

Female Pochard in winter. ◀

▲ **Female Ferruginous x Pochard hybrid.** Some hybrid Ferruginous Duck x Pochard can look very like genuine Ferruginous Duck. Carefully check bill pattern and shape, head shape, size, colour and body shape. Any grey scalloping will indicate Pochard influence.

Female Pochard in summer.

Male Pochard

Female Ferruginous Duck x Pochard hybrid.

Male Redhead

Male Ferruginous Duck x Pochard hybrid.

Male Ferruginous Duck x Pochard hybrid

Redhead, female (left) and male (right). North American vagrants. Male is darker grey than Pochard, note eye and bill colours. Female chocolate-brown.

Some **male Ferruginous Duck x Pochard hybrids** may resemble male Redhead, compare head shape mantle colour and bill pattern. ▼

identification problems. When faced with a potential Ferruginous, pay particular attention to the head and bill shape, overall size, bill pattern, undertail-coverts and wing-stripe. Hybrids reveal their Pochard parentage by a rather sloping forehead, a longer, deeper-based bill and, consequently, a more wedge-shaped head and bill profile; they are also larger and bulkier than Ferruginous. They may show faint grey vermiculations on the scapulars, possibly greyer flanks, whilst the undertail-coverts may have extensive dark feathering (juvenile Ferruginous should lose such markings by midwinter). Hybrids have a larger black bill tip, sometimes with a noticeable pale subterminal band curving back towards the bill sides; the wing-stripe usually shows a certain amount of grey. *Ferruginous Duck × Tufted* Genuinely rare compared to Ferruginous Duck × Pochard. A male in Somerset closely resembled Ferruginous in structure (but with a more pointed rear crown at rest), white eye colour, extensive white wing-stripe and white undertail-coverts (although the latter showed a small brownish triangle just above the waterline). The plumage, however, was distinctly dark brown and it had a larger black bill tip. A female also in Somerset resembled Ferruginous in its structure, reddish plumage tones, large white undertail-coverts and striking white wing-stripe, but it also showed obvious pro-Tufted features, such as a chocolate tone to the head and body, a dark yellow eye and a large black bill tip with a broad whitish subterminal band, as well as structural anomalies.

Pochard-like and Redhead-like hybrids

Pochard-like hybrids are also frequent. Their parentage is not always immediately apparent, but most appear to be crosses with Ferruginous Duck. They tend to resemble male Pochards but are slightly smaller with more of a domed head shape. They are darker grey on the upperparts and flanks and, most significantly, they often show a strong burgundy tone to the breast (inherited from Ferruginous). The eye is yellowish or orangey and the bill is paler and bluer grey than Pochard's, often crossed with a pale subterminal band. Such features strongly suggest Redhead *A. americana*, a North American species, recorded only once in Britain. Male Redhead is similar to Pochard but is darker grey, with a steep forehead, a very rounded head, a yellow eye and a different bill pattern (see below). Unlike female Pochards (greyish in winter plumage) female Redheads are light chocolate-brown all year, with a darker crown, a distinct pale eye-ring and whitish lores (reminiscent of female Ring-necked). Many females show white flecking on the nape. Juveniles are similar to females but show immaculate plumage. Pochard-like hybrids that resemble Redheads have more of a wedge-shaped head-and-bill profile (head often domed or with a strong peak on the forecrown), a reddish or orange tint to the eye, and a whitening of the wing-stripe. The bill pattern is important: male Redhead has blue-grey bill with a narrow white subterminal band and large black 'dipped-in-ink' tip (female's is similar, but with a blacker base); hybrids generally have a darker-based bill, a black tip that is curved on its inner edge, and a more U-shaped or irregular pale subterminal band.

Reference Hudson *et al.* (2012).

Eiders: females, immatures and eclipse males

Where and when Common Eider is a numerous but declining breeding bird around the coasts of Scotland, n. England (Cumbria and Northumberland) and Northern Ireland. In winter it is common in Scotland and much more widespread around English and Welsh coasts (but rather local in the south). It is extremely rare inland. King Eider is a regular visitor from the Arctic, mainly in winter. Almost 70% have occurred in e. Scotland, from Shetland to the Firth of Forth. It is very rare in England with just one record in Wales. About 80% of records have been males. Steller's Eider is a great rarity with 15 records, mostly in the Northern Isles (last in 2000).

Common Eider *Somateria mollissima*

Structure A large, bulky sea duck with a strikingly wedge-shaped head and bill profile. It dives with a characteristic open-wing action. It appears fat and heavy in flight with rather short, broad-based wings, a short neck and a large wedge-shaped head (often held low, contributing to a hump-backed impression). It usually flies low over the sea in lines or disorganised bunches.

Plumage *Adult female* In many ways suggests a female Mallard *Anas platyrhynchos*, being completely orangey-brown (including the belly) heavily barred black (quite fine on head and breast) and having a faintly paler eyebrow. Like Mallard, it has a dark speculum, narrowly bordered with white. The bill and frontal lobes are grey, the former with a dull green nail. *Juvenile* Similar to adult female but rather oily, dark chocolate-brown, immaculately patterned with dark buff feather fringes, producing a finely barred or mottled appearance (thus appears darker and more uniform than adult female). Note that the tips to the greater coverts and secondaries are dark buff and inconspicuous (obviously white on adult female). The bill and frontal lobes are dark greenish-grey and the head fairly plain. In flight, it lacks obvious pale borders to the black speculum, so the wing appears plain. *First-winter male* From their first autumn, juvenile males start to acquire variable amounts of adult-like feathering. Blackish appears on the head, back, scapulars and flanks (but a speckled brown belly is retained); the most distinctive feature, however, is a gradually developing white breast. The bill is dull lime-green and a pale stripe curves from the bill over the eye. By their first summer, they usually show a completely white breast and white feathering on the back and scapulars. They retain dark wing-coverts until their first-summer moult. *Second-winter male* Much more adult-like but usually retains slight traces of immaturity. *Eclipse male* Appears uniformly dull black (browner on breast) but with a variable thick, pale eyebrow or a large swathe of whitish-buff over the eye; also a pale greenish-grey bill and frontal lobes. It retains extensive white in the wing-coverts and, when not moulted, the tertials.

Northern Eider The northern race *borealis* occurs in the Arctic from ne. Canada east to nw. Russia and is a rare visitor to n. Scotland. Most are difficult to separate from the nominate but males from w. Greenland and Baffin Island can be distinguished by their bright orange-yellow frontal lobes, bill, legs and feet (these areas are usually dull greenish-grey on the nominate). They also show two, small triangular white 'sails' on their back (stiff modified scapulars, similar to those of male King Eider) as well as strongly decurved tertials. However, the identification of such vagrants may be clouded by the existence of intergrades.

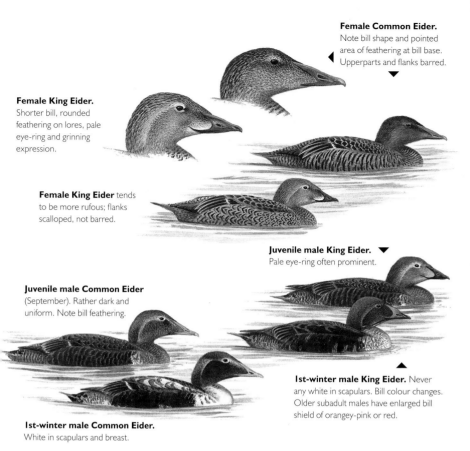

Female Common Eider. Note bill shape and pointed area of feathering at bill base. Upperparts and flanks barred. ▼

Female King Eider. Shorter bill, rounded feathering on lores, pale eye-ring and grinning expression.

Female King Eider tends to be more rufous; flanks scalloped, not barred.

Juvenile male King Eider. ▼ Pale eye-ring often prominent.

Juvenile male Common Eider (September). Rather dark and uniform. Note bill feathering.

1st-winter male King Eider. Never any white in scapulars. Bill colour changes. Older subadult males have enlarged bill shield of orangey-pink or red.

1st-winter male Common Eider. White in scapulars and breast.

Dresser's Eider Males of the race *dresseri* (North America from Labrador to Maine) have large, rounded frontal lobes that almost reach the eye (plus a line of green running horizontally below the black cap). The lobes vary from greyish, through greenish to bright orange. This form also has white sails, a shorter bill and a steep, highly peaked forehead (Garner 2008). A male was seen in Co. Donegal, Ireland, in 2011.

King Eider *Somateria spectabilis*

General Slightly smaller, shorter-billed, squarer-headed and more compact than Common Eider. Males have a distinctive large frontal shield over the bill (reduced in first-winters and in eclipse). In flight they appear shorter-necked and blunter-headed than Common.

Plumage *Adult female* Amongst a flock of Common Eiders, appears slightly smaller, more compact and more orangey-looking. The important area on which to concentrate is the head. Common Eider has a rather rounded head that merges into a long, wedge-shaped bill, creating a very distinctive 'long-nosed' profile, even at a distance. Female King Eider has a shorter bill and, when relaxed, the forehead slopes up more steeply from the bill to a peak on the forecrown, immediately above the eye. A flat crown then slopes gently backwards before turning and dropping vertically down the nape. Consequently, female King Eider has a 'square-headed/short-billed' profile. The key feature on which to then concentrate is the pattern of the feathering at the base of the bill. On female Common, a long pointed wedge of head feathering intrudes into the bill base. On female King Eider, this feathering is shorter and distinctly rounded at the tip (although, confusingly, King Eider can occasionally show a rather pointed wedge). The tip of the feathering is whitish, often giving the impression of a pale spot at the base of the bill. Two other features are also central to the identification. **1** EYE-RING There is a variable pale eye-ring that is often broader above the eye, producing an 'eyebrow' (some adult females show prominent white eye-rings with a conspicuous whitish 'tear-line' curving back behind the eye). **2** GAPE LINE The gape line curves upwards as it extends back into the facial feathering, producing a characteristic 'grinning' expression. Coupled with the eye-ring, this produces a pleasant and rather comical look. The body plumage is warmer brown than female Common, having a distinct orangey or fulvous tone. This is particularly strong on the head and scapulars (the latter have black feather centres with broad orangey fringes, forming U-shaped scallops). The flanks are similarly coloured but with thick black V-shaped barring; the breast is more faintly mottled with black. *Juvenile* Somewhat plainer, browner and less fulvous than adult female, and slightly darker-headed. It shows only a narrow white border to the front and rear edges of the speculum (adult female usually has a broader white border); since King Eiders open their wings to dive, this difference may be readily apparent before they submerge. First-winter plumage is gradually acquired during the course of the winter and, by late winter, females should show adult-like black-centred scapulars with broad orangey U-shaped fringes. *First-winter male* Like Common, juvenile male King Eiders moult almost continuously through their first winter but the rate at which they acquire adult-like features varies individually. They gradually show a dark chocolate-brown or sooty-black head and body, a white breast, a variable whitish rear-flank patch and an orange-yellow base to the pink bill (by spring, the base may start to swell into a yellow shield). Many also show white on the throat and foreneck, separated from the white breast by a blackish neck-ring. All such birds retain a heavily mottled brown juvenile belly and completely dark brown wing-coverts. Their post-juvenile moult ends when they commence their moult into eclipse in their first summer. *Second-winter male* Much more adult-like, but readily distinguished by the smaller shield, a darker grey crown, whitish rather than pale green cheeks, and the lack of a pink flush to the breast. They can be separated from exceptionally advanced first-winters by their black belly and at least partially white upperwing-coverts, which are not acquired until their second winter. By the end of the winter, they are even more adult-like, but they can still be distinguished by their slightly smaller shield, by the presence of a thin black vertical line down the centre of the nape and by the restricted and/or rather mottled white upperwing panel (Dawson 1994, Ellis 1994). *Eclipse male* In early summer, males show a great deal of variation as they enter eclipse. First-years moult first, followed by males that have failed to breed and then by breeding males, which commence their moult once their mates have started laying. In eclipse, males show a black body, a messy blackish head and a dingy whitish breast. Adults show a white wing-covert patch (usually obvious when diving) and retain a dull red bill (shorter than Common) with a somewhat shrunken orange shield.

Steller's Eider *Polysticta stelleri*

Being only about two-thirds the size of Common Eider (and only the size of Goldeneye *Bucephala clangula*) this eider is not confusable with the other two. It also has a squarer head, a more spatulate grey bill and a long, pointed tail that is cocked-up at rest like a Common Scoter's

Melanitta nigra. Males in full plumage are stunning, but females are dark chocolate-brown with a pale eye-ring, curved blue tertials (tipped white) and a white-bordered blue speculum, like a Mallard. Juveniles have a weaker white border to the blacker speculum and lack the curved tertials. First-year males gradually acquire subdued male-like features by their first summer. Adult males in eclipse show a white forewing.

References Cramp & Simmons (1977), Dawson (1994), Ellis (1994), Garner (2008).

Scoters

Where and when Outside the breeding season, essentially marine. Only Common Scoter breeds here (small numbers in Scotland, also Ireland) but from late summer to spring it is locally common around coasts, with huge concentrations in Liverpool, Carmarthen and Cardigan Bays. Velvet Scoter is much rarer and is declining, so much so that it has now been listed as Endangered on the IUCN Red List of Threatened Species. Often found in Common Scoter flocks, the largest concentrations being in e. Scotland (particularly the Moray Firth and the Forth Estuary). Small numbers of Common Scoters occur inland on migration, particularly in March/April, mid June to August (latter mainly males on moult migration) and in November. Velvet is very rare inland, but may appear in late autumn, winter and spring, particularly during freezing weather. Surf Scoter, a very rare North American visitor, currently averaging 19 records a year, has occurred on most coasts (sometimes in small parties) often in Common Scoter flocks, with a few inland records. Two other N. American and E. Siberian species have recently been split: Black Scoter is the equivalent of Common Scoter (ten records to 2011) and White-winged Scoter is the equivalent of Velvet. The latter species has two races: American *deglandi* (one record in Scotland in 2011) and E. Asian *stejnegeri* (one record in Ireland, also in 2011).

Ageing and sexing Scoters vary in size and males average distinctly larger than females. Identification of adult males is straightforward, but females and immatures are more similar. Juveniles resemble adult females but at close range show immaculate plumage. Unlike most ducks, juvenile scoters are distinctly whiter on the belly than adults, not darker, showing a large ill-defined but obvious buffy-white belly patch (delicately speckled at close range). Young males gradually acquire adult bill colours during their first winter, so the bill often appears much duller than the adult male's. Black feathering also appears during the first winter, but the progression varies individually; a whitish belly patch is retained until a complete moult in late summer and this allows easy ageing of distant birds, even in flight. Eclipse males are duller than full males.

Common Scoter *Melanitta nigra*

General features Often gathers in large rafts and may migrate low over the sea in tight bunches, straggling flocks or long lines. Generally the smallest and slightest scoter, compact with a round head (but females often with a steep forehead), a thinner neck and a smaller bill than the other species. The tail is relatively long and pointed, and is held cocked when asleep. In flight, appears stocky and relatively short-necked. The wings are plain and dark but, on the underwing, the remiges and greater coverts appear pale silvery-grey; both sexes show this feature but, because of the contrast with the black underwing-coverts and belly, it is especially obvious on adult males, even at a distance. It usually dives with closed wings (Velvet and Surf usually dive with open wings). When wing-flapping, it has a unique downward S-shaped head movement.

Plumage *Adult male* Entirely black. The bill has a prominent strip of bright orangey-yellow on the upper surface, just extending onto a round black knob at the base, which creates a very distinctive profile, even at a distance. *Female and juvenile* Readily distinguished from other scoters by the well-defined pale, brownish-white face. Juveniles may also show a faint, ill-defined vertical greyish line (or 'strap') around an otherwise paler, whitish face. *First-winter male* See 'Ageing and sexing' above. When moulting out of juvenile plumage, they can show dusky patches on the head-sides, which can suggest Velvet and Surf Scoters.

Velvet Scoter *Melanitta fusca*

General features The largest, bulkiest scoter with a long, wedge-shaped head and bill profile. The tail is proportionately shorter than Common Scoters. Unlike other scoters, it has stunning white secondaries in all plumages. However, these may not be visible at rest but instead may appear as a small white triangle or as a narrow crescent on the rear of the wing. When diving, it usually (but not always) opens its wings like a Common Eider *Somateria mollissima* (but Common will also occasionally do this). Velvets tend to be more approachable than Common and, on occasion, exceptionally tame.

Plumage *Adult male* Most of the bill is bright orangey-yellow, and it has a white crescent-shaped patch under the white eye. When visible, the legs and feet are pink, becoming bright red in breeding condition (black in Common). *Female and juvenile* Unlike Common, they have two large whitish facial spots, the one before the eye usually larger, more diffuse or buffier; the one behind usually smaller, rounder, whiter and more sharply defined. However, the exact pattern varies individually; the spots may even run together, whilst some females, particularly first-winters, virtually lack them. The legs are pink, duller than the adult male's. *First-winter male* See 'Ageing and sexing' above. By January, advanced first-winter males appear uniformly dull black (with variable amounts of brown feathering) but they lack the adult's white eye-crescent and have a duller bill.

Surf Scoter *Melanitta perspicillata*

General features Rare and should be identified with caution. Averages larger and bulkier than Common, being stockier with a short, thick neck. It is superficially similar to Velvet but, in direct comparison, distinctly smaller (about four-fifths the size) with a shorter body and a more rounded back. A heavy, deep-based bill produces a wedge-shaped head and bill profile (recalling Common Eider) with a sloping forehead, flat crown and fairly vertical nape. In direct comparison with Velvet, the head shape is somewhat squarer and more angular. Like Velvet, it usually opens its wings to dive (check for all-dark wings).

Plumage *Adult male* Easily identified by its large white patch on the nape and a smaller one on the forehead. The large bill is intricately patterned with red, white and yellow, appearing mainly orange at a distance. Like Velvet, the male has pink or red legs. *Female and juvenile* Dark crown contrasting with a slightly paler face, particularly on juveniles, produces a slight

Adult male Common Scoter

Diagnostic head jerking flap.

1st-winter male Common Scoter

Adult female Common Scoter

Compare head and neck shapes, and bill, tail and body lengths.

1st-winter male Velvet Scoter in advanced plumage. No white fleck behind eye as in adult male. Wing-bar not always visible at rest.

Female Velvet Scoters. Note variable head patterns.

◀ **1st-winter male Surf Scoter.** Note head pattern and bill colour.

Juvenile Greater Scaup. Sits higher on water than scoters. Note short tail.

▲ **1st-winter female Surf Scoter**

capped effect. Juveniles and most adult females show two large whitish spots on the face (the rear spot usually tear-shaped and on some juveniles these may run together). Some adult females show a large whitish patch on the nape. *First-winter male* See 'Ageing and sexing' above. May start to show adult bill patterning by December (brightening towards spring) when advanced birds also show the diagnostic white nape patch. However, the white forehead is not attained until the second winter.

Other confusion species Beware of confusing Velvet and Surf with immature Eider and juvenile Scaup *Aythya marila* (see pp. 70 and 59) or inland female and juvenile Common with female and juvenile Red-crested Pochard *Netta rufina* (latter paler with huge white wing-stripe).

Black *Melanitta americana* and White-winged Scoters *M. deglandi*

Both species require good views and a *detailed* description of the bill for acceptance; expert advice should be sought.

Black Scoter Similar to Common but, instead of Common's black knob on the bill, Black's base is gently swollen and extensively bright orange-yellow. Consequently, it stands out like a Belisha beacon amongst accompanying male Commons. In direct comparison, Black is also slightly but distinctly larger, bulkier and larger-headed. Realistically, only males are likely to be separable in the field but Garner (2008) indicates that female and juvenile Black have a gently swollen bill base, an arched nail in a side-on view and the brown on the nape is squared-off in a back-on view (pointed on Common).

American White-winged Scoter *deglandi* Adult male is similar to Velvet but has a swollen base to the bill with a distinct 'step' between the base and the tip. The bill colour is a striking mixture of bright orange-yellow on the tip, white on the top and orange-yellow and bright pink on the sides. It has a longer and more prominent white eye crescent than Velvet, and brown flanks contrast with the rest of the plumage. Female and juvenile identical to Velvet but show a slightly swollen bill base at very close range.

Asiatic White-winged Scoter *stejnegeri* As American, but bill reddish-pink with yellow around the tip and a peculiar forward-pointing knob at the base. Flanks black.

Reference Garner (2008).

Goosander and Red-breasted Merganser

Where and when Both species breed in Scotland and N. England. Red-breasted Merganser *Mergus serrator* also breeds in N. Wales and in Ireland, while Goosander *M. merganser* now also breeds in central Wales and Devon. In summer, both occur on freshwater rivers and lakes, but mergansers also commonly breed in coastal areas. In winter, mergansers are widespread but mainly coastal, whereas Goosanders occur mainly on fresh water (although in Scotland, where it is most numerous, large concentrations are estuarine). Goosander generally becomes scarcer in S. Britain and is rare in Ireland. The ecological split is a good clue to their identity, but note that exceptions occur: migrating mergansers sometimes occur inland, particularly in late autumn and early winter, and also during periods of severe freezing, while even in southern areas the occasional Goosander may appear on the coast.

Structure Both are long, slim, thin-billed, fish-eating ducks. Goosander is much the larger and bulkier (10–20% bigger) with a large head, a thick neck and a large, full crest. Compared to Goosander, merganser is distinctly smaller, slimmer and thinner-necked; its crest is thinner and wispier, generally forming two distinct tufts on the back of the head (Goosander has a fuller, more even 'lump'). Despite its smaller size, merganser has a longer, thinner bill, which shows more of a gentle up-curve.

Plumage Full-plumaged males are easily separated. The best way to separate females and immatures is by the lower border of the reddish-brown of the head: on Goosander, it ends

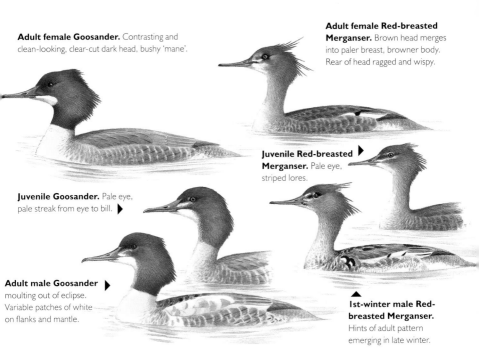

Adult female Goosander. Contrasting and clean-looking, clear-cut dark head, bushy 'mane'.

Adult female Red-breasted Merganser. Brown head merges into paler breast, browner body. Rear of head ragged and wispy.

Juvenile Red-breasted Merganser. Pale eye, striped lores.

Juvenile Goosander. Pale eye, pale streak from eye to bill. ▶

Adult male Goosander ▶ moulting out of eclipse. Variable patches of white on flanks and mantle.

1st-winter male Red-breasted Merganser. Hints of adult pattern emerging in late winter.

abruptly, forming a clear demarcation with the obviously whitish breast. On merganser, it gently merges to create a dingier, more uniform impression to the front of the neck. The upperparts and flanks of Red-breasted Merganser also show a distinct brownish cast, compared to the clean grey plumage of Goosander, rendering the merganser a much duller bird in direct comparison. Also, Goosander shows a clear-cut white throat patch, whereas merganser has a more pronounced whitish loral line, a variable pale eye-ring and a dull throat (although note that juvenile Goosander also shows a whitish loral line, which may persist into the winter). Whereas 'brownhead' Goosanders have brown or dull yellow eyes (see below) adult female mergansers have red eyes, often with variable amounts of black surrounding them (as well as on the lores). Eclipse males of both species resemble females but retain the large areas of white on the forewings (which females lack) and so are easily sexed; eclipse male Goosanders show a diffuse white line down the front of the otherwise reddish-brown neck. Some early migrating males arrive in the winter quarters still in eclipse plumage, so males showing a mixture of 'full' and 'brownhead' plumage in October/November will be adults moulting out of eclipse. *Juveniles* Both species have a noticeable whitish line from the bill back below the eye, with a whitish ring above the eye. These are gradually lost by late autumn but may be detectable into midwinter. Both initially lack a crest, and some do so even into the New Year. Goosanders show dull yellow eyes (brown on adult females) while juvenile mergansers show dull orange or brown eyes (red on adult females, like males). Juvenile/first-winter males of both species gradually acquire traces of adult plumage during their first winter, but the amounts attained are generally small (particularly with Goosander) even by late winter and early spring.

Diving action Mergansers *may* dive with a more energetic, hump-backed action.

Divers

Where and when In winter, Red-throated Diver is the most widespread and, generally, the most plentiful diver, particularly along the east coast. Black-throated Diver is the least numerous and most localised (mainly in Scotland but with a large concentration in Gerrans Bay, Cornwall). Both species breed in Scotland, Red-throated being by far the commoner. Great Northern Diver is most numerous in the north and west, particularly in Scotland and Ireland. All three occasionally occur inland, Great Northern being the most frequent, from November onwards. White-billed Diver is a very rare winter visitor, mainly in Scotland, with a regular early spring gathering on the Outer Hebrides, principally off Lewis, and another off the north coast of Aberdeenshire. Pacific Diver was first recorded in 2007 but there have been another four records since (to 2009).

Diving actions A very easy distinction from Cormorant *Phalacrocorax carbo* and Shag *P. aristotelis* is that divers do not leap out of the water to dive; neither do they have a long tail. Instead, they gently lurch the head forward and submerge gracefully with a smooth action. Diving may follow a period of underwater surveillance (or 'snorkelling') with the bill and head partly submerged in the water.

Flight identification All species have a very distinctive shape in flight: a 'humpbacked' appearance with the head and neck held slightly below the horizontal and the feet trailing behind. They often fly well above the horizon, sometimes in small groups (although Great Northern tends to be more solitary). Unlike most waterbirds, divers (and grebes) do not push their legs forward when landing; instead they drop belly-first onto the water, often skidding across the surface before settling. All have plain wings, which instantly separates them from grebes. Separating diver species in flight can be difficult and distant or doubtful individuals should be logged as 'diver sp.'.

Summer plumage Note that newly arriving late autumn/early winter adults and spring adults may show variable amounts of summer plumage, very useful with migrating birds on seawatches. With the exception of Red-throated, first-summer divers do not show significant amounts of summer plumage, often becoming bleached.

Red-throated Diver *Gavia stellata*

Size, structure and bill The smallest diver, although it overlaps in size with Black-throated. In direct comparison, it is not much larger than Great Crested Grebe *Podiceps cristatus* (although *c.* 50% heavier). Its shape is usually distinctive, with a rather rounded back, shallow breast, full throat and rounded head. The bill is slender and usually held upwards, forming a continuous line with the throat and accentuating an upcurved lower mandible (although the upcurve can be slight). Bill generally pale in winter, appearing greyish at any distance, but may darken towards spring.

Plumage Much browner than the other species, with the crown and nape often appearing a rather dusky grey-brown; however, the upperparts can look dark at any distance. Has diagnostic white upperpart flecking (also difficult to see at any distance). The extent of white on the face is variable, but it generally extends up and around the eye (usually most obvious before the eye); classic individuals, with the eye isolated in the white face, appear rather white-headed from a distance. Compared with Black-throated, the demarcation between the dark nape and the white neck is further to the rear, producing a narrower nape line when viewed from behind. The amount of white visible on the flanks depends on the bird's attitude on the water; when at ease, it generally shows a strip of white along the entire length of the flanks. *Juvenile* Before moult (generally in midwinter) juveniles are fresh and immaculate with variable amounts of fine streaking on the sides of the head and neck, obscuring the line of demarcation. On extreme examples, the whole head can look greyish from a distance, contrasting with the darker body. Juveniles may also show a dark chestnut patch on the upper foreneck, mirroring the adult's summer plumage. They become difficult to age once juvenile plumage is lost in late winter. However, adults begin a complete post-breeding moult in early winter, so any Red-throated in active primary moult at that time will be an adult.

In flight Appears small, slim and slight with small feet producing a rather tapered rear end (foot projection about half the head and neck length). Compared to Black-throated, winter birds have browner upperparts, usually with obvious and extensive white on the face (but juveniles are often dusky-headed). It has a peculiar yet distinctive habit of intermittently raising its head and neck in flight.

Pitfalls If seen well, typical individuals are readily identifiable, but not all are so straightforward. Specific points to remember are: **1** Red-throated not infrequently holds its bill horizontally; **2** its head shape may, in certain postures, appear more angular; **3** in certain lights the upperparts can look very dark and contrast markedly with the underparts (provoking confusion with Black-throated) and **4** in certain positions, it can show an isolated white patch on the rear flanks, again suggesting Black-throated. Identification of distant individuals demands caution.

Black-throated Diver *Gavia arctica*

Size, structure and bill Intermediate in size between Red-throated and Great Northern (but measurements overlap with both). Very sleek, streamlined and graceful, but fuller-breasted than Red-throated. Unlike Great Northern, the head is gently, smoothly and evenly rounded, although in certain postures it can show a more abrupt forehead. The bill is usually held horizontally (occasionally at an upward angle) and is slender, pointed and straight, with the upper mandible gently downcurved towards the tip; the shape, however, is variable, some being slightly thicker-billed. The bill is grey with a dark culmen and cutting edges (less obvious than on Great Northern) but it blackens towards spring.

Plumage A crisply clean, almost auk-like black-and-white diver. The dark of the head is tinged velvety grey at close range (in late spring they can look quite grey-headed) but they usually look strikingly black and white, with a sharp and even line of demarcation running below the eye and down the middle of the neck-sides, thereby producing a larger dark nape area than Red-throated (note that Great Northern is rather dusky about the head with a dark half-collar on the sides of the lower neck). The eye-ring is thin so that, unlike most Great Northern, the eye does not stand out within the dark head. The flank pattern is a useful feature: the fore-flanks are blackish but the rear flanks are white, depending on posture often standing out as a distinctive isolated white patch, obvious even at a distance; this can be reduced to a small round spot (Red-throated may show a similar pattern in certain postures, so this feature should not be used in isolation). *Juvenile* Has noticeable pale mantle and scapular fringes until it moults in the New Year (adult much plainer). After moulting, ageing is less easy but adult (unlike Red-throated) has a complete *pre*-breeding spring moult, so any Black-throated in active primary moult in spring will be an adult.

In flight Compared to Red-throated, winter birds look much more black and white and tend to look 'long and straight' in flight. Note that the feet look long and large (Red-throated has small feet that produce a short, tapered rear end). In spring, white patterning on summer-plumaged scapulars may be surprisingly obvious, even at a distance.

Great Northern Diver *Gavia immer*

Size, structure and bill A large, heavy, cumbersome diver. Inexperienced observers may confuse it with first-winter Cormorant, but note the latter's large, full tail as it leaps to dive. Small Great Northerns do occur, provoking confusion with Black-throated, so head structure is important for separation. Great Northern has a large head with a steep forehead (often with a distinct 'bump' where the forehead meets the crown), a rather flat crown and another angle where the crown merges into the nape. However, when alarmed or in diving, feather-sleeking may produce a much more streamlined appearance. Its heavy-headed effect accentuates a heavy, deep-based bill, which usually shows a stronger gonydeal angle than Black-throated; the colour is pale grey, with a dark culmen and cutting edges.

Plumage Essentially black and white, but messier-looking than Black-throated, lacking the latter's smart contrasts: demarcation between the crown/nape and the throat/foreneck/breast is duskier and less clear-cut, showing a dark half-collar on the lower neck with a white indentation above. A pale eye-surround produces a more isolated and more obvious eye than Black-throated. Great Northern shows a long, rather uneven flank line, invisible if the bird sits low in the water. Newly arrived late autumn adults usually retain obvious traces of summer plumage (almost complete on some) and this may persist well into winter. *Juvenile* Shows prominent pale mantle and scapular fringes, producing a scalloped pattern (adult much plainer). After moulting in mid to late winter, ageing is more difficult, but moulting adults in late winter lose their flight feathers.

In flight Although size may be difficult to judge on distant flying birds, Great Northern usually looks distinctly larger, heavier and significantly more substantial than the two smaller species, with its heavy bill and large feet being prominent. The extensive dark on the head and the dark 'half-collar' on the sides of the neck may also be apparent. A useful tip is that, on spring migration, Great Northern Divers in the English Channel invariably fly high to the west towards their Canadian breeding grounds, whereas Red and Black-throated usually fly east towards Scandinavia.

White-billed Diver *Gavia adamsii*

Size, structure and bill A large, heavy diver, slightly bigger than Great Northern, with a thicker and heavier head and neck. It usually shows a distinct 'lump' where the forehead meets the forecrown. The bill is large, very long and ivory-coloured, with a straight upper mandible but a prominently uptilted lower (appears almost banana-like); it is usually held upwards, suggesting a gigantic Red-throated. Unlike Great Northern, the *bill lacks any dark along the culmen and the distal cutting edges*, although dark feathering protrudes into the base of the upper mandible to the nostril. Note: pale-billed Great Northerns occur, but they always show a dark culmen ridge and cutting edges.

Winter Black-throated Diver. Pale nape and dark sides of neck, rather black and white overall. Long legs/feet.

Summer Red-throated Diver. Head dark, neck often drooping.

Winter Red-throated Diver. Rapid flight, pale head often moved up and down. Small feet.

Summer Black-throated Diver. Pale nape and white chequering on scapulars.

Summer Great Northern Diver. White chequering on scapulars.

Winter Great Northern Diver. Large, huge feet and distinct half collar.

Adult winter Red-throated Diver. White 'face', uptilted bill, upperparts spotted white.

Juvenile Red-throated Diver. Duskier head and neck than adult, often shows reddish patch on throat. Note continuous white flank line.

Adult winter Black-throated Diver. Black and white, very contrasting, head greyer.

Slender, straight bill.

Juvenile Black-throated Diver. Pale feather edges to upperparts, grey crown and hindneck. Note white patch to rear of flanks.

Adult winter Great Northern Diver. Plain above, note heavy bill and half collar.

1st-winter Cormorant may resemble a diver. Note angular head, bill held upwards, orangey face and long tail (obvious when diving).

Juvenile White-billed Diver. Note all pale, uptilted bill (no dark along culmen). Greyish neck sides with a darker patch on ear-coverts, and half collar.

Juvenile Great Northern Diver. Note heavy, pale bill with dark culmen, and a large angular head. Extensive black on head, and dark half collar. Upperparts show fine pale feather edges.

Plumage Similar to Great Northern, but the head is rather paler, greyer and duskier (contrasting more with the dark upperparts) and it has a noticeable pale surround to the eye. As with Great Northern, winter adult White-billed often shows traces of white spotting on the upperparts. *Juvenile* Even duskier-headed than adult, often showing a diffuse dark patch on the ear-coverts (but adults may show a similar dark area); prominent pale mantle and scapular fringes, more pronounced than on juvenile Great Northern.

In flight Identification should be attempted with great caution. Like Great Northern, a large, heavy diver (thick neck and huge feet) but with a whitish 'upturned banana' bill. However, bill details are difficult to discern in flight. Summer-plumaged birds are much more distinctive as the whitish bill contrasts with the black head, but beware of Great Northerns looking pale-billed in certain light conditions.

Pacific Diver *Gavia pacifica*

Despite its name, this North American species breeds as far east as Baffin Island and it seems likely that it will to prove to be a fairly regular vagrant here. Very much the North American equivalent of Black-throated Diver. The most important difference is that *the flanks are entirely black, lacking Black-throated's large white patch on the rear flanks.* However, since the flank pattern varies according to the bird's posture, prolonged and careful observation is necessary to confirm this. In direct comparison, size and structural differences are also obvious (although remember that male divers are larger than females). A Cornish adult appeared *c.* 20% smaller than an accompanying Black-throated, with a shorter and much slimmer 'stuck-on bill'. It also showed a steeper forehead, a more rounded and blacker head and a more rounded back. Juvenile Pacific, however, has a greyer head with dusky ear-coverts (which are cleaner and whiter on juvenile Black-throated). A black 'strap-line' at the juncture of the throat and the upper neck is an important confirmatory feature (but often indistinct or even absent on juveniles). A thick black band (or 'vent strap') behind the legs, isolating the white undertail-coverts, may be noted, but this can also be shown by some Black-throated (as this is below the surface of the water, the bird will need to be seen rolling onto its side). For further information see Mather (2010).

References Appleby *et al.* (1986), Burn & Mather (1974), Mather (2010).

Cory's and Great Shearwaters

Where and when Great Shearwater is a summer visitor from its breeding islands in the South Atlantic (principally Tristan da Cunha) reaching British and Irish waters from July to October, with most in August to September; although numerous offshore, it is rarely seen from land, even during strong westerly gales. The best way to see them is by joining an organised Atlantic pelagic trip, when they can often be found following trawlers. Cory's Shearwater occurs from April to September, with most in July and August, principally off Cornwall and s. Ireland. Annual totals vary considerably: more than 5,000 in 1998 but just 87 in 2007.

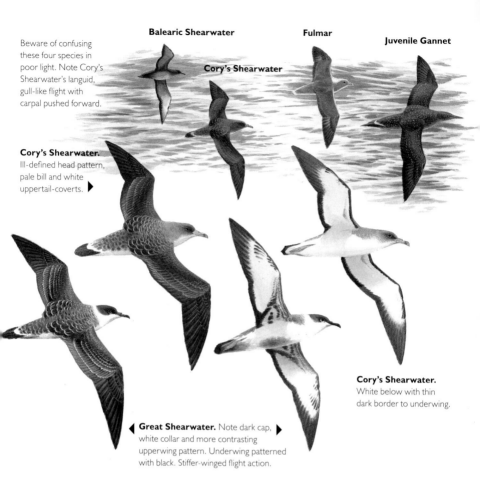

Balearic Shearwater

Cory's Shearwater

Fulmar

Juvenile Gannet

Beware of confusing these four species in poor light. Note Cory's Shearwater's languid, gull-like flight with carpal pushed forward.

Cory's Shearwater. Ill-defined head pattern, pale bill and white uppertail-coverts. ▶

Cory's Shearwater. White below with thin dark border to underwing.

◀ **Great Shearwater.** Note dark cap, ▶ white collar and more contrasting upperwing pattern. Underwing patterned with black. Stiffer-winged flight action.

Cory's Shearwater *Calonectris borealis*

Size and shape About the size of Lesser Black-backed Gull *Larus fuscus*, Cory's is the largest European shearwater, being slightly larger than Great with slightly longer wings in direct comparison.

Plumage Rather featureless, leading to its epithet 'the Garden Warbler *Sylvia borin* of the oceans'. Brown above and white below with a narrow brown border to the underwings. The two main differences from Great are the dusky-grey sides to the head, which gradually merge into the white underparts, and the pale yellow bill, which in good light is obvious at considerable distances. Other features include an inconspicuous narrow white horseshoe-shaped uppertail-covert patch, immediately in front of the tail, and it can also show a faint dark 'W' across the upperwings, a result of slightly darker primaries and rear wing-coverts.

Flight Much more distinctive is its shape and flight. It flies on distinctly down-bowed wings with the carpals pushed forward and the primaries angled backwards. It has a slow, lazy and rather languid flight, rising up over the waves then gliding low over the surface, which it tends to hug, rather than shearing from side to side.

Scopoli's Shearwater *Calonectris diomedea*

There is a single accepted record of Scopoli's Shearwater, which replaces Cory's in the Mediterranean. This species, split from Cory's in 2012, may be overlooked. In direct comparison it is smaller and slighter than Cory's with a slimmer bill; most importantly, the white of the underwing intrudes extensively into the primaries (see Fisher & Flood 2010).

Great Shearwater *Puffinus gravis*

Plumage Although similar to Cory's in size and general appearance, Great has much more definite plumage characters that are easily seen in a reasonable view. Most distinctive is a somewhat skua-like head pattern with a well-defined black cap that appears tipped forward, contrasting sharply with the gleaming white face and collar, the latter extending up around the sides of the neck. Unlike Cory's, it has a black bill. The underparts show a dark shoulder patch, recalling the 'breast peg' of non-breeding Black Tern *Chlidonias niger*; there is a narrow but variable diagonal black bar across the underwing-coverts and a small black belly patch – or 'oil stain' – that may be apparent in closer views. Like Cory's, it has a white uppertail-covert crescent but this tends to stand out more strongly against the darker brown upperparts and the blacker tail. Note that these plumage differences are strongly influenced by the light. In good light, Great looks more 'two-toned' than Cory's, the blacker outer wings contrasting more strongly with the browner inner wings. This effect may be further emphasised by the fact that the rear of the wing may look pale as a consequence of the light reflecting off pale wing-covert fringing. However, Great's upperwings look more uniform in dull light and, if seen against the sun, the differences between the two species become less obvious.

Flight Great does not have Cory's lazy flight; although it too flies on bowed wings, it instead has a stiffer-winged action that is more reminiscent of smaller shearwaters, such as Manx *Puffinus puffinus*. It has slightly quicker wingbeats than Cory's with four to five flaps followed by a glide and, like Manx, it tends to shear more over the surface of the water. However, variation exists depending on what the birds are doing (e.g. feeding or purposefully migrating) and on the prevailing weather conditions, particularly wind strength.

Pitfalls Undoubtedly the greatest identification problems occur not with separating them from each other, but with separating them from other species. Being a rather featureless shearwater, Cory's is a notorious 'beginner's bird' and the less experienced seawatcher is urged to exercise caution and to claim only those individuals that are seen well. Fulmar *Fulmarus glacialis* provides the greatest pitfall, particularly when silhouetted at long range or in inclement weather. In these circumstances, Fulmar's characteristic white head and pale primary bases may be impossible to detect, while 'blue-phase' Fulmars would not show these features anyway. In strong winds, Fulmars also bound in high arcs over the waves, so pay attention to shape and flight: Fulmars have fat, cigar-shaped bodies and fly with a flap-flap-glide on stiff but slightly bowed wings, interspersed with periods of banking and gliding. Another pitfall is provided by Balearic Shearwater *Puffinus mauretanicus*, which is slightly larger and bulkier than Manx. Paler examples are brown above and dusky-white below but their shape is more similar to Manx (see p. 85) like Manx, they also fly with rapid, shallow, stiff-winged strokes (but tend to shear slightly less, producing a more direct flight). The underwing is dusky-white with a brownish tip, trailing edge and axillaries. At close range, the small bill is blackish. Silhouetted

Gannets *Morus bassanus* can also be confused, but are easily eliminated by their long, narrow, sharply pointed wings, long head and bill profile, and long, pointed tail. Immature Herring *L. argentatus* and Lesser Black-backed Gulls also need to be considered, bearing in mind the rather languid, gull-like flight of Cory's.

Reference Fisher & Flood (2010).

Manx, Balearic and Sooty Shearwaters

Where and when Manx Shearwater is an abundant breeding species at selected sites in n. and w. Britain (and Ireland). It is likely to be seen off all coasts, mainly from March to October. Balearic Shearwater is mainly a summer visitor in varying numbers, traditionally encountered off s. England in July and August. Listed as Critically Endangered but, paradoxically, it has recently increased in the English Channel, apparently as a result of its non-breeding range shifting northwards in response to global warming. It is now also seen more widely throughout the year, particularly from Dorset to Cornwall but also penetrating the Irish and North Seas, with a few even reaching Scotland. Sooty Shearwater is usually an uncommon but widespread summer visitor from the Southern Hemisphere, reaching British and Irish waters from July to October (large numbers may be seen off w. Ireland after autumn gales).

Manx Shearwater *Puffinus puffinus*

Easily identified: strikingly black above and white below, the demarcation being sharply defined. It flies with rapid, stiff-winged strokes, followed by a period of shearing, gliding and banking low over the waves, alternately revealing the upperparts and then the underparts. The long wings and shearing flight instantly separate it from auks, which can appear similarly patterned at a distance.

Balearic Shearwater *Puffinus mauretanicus*

Size and structure Although similar to Manx in size and structure, it averages *c.* 10% larger. It is also distinctly heavier and thicker-set, particularly about the head and neck, creating a front-heavy appearance. These differences are best appreciated in direct comparison with Manx, when Balearic appears positively thickset and stocky. It is also shorter-tailed but, unlike Manx, the feet project quite noticeably beyond the tail. The wings are *proportionately* slightly shorter than Manx, contributing to a slightly flappier flight with an upward lift followed by a low glide over the water. Flight differences are otherwise subtle, although it tends to shear slightly less than Manx, producing a slightly more direct flight.

Plumage Surprisingly variable but, in direct comparison with the smart and contrasting black-and-white Manx, most Balearics appear brown and rather dusky (note, however, that in bright light Manx's black upperparts can momentarily appear very brown at certain angles). The head, flanks and undertail-coverts are brown, merging with the rest of the underparts,

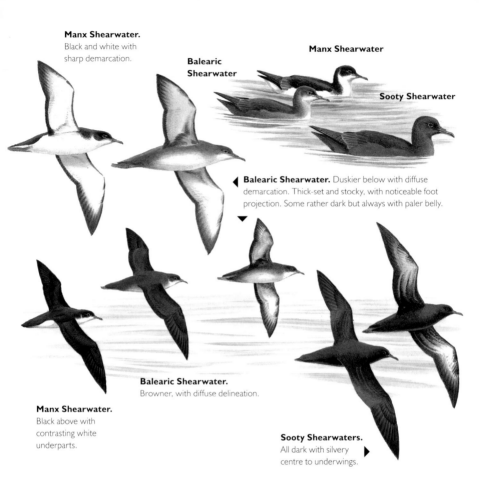

Manx Shearwater.
Black and white with
sharp demarcation.

**Balearic
Shearwater**

Manx Shearwater

Sooty Shearwater

◀ **Balearic Shearwater.** Duskier below with diffuse
demarcation. Thick-set and stocky, with noticeable foot
projection. Some rather dark but always with paler belly.

Balearic Shearwater.
Browner, with diffuse delineation.

Manx Shearwater.
Black above with
contrasting white
underparts.

Sooty Shearwaters.
All dark with silvery
centre to underwings. ▶

which are a rather dusky whitish. The underwings too are dusky (with a dark trailing edge).
Brown 'armpits' (white on Manx) and a brown band across the base of the underwings may
also be noticeable. Late summer adults may be worn and bleached, some showing a pale collar
around the back of the neck and/or patchy plumage on the upperparts (Manx apparently
has a more stable black pigment; Yésou *et al.* 1990). More confusingly, some Balearics are
peculiarly dark, with a limited dusky patch confined to the breast or central belly (and thus
invisible when sitting on the water). Such birds may be confused with Sooty Shearwater and
size and structural differences then become important (see p. 87). A minority of Balearics are
more extensively white below, resembling Yelkouan Shearwater *P. yelkouan*, which replaces
Balearic in the e. Mediterranean. Also bear in mind that some worn Manx can appear
distinctly brown-toned above, particularly in bright sunlight.

Ageing Although variable in their plumage tones, from July to September, juveniles can
be distinguished from adults by their darker, fresh and immaculate plumage at a time when
adults are often worn and showing traces of moult. Pale juveniles are apparently scarce.
Ageing becomes increasingly difficult as the adults approach the end of their moult, which is
completed in October (Yésou *et al.* 1990).

Sooty Shearwater *Puffinus griseus*

On the face of it, easily identified: an all-dark shearwater which, in good light, shows a distinctive silvery centre to the underwing. The problem, however, is dark Balearic Shearwaters (see p. 86). The easiest way to separate them is by shape: Sooty is a rather fat, bulky-bodied shearwater, with long, narrow, pointed wings that are usually *angled back from the carpals* and slightly bowed, creating a 'mini-albatross' shape (at distance, they may vaguely suggest dark-phase Arctic Skua *Stercorarius parasiticus*). Balearic is closer to Manx in shape, so it has proportionately broader, shorter wings that are held more stiffly, less angled back. When flying with a strong tail wind, Sooties tend to 'bound' over the sea in a series of high arcs, but the flight is more languid in calm conditions. Balearics typically fly closer to the surface with a more direct flight on stiffer wings and a flappier action. Although size comparisons are particularly difficult at sea, Sooty is about 15–20% larger than Balearic.
Reference Yésou *et al.* (1990).

Storm, Leach's and Wilson's Petrels

Where and when Storm Petrel is a numerous breeder on northern and western coasts, but is usually seen from shore only after gales. Leach's Petrel breeds in small numbers off w. Scotland and Ireland, as well as in the Faeroe Islands and Iceland, but migrants thought to originate from the larger North American colonies occur in late autumn and winter. They can be found from early September to February, but the peak month is usually November. Large numbers are occasionally 'wrecked' and Merseyside has proved to be the area where they occur most regularly, sometimes in considerable numbers during north-westerly gales. Leach's are more likely to be seen inland than Storm. Wilson's Petrel was once regarded as a very rare vagrant from the Southern Hemisphere but is now known to be reasonably numerous in the western approaches in July–September (as many as 103 were recorded in 1988). However, to see this species it is essential to book onto an organised pelagic trip, the most reliable ones being from the Isles of Scilly or off w. Ireland.

General approach Given a good view, the three species are not difficult to identify but, since many sightings are brief, distant and during inclement weather, caution is recommended. Although seawatching experts can separate them by flight alone, less experienced birders should rely more on their shape and plumage; also bear in mind the difficulties of interpreting and accurately describing flight actions. Particular caution is demanded when identifying Wilson's but on a pelagic trip they can often be attracted close to the vessel by the use of 'chum'.

Storm Petrel *Hydrobates pelagicus*

Size and shape A small petrel (about two-thirds the size of Leach's) with a square tail and relatively short, quite pointed, rather triangular wings that lack a definite angle at the carpal joint. The wing-tips appear slightly more rounded when feeding. Unlike Wilson's, the legs are short and are not normally visible except when foot paddling.

Plumage Appears black, with a prominent contrasting square white rump that extends onto the sides of the ventral area (unlike Leach's). Unlike Leach's and Wilson's, it has plain upper-wings, lacking a prominent pale grey panel across the coverts. At close range, however, adults show slightly browner greater coverts and juveniles a more clear-cut but *narrow* whitish greater covert bar. The most distinctive feature is a prominent *white line on the underwing*, lacking on both Leach's and Wilson's.

Flight Usually seen flying low over the water like a bat or a big black House Martin *Delichon urbicum* (unlike Leach's, inland Storm Petrels can be quite difficult to pick out from hirund-ines). Normal flight is fast and fluttery, with quick, flappy wingbeats and short glides on quite bowed wings. It shears over the waves when travelling more purposefully. When feeding, it flies into the wind and then hangs in one place, either foot paddling or sitting on the water with the tail splayed and the wings raised 10–20 degrees above the horizontal.

Leach's Petrel *Oceanodroma leucorhoa*

Size and shape Noticeably larger than Storm Petrel; in size and shape resembles a small Black Tern *Chlidonias niger*. Colour and shape may also suggest a miniature Arctic Skua *Stercorarius parasiticus* (it is sometimes mobbed by Black-headed *Larus ridibundus* and Common Gulls *L. canus*). Long, rather pointed wings, noticeably angled back from the carpals, and a longish forked tail (adults moult in late autumn and winter and can at times look slightly round-er-winged, while the tail fork can be very difficult to see at any distance). When sitting on the water, it looks long and horizontal with a long rear end.

Plumage Slightly paler than Storm Petrel, appearing dark brown. The most distinctive feature is a broad, dirty grey band across the upperwing-coverts, also visible at rest; this can, however, largely disappear on moulting adults. The white rump is narrower than on Storm Petrel and is V-shaped when seen from above; a narrow dark central bar may be visible at close range, but is often difficult or impossible to see at a distance, when the rump itself is less obvious than on Storm.

Flight Buoyant, effortless, and also slower and lazier than Storm, with an easy action recalling Black Tern or European Nightjar *Caprimulgus europaeus* (it is possible to count each wingbeat). In travelling flight, it shears up from the surface to bound forward, often with sudden changes in speed and direction (again recalling European Nightjar) but its wing-beats are deeper and 'flappier' when flying into the wind. When feeding, it hangs over the surface with swept-back wings, slightly bowed and flatter than both Storm and Wilson's; it then half-flies, half-walks across the surface.

Wilson's Petrel *Oceanites oceanicus*

Size and shape When seen well, not difficult to identify, but most observers unfamiliar with the species initially find it difficult to pick out from Storm Petrel, with which it associates. Distinctly larger than Storm and, although basically similar in shape, the longer and broader wings are more rounded and hence rather paddle-shaped. Also, the rear edge of the wing is very straight. However, in travelling flight the wings look more pointed, with the primaries more angled back. In late summer (when most occur) many are moulting and the old outer primaries often project beyond the still-growing new inner feathers, producing a hooked effect to the wing-tips (most Storm Petrels seen at this time are fresh

Storm Petrels, three feeding. Note square tail, white rump which extends around ventral area, rounded wings with white stripe on underwing.

Wilson's Petrels, three feeding. Note long legs, lack of pale underwing stripe. 'Bounces' when feeding and often dangles legs when moving from one feeding area to another.

Adult Wilson's Petrels are often still moulting in late summer.

Leach's Petrel. When feeding half flies, half walks with bowed flatter wings.

Wilson's Petrel. Longer, broader wings, pale covert bar and feet extend beyond tail tip.

Storm Petrel. Compact, rounded wings, square tail and rump patch, plainer upperwing.

Leach's Petrel. A large petrel, long pointed wings and lazy nightjar-like flight action. Grey covert bar, V-shaped white rump and forked tail.

juveniles). Of particular importance is leg length: the long legs are prominent when feeding and are often trailed at 45 degrees when flying short distances; in full flight, the legs project quite prominently beyond the tail (but the diagnostic yellow webs to the feet are incredibly difficult to see in the field).

Plumage Paler than Storm Petrel, looking more faded, but the best feature is a broad pale grey upperwing panel, obvious at closer ranges (similar to Leach's and unlike the plain-winged Storm). The rump is more obvious than on Storm, extending further onto the sides and appearing to be continuously on view. It may show a diffuse pale stripe on the underwing, but it lacks the prominent white underwing line of Storm Petrel (being a 'negative' feature, the lack of this character is not always immediately apparent).

Flight Full flight is vigorous and direct, with rapid wingbeats followed by short glides, often several metres above the surface (recalling Swallow *Hirundo rustica*). When feeding, the wings are held in a shallow V while foot paddling with its long legs. The feeding flight is slower, easier and more butterfly-like than Storm Petrel, tending to glide, skim and skip over the surface of the waves with the legs dangling and the wings slightly bowed.

References Flood (2010), Harrison (1983).

Slavonian, Black-necked and Red-necked Grebes

Where and when Both Slavonian and Black-necked Grebes are very rare breeding species (the former rapidly declining in Scotland, the latter at scattered sites in England) but are more widespread as coastal winter visitors and passage migrants. In Britain, Slavonian is locally common from September to April, with Icelandic immigrants increasing in n. Scotland but Continental birds declining around e. and s. England (Harvey & Heubeck 2012); it is scarce and local in Ireland. Slavonian frequents sheltered bays, estuary mouths and harbours, and sometimes occurs inland, mainly in late autumn and winter, often during spells of severe freezing weather. In winter, Black-necked Grebe is distinctly rarer than Slavonian, with a more southerly distribution, occurring locally off s. England (mainly Dorset and Hampshire). It is rare in Ireland. It prefers more sheltered environments than Slavonian and is less likely to be seen riding out rough seas. It also turns up inland, particularly on reservoirs, but mainly from July to November and again in early spring; it is rare inland in midwinter. Red-necked Grebe is the rarest and is mainly an uncommon and apparently declining winter visitor from September to April (current winter population estimated at 55). It is most frequent off coasts of e. and s. Britain; it, too, is rare in Ireland. There is an annual late summer moult gathering in Gosford Bay in the Firth of Forth (recently around 50, but now reduced). It may be much more widespread in severe winters, even inland (*c.* 500 were recorded in 1978/79).

Slavonian *Podiceps auritus* and Black-necked Grebes *P. nigricollis*

Size and Shape Both are small black-and-white grebes closer in size to Little Grebe *Tachybaptus ruficollis* than to Great Crested *P. cristatus*. The simplest way of separating them is by overall shape and by their head patterns. A useful *aide-mémoire* is to think of Slavonian as a miniature black-and-white Great Crested and Black-necked as a large black-and-white Little Grebe. In shape, Slavonian suggests Great Crested Grebe in that it has a rather low forehead, a flat crown and a peak at the rear of its head. The back, too, is rather flat with the rear end tapering off towards the water's surface; it may, however, appear more fluffed up and rounded when resting. Compared to Slavonian, Black-necked is altogether a more rounded, fluffier-looking bird, with a high, steeply sloping forehead with a distinct peak either at the front or in the centre of the crown. It has a rather thin neck, the back is rather more rounded than Slavonian and the rear end tends to be held higher and more fluffed out, the whole shape being strongly reminiscent of Little Grebe, with which it often associates.

Bill Slavonian has a straight, pale-tipped bill, whereas Black-necked has an uptilt to the lower mandible, but this can be surprisingly difficult to detect at any distance, particularly on juveniles.

Plumage *Slavonian* Even at a distance, Slavonian is a small, smart, black-and-white grebe, again suggesting a miniature Great Crested. The black cap extends straight back from the bill and through (but not below) the eye. It is thus sharply demarcated from the gleaming white face and cheeks. At close range, it often shows small pale patches on the lores (usually lacking or very tiny on Black-necked). The breast is white but variable amounts of dark shading may extend around the upper foreneck (perhaps strongest on first-winters). Adults moult in late summer/early autumn so they may retain traces of summer plumage into October. Early autumn juveniles are similar to winter adults, but rather less clean-cut and they may retain traces of head striping behind the eye and on the lower cheeks. Post-juvenile moult continues throughout early winter, so that first-winters and adults are soon indistinguishable (the adult's brighter red eye may be the best difference). ***Black-necked*** Unlike Slavonian, in winter Black-necked the black of the head extends *below the eye with a prominent lobe of white extending up behind the eye into the black of the rear crown* (the black cap bears an uncanny resemblance to a showjumper's hat). The whitish ear-coverts and throat are generally dingier than Slavonian but the face nevertheless contrasts with the dull greyish foreneck, which in turn contrasts with the white breast. Thus, the overall effect is of a drabber, messier bird than Slavonian. This dingier appearance is especially true of autumn juveniles which, as well as being slightly browner than winter adults, show a dull orangey tint to the ear-coverts (often misinterpreted as remnants of summer plumage) as well as a dark line behind the eye. Also, juveniles have dull orange eyes (brilliant red on adult). By late autumn, all ages look much more black and white and some smarter individuals can suggest Slavonian, especially at a distance. Eye colour is probably the best ageing feature by then. Note that, in March/April, winter plumage differences often 'break down' as both species start to moult into summer plumage. Of interest, Black-necked Grebes may call in both spring and autumn: a high pitched upslurred *oo-ee* or *pee-eep*.

Flight identification Slavonian's wing pattern is similar to Great Crested's with variable amounts of white on the inner forewing and a large white patch confined to the secondaries. Black-necked lacks white on the forewing but the white patch on the rear of the wing is larger, extending onto the inner four or five primaries (although differences may be difficult to evaluate on the rapidly moving wing).

Other confusion species Beginners may confuse Slavonian and Black-necked Grebes with other species. Particularly pale or contrasting winter-plumaged Little Grebes may be mistaken for juvenile Black-necked, but note Little's essentially *brown* upperparts and *buff* underparts. Great Crested Grebe may be confused with Slavonian, but the former is larger, longer-necked, browner above, has a prominent white line above the lores and, most obvious of all, a long pink bill. Also, although Great Crested has red eyes, they do not stand out and 'glare' as much as Slavonian's. Distant winter Slavonian may be confusable with auks, but the latter have short, thick necks, longer bodies, a longer tail (only Razorbill *Alca torda*) and distinctive open-wing diving actions.

Red-necked Grebe *Podiceps grisegena*

Size and shape Intermediate in size between Slavonian and Great Crested, but distinctly shorter-necked, stockier and more compact than the latter.

Bill Rather dagger-like, on juveniles and first-winters pale yellow (sometimes *greenish*); on adults black with a *prominent bright yellow patch at the base*, sometimes extending towards the tip, particularly on the lower mandible (bill pale pink on Great Crested).

Winter plumage Remember that the pattern of juvenile and winter plumages reflects that of summer adult: pale face, dark neck and pale breast. Consequently, compared to Great Crested (which is gleaming white below) Red-necked looks dull and scruffy. Whereas Great Crested has an obvious white line above the lores, Red-necked has a black cap that *extends down to the*

Winter Slavonian (right) and **Black-necked Grebes** (left). Note differences in head shape and extent of black on head. Black-necked Grebe puffs up body like Little Grebe. ▼

Winter Little Grebe can be pale below, but essentially brown-and-buff, never black-and-white.

Black-necked (left) and **Slavonian Grebes** when alert. Again note head shape and head patterns. ▼

Black-necked and Slavonian Grebes can be confused with female or immature **Smew** (left) and winter male **Ruddy Duck** (right). ▲

Note up-tilted bill of Black-necked Grebe (left) and pale-tipped bill of Slavonian Grebe (right). Both features can be difficult to see.

◄ **Winter Great Crested** (right) and **adult winter Red-necked Grebes** (left). Note prominent yellow base to bill of Red-necked, and dark cap down to eye. Neck dusky, blackish back and white flank 'flash'.

Great Crested Grebe is much whiter below, has all pink bill and white between bill and eyes.

Great Crested (left) and **Red-necked Grebes**. Note latter's rounded dark cap and dark upperparts. Great Crested Grebe's head is peaked at the rear and shows white before and above eye. ▶

92

Juvenile Red-necked Grebe.
Rusty neck, remnants of head stripes.

1st-winter Great Crested Grebe, early winter. Pink bill, extensive white before eye. Retains traces of juvenile head stripes.

Juvenile Black-necked Grebe
has yellowy tint to ear-coverts. Generally buffy on neck and breast and may recall Little Grebe.

1st-winter Red-necked Grebe.
Yellow extending towards bill tip. May show some head stripes.

Juvenile Slavonian Grebe
shows traces of head stripes.

Winter Black-necked Grebe.
More clear-cut than juvenile, with brighter eye. Note shapes of head and bill and extent of black on head.

Winter Slavonian Grebe.
Note angular head and clear-cut black cap down to eye. Straight pale-tipped bill.

Winter Little Grebe.
Very buff and brown. Note stubby bill.

eye. Dingy whitish cheeks then contrast with a darker neck, which varies from buffy-brown through brownish-grey to quite a dark purplish-brown. Blacker above than Great Crested and usually shows a striking white flash along the upper flanks. Distant birds are perhaps more likely to be confused with Slavonian Grebe, particularly those with dark shading on the neck, or transitional ones in spring that have started to acquire orange summer neck feathering. If in doubt, check the bill colour.

Juvenile In early autumn, juveniles are surprisingly similar to summer adults, with a strikingly chestnut neck that contrasts with the whitish face; however, young birds often show prominent black head striping. This plumage is lost after a late autumn/early winter post-juvenile body moult, although vestigial head striping can persist well into the winter.

Flight On autumn and winter seawatches, Red-necked appears similar in size and wing pattern to Great Crested, but is somewhat stockier and shorter-necked. Its dusky head and neck are in marked contrast to winter Great Crested, which always looks very white in these areas.

Diving actions When feeding in deep water, Red-necked often leaps well clear of the surface, rather like a Shag *Phalacrocorax aristotelis*. Unlike Great Crested, fish are often brought to the surface and manipulated before swallowing. Great Crested usually submerges smoothly, without leaping. Slavonian and Black-necked tend to spring forward with a quick, dapper action. However, such differences are not diagnostic and, in general, the deeper the water the more energetic the diving action.

Reference Harvey & Heubeck (2012).

Bittern and Night Heron

Where and when Bittern is a localised breeding species, with currently *c.* 100 pairs, mostly in e. England and Somerset. It is more widespread in winter (mainly November to March) when numbers are augmented by continental immigrants, particularly in cold weather. They breed in extensive reedbeds but are often found in much smaller ones in winter. Night Heron is a vagrant mostly in spring (late March to June) but small numbers also occur in late autumn and occasionally in winter. It currently averages 14 records per year but is prone to irregular spring influxes, with as many as 61 in 1990.

Bittern *Botaurus stellaris*

Bitterns are habitat specialists, associated almost exclusively with reeds. They are most easily detected by the male's low, far-carrying booming call, given in late winter and spring, not only during the day but also throughout the night during the peak of their activity. They are usually seen flying low over reedbeds (occasionally at some height) and in winter tend to be most active at dusk, as they fly to roost. They are, however, most easily seen in late spring, when the female has young and makes regular feeding flights. Severe winter freezes will also induce them into the open. Bittern is a large, owl-like heron, rich buff in colour, densely and intricately patterned with black. The most obvious and significant plumage features are the black crown, the black moustache extending back below and beyond the eye, thick black-and-brown lines down the foreneck and onto the breast, and owl-like black barring across the rich orangey-brown primaries and secondaries. It has a dull lime-green bill and legs, the latter longer than those of Night Heron, with the feet projecting well beyond the tail (note also Bittern's remarkably long claws). They often fly with their neck stretched out, particularly after take-off.

Calls Occasionally calls in flight: usually a low *kwok* but in spring also a loud, gull-like *caw*.

Night Heron *Nycticorax nycticorax*

Unlike Bittern, Night Heron is not a reedbed specialist but instead is likely to be found roosting in trees, particularly willows and alders, around the edges of lakes and marshes. As its name suggests, it is markedly crepuscular, flying out to feed at dusk (often at some height) at a time when other herons are going to roost. The first indication of its presence may be a loud, deep *cok* call, strongly reminiscent of a Raven *Corvus corax*. It is rather stocky and thickset, the head merging into the body when roosting. In flight it is compact with relatively short rounded wings. It has a thick pointed bill, a rather neck-less appearance and short legs, with only the feet projecting beyond the tail in flight (producing a short rear end). *Juvenile* With their black crown, mantle and scapulars, and pale grey wings and underparts, adults are easily identified, but juveniles are brown and much more Bittern-like. However, they are slightly but distinctly smaller and more compact than Bittern, with shorter legs. Their plumage is dark chocolate-brown, the *upperparts profusely spotted buffy-white* (each feather on the back, scapulars and wing-coverts having a buffy-white tear-shaped spot at the tip). The underparts are heavily but diffusely streaked. In flight, Night Heron lacks Bittern's heavily barred, owl-like primaries; instead the primaries and secondaries are plain brown, with a white trailing edge and there are lines of buff-white spots across the tips of all the major feather tracts. Although adults have a black bill, the juvenile's bill is initially dull creamy (with dark along the culmen ridge and cutting edge) gradually turning to yellow-green and then yellow. It has large orange eyes and blue-green or lime-green lores. The legs vary from dull green to bright greenish-yellow. *First-summer* A post-juvenile body moult usually begins in late winter, producing a plain brown or grey-brown back. By their first-summer, Night Herons have an intermediate pattern mirroring that of the adult's, with the crown rather adult-like (but subdued), the back plain grey and diffuse brown streaking on the breast and neck-sides (they may even grow adult-like white plumes on the back of the head). However, the wings remain juvenile (spotted with white). The bill and legs begin greenish-yellow but the bill may start to turn black. *Second-summer* It seems that during the second half of their second calendar year (i.e. when just over one year old) they moult into a second immature plumage that is more adult-like, but uniformly brownish-grey across the back and wing-coverts with diffuse streaking on the neck (van Duivendijk 2011).

American Bittern *Botaurus lentiginosus*

Covered here for the sake of completeness, American Bittern is a very rare vagrant (41 records, but just nine in 1950–2010). It usually occurs in western areas in November, sometimes remaining into winter. Slightly smaller than 'Eurasian' Bittern and easily identified in flight by *plain* black primaries and secondaries with an obvious broad chestnut-buff trailing edge to the wing. The wing-coverts appear contrastingly sandy. The crown is brown (not black), with a noticeable pale buff supercilium, a dark orangey-brown rear face, a long black moustachial stripe (reaching down to the neck-sides) and long, heavy, rich chestnut stripes down the foreneck and onto the breast.

Reference van Duivendijk (2011).

Purple Heron

Where and when Purple Heron *Ardea purpurea* is a vagrant with records from March to November, but mostly in April and May. It currently averages 22 records a year with a peak of 35 in 1987. It occurs mainly along coasts of e. and s. England. A pair bred in Kent in 2010.

Habitat and behaviour Much more dependent on reeds than Grey Heron *A. cinerea*. Secretive and often very difficult to detect as it stands motionless at the edge of the reeds. Unlike Grey, it often lands on the tops of reeds or in trees and bushes within reedbeds.

Size and structure Distinctly smaller, slighter and 'scrawnier' than Grey Heron, lacking its obvious bulk (about 10% smaller in size and about 20% smaller in weight). At rest, it has a remarkably thin, snake-like neck as well as a strikingly long dagger-like bill that seems to merge seamlessly into its rather small head; its bill is much longer than that of Grey Heron. In flight, the folded neck forms a prominent low-slung 'neck pouch' that is obvious even in a back-on view. Befitting a lighter bird, its wings are slimmer and less rounded, resulting in a lighter, less lumbering flight with a tendency for vertical body movement during the down stroke. Also obvious are very large and conspicuous orangey-yellow feet, with very long toes reaching conspicuously beyond the tail.

Plumage *Adult* Easily identified by its solid black crown and striking reddish-brown neck, with a long thin black stripe that extends from below the eye to the bend of the wing. There are further black stripes on the front of the neck, but most distinctive is a narrow black 'connecting stripe' from below the eye to the upper nape, where it meets the crown and nape plumes. The upperparts are a darker, slatier-grey than Grey Heron, the contrast between the wing-coverts and the primaries/secondaries being much less obvious. There is a dark purple patch at the bend of the wing and a dark purple belly and thighs. Elongated brown scapulars and a brown cast to the wing-coverts combine with the chestnut-brown neck and dark purple areas to create a much browner impression that is particularly apparent in flight. A dark chestnut panel on the underwing-coverts may be particularly obvious if a bird is flushed or seen overhead. The long yellow bill and yellow legs add to the more colourful impression in flight (most obvious on adults in full breeding condition). When flying head-on, it shows deep buff 'landing lights' at the bend of the wing (white on Grey Heron). *Juvenile* Distinctive in its own right, being strikingly sandy-brown over the neck and upperparts, lacking the adult's prominent black neck-stripes and its dark chestnut and purple areas. Closer views reveal broad buff feather fringes to the upperparts, narrow broken stripes on the foreneck and a faint shadow of the adult's facial pattern. Juvenile Purple Heron is so brown that it may suggest a Bittern *Botaurus stellaris*, but it has a typically heron-like shape, with the prominently bulging neck, long bill, large feet and black primaries and secondaries. Bittern looks much more 'owl-like' in flight. *First-summer* Juvenile plumage is retained until the birds arrive in Africa where a partial moult takes place in their first winter, involving the body and some wing-coverts. It seems that at least some immatures remain there during their first summer (Cramp & Simmons 1977). Those that return north are adult-like but retain variable amounts of buff feathering, particularly across the upperwing-coverts (obvious in flight). Some are more juvenile-like, the neck stripes being browner and less well developed; others show only a 'ghost' of the adult's face stripe and lack the distinctive black stripe on the side of the neck. They may retain at least some old juvenile secondaries, which appear contrastingly browner and also slightly shorter than the new black feathers. *Older immatures* From the autumn of their second calendar year, young Purple Herons become much more adult-like, although it seems that second-summers may still retain signs of immaturity, such as brown feathering in the crown, duller plumage and the retention of some broad pale feather fringing on the wing-coverts (Cramp & Simmons 1977). Unfortunately, these immature plumages do not appear to have been well studied and there is also likely to be significant individual variation.

Call Higher-pitched, shorter, quieter and gruffer than Grey's.

Melanistic Grey Heron Dark Grey Herons (melanistic or stained) are not infrequent. Their identification should be relatively straightforward, particularly if careful attention is paid to the structural and plumage differences outlined above (but beware of distant birds or those seen briefly).

Reference Cramp & Simmons (1977).

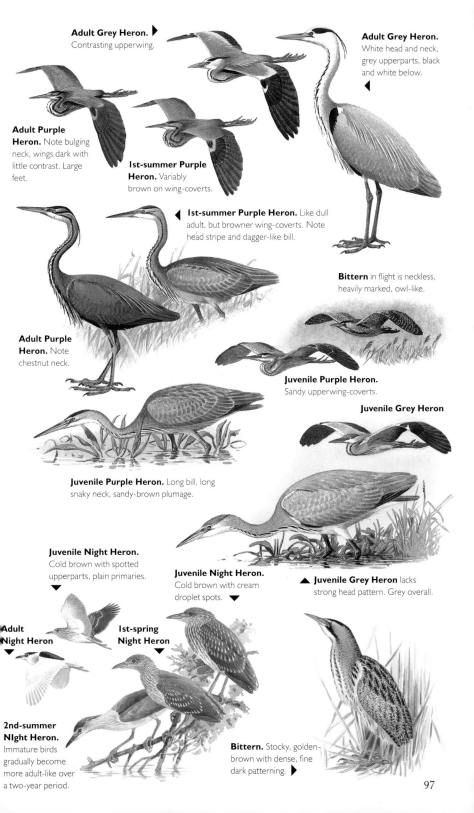

Adult Grey Heron. ▶
Contrasting upperwing.

Adult Grey Heron.
White head and neck,
grey upperparts, black
and white below. ◀

**Adult Purple
Heron.** Note bulging
neck, wings dark with
little contrast. Large
feet.

**1st-summer Purple
Heron.** Variably
brown on wing-coverts.

◀ **1st-summer Purple Heron.** Like dull
adult, but browner wing-coverts. Note
head stripe and dagger-like bill.

Bittern in flight is neckless,
heavily marked, owl-like.

**Adult Purple
Heron.** Note
chestnut neck.

Juvenile Purple Heron.
Sandy upperwing-coverts.

Juvenile Grey Heron

Juvenile Purple Heron. Long bill, long
snaky neck, sandy-brown plumage.

Juvenile Night Heron.
Cold brown with spotted
upperparts, plain primaries.
▼

Juvenile Night Heron.
Cold brown with cream
droplet spots. ▼

▲ **Juvenile Grey Heron** lacks
strong head pattern. Grey overall.

**Adult
Night Heron**
▼

**1st-spring
Night Heron**

**2nd-summer
Night Heron.**
Immature birds
gradually become
more adult-like over
a two-year period.

Bittern. Stocky, golden-
brown with dense, fine
dark patterning. ▶

97

Egrets

Where and when Little, Great White and Cattle Egrets were formerly vagrants to Britain. Little Egret first bred in 1996 and, by 2010, the breeding population was approaching 1,000 pairs, with at least 4,500 wintering. It is commonest in England and Wales, but is scarce in Scotland. Great White Egret remains rare, but is increasing, with breeding in Somerset from 2012. Cattle Egrets have also increased, with as many as 168 recorded in 2008, and a pair bred in Somerset in 2008–09. Since then, numbers have declined due to a series of cold winters. It remains to be seen whether the species will eventually colonise.

Little Egret *Egretta garzetta*

By far the commonest egret, occurring in a wide variety of wetland habitats, both fresh and saline. It feeds in shallow water, either by stalking or energetically chasing mobile prey. Like Cattle Egret, it will also feed on earthworms in damp fields, sometimes associating with cattle. It nests in trees, often with Grey Herons *Ardea cinerea*.

Structure Medium-sized, slim and elegant; about two-thirds the size of Great White Egret. Long, thin neck and a long, slim, dagger-like bill. Longish black legs project well beyond the tail in flight, with the adults' *bright yellow feet* being both obvious and diagnostic (grey-green on juveniles). It fails to show an obvious neck pouch in flight (appearing rather 'flat-necked' compared to Great White). Its flight is rather flappy and energetic (*cf.* Great White).

Plumage *Adult* In late winter, spring and summer two long plumes hang down the nape and it also has full plumes on the lower breast, back and scapulars, the latter forming a 'fluffy train' that overhangs the tail. To all intents and purposes, it loses the plumes in autumn (maybe just a vestige on the nape, back and scapulars and a slight ruff on the breast). *Juvenile* Weaker and fluffier looking plumage than the adult, often appearing distinctly 'textured' on the breast. It lacks long plumes but, when the head feathers are raised, it often shows a slight tufted effect to the fore-crown and a ragged impression at the rear, with an inconspicuous short plume often visible, and also a slight ruff on the breast (young birds may also retain traces of down on the head). Post juvenile moult takes place from August to November (*BWP*).

Bare parts *Adult* The bill is black, sometimes with grey or pink along the base of the lower mandible. The bare lores vary from dull blue-grey to lime green and dull yellow, but for a short period in spring, when in full breeding condition, both the bill and legs become intensely black and the lores a vivid purplish-pink, with deep blue nearer the eye. *Juvenile* Easiest to age by foot colour: grey-green or greyish-yellow. Initially, the legs are also green, gradually turning black, but with variable amounts of grey-green extending up the legs to the 'knees' and thighs. Initially, the bill has a dark tip contrasting with an extensive pinkish base, but it gradually darkens, becoming a rather messy pale grey and then black, often with pink retained along the basal two-thirds of the lower mandible. The lores are an insipid grey, blue-grey, lime green or even pinkish.

Call A distinctive, deep, guttural *arggh arggh*, sometimes more of a blood-curdling, drawn-out rasping *kaaaaa*.

Great White Egret *Ardea alba*

Structure, posture and feeding action A very large egret, similar in size to Grey Heron (may be even taller in direct comparison). It is about one-third as big again as Little Egret and,

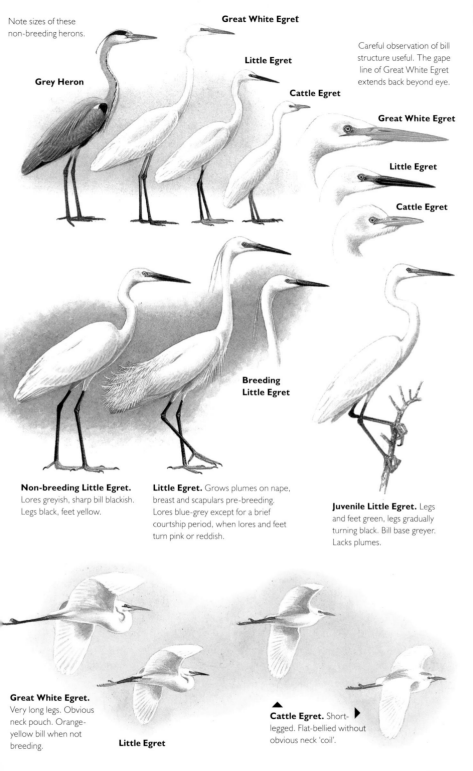

Note sizes of these non-breeding herons.

Great White Egret

Little Egret

Grey Heron

Cattle Egret

Careful observation of bill structure useful. The gape line of Great White Egret extends back beyond eye.

Great White Egret

Little Egret

Cattle Egret

Breeding Little Egret

Non-breeding Little Egret. Lores greyish, sharp bill blackish. Legs black, feet yellow.

Little Egret. Grows plumes on nape, breast and scapulars pre-breeding. Lores blue-grey except for a brief courtship period, when lores and feet turn pink or reddish.

Juvenile Little Egret. Legs and feet green, legs gradually turning black. Bill base greyer. Lacks plumes.

Great White Egret. Very long legs. Obvious neck pouch. Orange-yellow bill when not breeding.

Little Egret

Cattle Egret. Short-legged. Flat-bellied without obvious neck 'coil'.

99

when seen together, Great White towers over its smaller relation. Despite this size difference, it is surprisingly easy to confuse with Little Egret, especially at a distance and/or when direct size comparisons are not possible. Particular caution should be exercised with distant birds in flight. It has a long, slim, dagger-like bill and long legs but the most obvious structural feature is the remarkably long, thin snake-like neck. When feeding, it is often held in a distinctive shape, with the upper neck curved but the lower neck straight (like a question mark or the metal hook of a coat hanger). Also when feeding, it often stretches its neck out sideways at a bizarre angle, *c.* 45 degrees to the body, creating a very distinctive posture as it searches for food. It is a slower, more methodical and less energetic feeder than Little Egret (more like Grey Heron) and it does not 'gallop' after prey like the smaller species. Despite its large size and pure white plumage, it is surprisingly adept at disappearing into waterside vegetation.

In flight Appears very large (similar in size to Grey Heron) with a similar slow, easy flight action, but with shallower wingbeats and less bowed wings (back-on, it could be mistaken for a Mute Swan *Cygnus olor*, which is unlikely with Little Egret). Its wings are slightly but distinctly narrower, less rounded and more tapered than Grey Heron's, most obvious in direct comparison. It tends to glide and soar more than Little Egret. Its long, pointed bill and *strikingly long legs* are obvious and it also shows an obviously rounded neck pouch, reminiscent of Purple Heron *Ardea purpurea* (flatter on Little Egret).

Plumage *Adult* In summer, it lacks Little Egret's plumes on the rear head but has long fluffy ones on the scapulars, which often hang down over the tail (often in strands when wet, but sometimes stuck together in a single long, thin strand). The bill is deep black, with contrasting blue-grey or lime-green lores, and the legs are also black (including the feet) but with *yellow on the tibia*, down to and including the 'knees'. For a short period when in full breeding condition, the bill becomes an intense black, the base of the lower mandible and the lores a vivid turquoise-green and the legs pale red (this odd combination of colours creating an extremely attractive appearance). By late summer, orange or bright primrose-yellow starts to appear on the bill base, gradually increasing in size so that, by autumn, the *bill is completely orangey-yellow*, remaining this colour throughout the winter. Unlike Little Egret, the gape line extends a short distance back below and beyond the eye. *Juvenile* Similar to winter adult, with bright orangey-yellow bill, lime-green lores and black legs with dull yellow above the 'knees' (variable in extent).

Cattle Egret *Bubulcus ibis*

Cattle Egret is a bird of grasslands, where it hunts for insects and other prey around the feet of grazing herbivores (but Little Egret may also feed in this way). It has adapted this behaviour to take advantage of the spread of pastoral farming and is now one of the world's most successful birds.

Structure Although about the same size as Little Egret (slightly smaller) it has a shorter, thicker bill, a shorter neck, a stockier body and shorter legs. It also has a distinctive 'jowl' (the feathering of the chin extending below the base of the lower mandible) and a rather ragged forecrown.

In flight More compact than Little Egret, with slightly shorter wings and shorter all-dark legs and feet. It should be noted, however, that flying Cattle Egrets can be surprisingly difficult to distinguish from Little, even in mixed flocks.

Plumage *Adult* In breeding plumage, easily identified by its *pale orange crown, breast and lower back*. It lacks Little Egret's long head plumes, but has a short mane of longer feathering on the nape and also longer buff plumes on the mantle, hanging down over the tail. The most obvious difference, however, is the relatively *short, thick, pale orange or yellow bill*, which is visible at some distance, even in flight. In full breeding condition, the bill briefly becomes red with a yellow tip, the lores an intense purple, the eye turns from yellow to deep red and the legs become deep pink (although initially they may be yellow). Non-breeding plumage is completely white. *Juvenile* Also completely white, but easily aged by its black bill (with pale yellow lores). As a consequence of its bill colour, the juvenile is much more likely to be passed off as a Little Egret. However, the bill gradually starts to turn yellow from about August onwards.

Coromandus **Cattle Egret** The race *coromandus* occurs in s. Asia, from Pakistan east to Australia. This form is kept in captivity and examples have escaped. It differs from the nominate in being perceptibly longer-billed and longer-necked but most distinctive is that, in summer plumage, it has more extensive orange on the head, neck and breast (with a white forehead and supercilium, and white down the centre of the throat, ending in a point on the upper breast, within the orange). It also shows paler orange on the back (with orange plumes). A pale buff wash to the forehead, crown, sides of the neck and breast may persist throughout the winter. One escaped bird in Somerset was in wing moult in mid February.

Squacco Heron *Ardeola ralloides* Although distinctly different from Cattle Egret, distant Squacco Herons are confusable in flight, particularly if seen below eye level, when the orange or buff 'saddle' (back and scapulars) may not be visible.

Cormorant and Shag

Where and when Cormorant is common around all British and Irish coasts, as well as on inland waters, larger rivers and canals, even in cities. Note that the Continental race *sinensis* has colonised Britain and is spreading (see p. 102). The less widespread but more numerous Shag is essentially marine, commonest around rocky coastlines in the north and west, but rare inland (mostly in autumn and winter). Separating them is not always easy, particularly at a distance. Difficulties often arise with out-of-context Shags, particularly inland. A detailed description would then be essential for acceptance.

Cormorant *Phalacrocorax carbo*

Size and structure Most birders separate Cormorant and Shag by a combination of size, structure and 'jizz'. Cormorant is larger and towers above Shag at rest, although the height difference is emphasised by its tendency to hold its head and neck higher. However, Cormorants exhibit considerable size variation and particularly small females can look Shag-like if seen in isolation. Cormorant is bulkier, heavier and more angular than Shag, particularly when out of the water, with a thick neck and a heavy, angular head. A lower forehead and more tapered bill produce a more wedge-shaped head and bill profile. In flight, Cormorant looks heavy and massive, with a larger head, a thicker, slightly kinked neck and slow, rather goose-like wingbeats. **Facial pattern** At close range, pay particular attention to the fact that Cormorant has an extensive area of bare skin on the lores, face, chin and throat, extending narrowly above and behind the eye; this is often referred to as the 'gular patch'. On adults, this area is yellow in winter but, in breeding condition, the portion below the gape line becomes dark green, while the area immediately below the eye brightens from cream to bright yellow or orange and then to bright red during mating and egg laying. Note, however, that the bare skin on adult Cormorant's lores and chin is often obscured by minute black feathering, so some are less distinctive than others. On immatures, the whole of this facial area is usually yellow or pale orange and, even at a distance, individuals with extensive patches can be safely identified as Cormorants (Shags never show orange on the face).

Plumage *Adult and second-year* Adults in breeding plumage are readily identifiable. As well as the yellow or orange face, Cormorant has large white facial and flank patches, a purple-blue sheen to the head and underparts, and a bronze gloss to the upperparts (with broad dark feather fringes). White filoplumes on the head may be extensive and conspicuous (see below). Both species start to acquire winter plumage soon after they finish breeding and both become duller. Cormorants then lose their white flank patches, have a much duller white facial patch and lack the white filoplumes on the head; the bare facial skin may turn cream. Breeding plumage may be acquired again as early as late December (see below for racial differences). *Juvenile/first-winter* Both species are very variable, compounded by the fact that they undergo an almost continuous body moult from their first autumn until they acquire adult plumage two years later. Cormorant is blacker than Shag and *usually much whiter* below, many showing strikingly white underparts. Some are brown below, while on others the white is confined mainly to the belly (often looking mottled and 'moth-eaten' when moulting). Conversely, a few are the opposite, being blackish on the belly and brown on the neck and breast. Juvenile/first-winter Shag is distinctly paler and browner than Cormorant, normally showing pale brown feathering on the underparts, and lacking the obvious whiteness of most young Cormorants (but note that juvenile/first-winter Shags with very white underparts are occasionally recorded; see p. 104). *Older immatures* Both species become darker with age, with second-years similar to winter adults, but duller and browner.

'Continental Cormorants' *Phalacrocorax carbo sinensis*

The race of Cormorant that has traditionally bred in Britain is nominate *carbo*, which usually nests on sea cliffs, mainly in northern and western areas. However, since the early 1980s, birds of the Continental race *sinensis* have colonised, nesting in trees and bushes on inland waterbodies. They have subsequently spread west and are now common in many areas. It has been established, however, that *carbo* and *sinensis* are now interbreeding in some areas (Ekins unpubl.). *Shape of the gular patch* The only way to separate the races with any degree of confidence is by the shape of the bare gular patch. In *carbo*, the bare skin extends downwards and backwards from the eye and forms a fairly acute angle at the base of the gape before sloping down and then *forwards* again towards the chin. On *sinensis*, the bare skin also extends downwards and backwards from the eye but, instead of forming an acute angle at the gape line, it drops vertically towards the throat (see illustration). In other words, *carbo* has an 'acute' angled patch, *sinensis* a 'square' one. *Other differences* Other 'on average' differences are much more difficult to evaluate and bear in mind that intergrades between the two forms will confuse the issue. *Carbo* averages 25% larger than *sinensis*, tending to make the latter

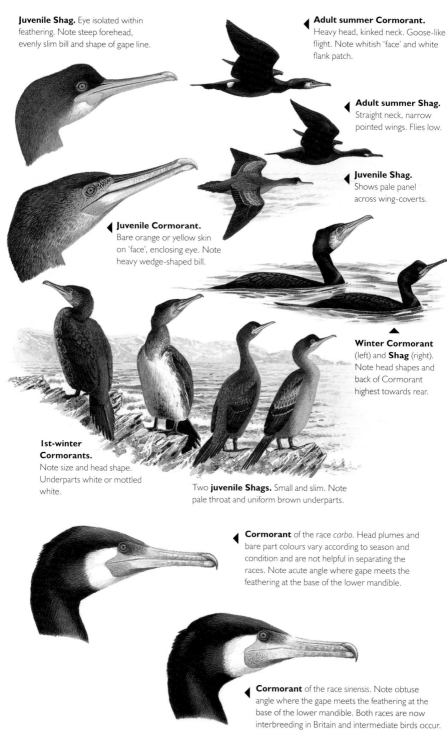

Juvenile Shag. Eye isolated within feathering. Note steep forehead, evenly slim bill and shape of gape line.

Adult summer Cormorant. Heavy head, kinked neck. Goose-like flight. Note whitish 'face' and white flank patch.

Adult summer Shag. Straight neck, narrow pointed wings. Flies low.

Juvenile Shag. Shows pale panel across wing-coverts.

Juvenile Cormorant. Bare orange or yellow skin on 'face', enclosing eye. Note heavy wedge-shaped bill.

Winter Cormorant (left) and **Shag** (right). Note head shapes and back of Cormorant highest towards rear.

1st-winter Cormorants. Note size and head shape. Underparts white or mottled white.

Two **juvenile Shags.** Small and slim. Note pale throat and uniform brown underparts.

Cormorant of the race *carbo*. Head plumes and bare part colours vary according to season and condition and are not helpful in separating the races. Note acute angle where gape meets the feathering at the base of the lower mandible.

Cormorant of the race *sinensis*. Note obtuse angle where the gape meets the feathering at the base of the lower mandible. Both races are now interbreeding in Britain and intermediate birds occur.

look distinctly smaller and slighter in direct comparison, but this difference is obfuscated by the fact that male Cormorants of both races average 20–25% larger than females (males also have a longer bill with a more prominent hook at the tip). Adult *sinensis* can acquire breeding plumage by late December and most are in full breeding plumage by late January. Adult *carbo* typically acquires breeding plumage later, although some can be in full plumage by early February. Despite assertions to the contrary, in breeding plumage there is no difference between the two races in the extent of the white on the head. Juvenile *sinensis* tend to look smaller, shorter-billed, darker-bellied and longer-tailed than *carbo*. Classic birds show much less white on the belly, with just a few paler feathers confined to the central belly and upper breast, forming a paler 'T'-shaped area on the breast.

Shag *Phalacrocorax aristotelis*

Size and structure Distinctly smaller, slimmer and slighter than Cormorant with a smaller head (more rounded at the rear), a steep forehead (but flattened when diving) and a narrow, parallel-sided bill. In flight, it looks slimmer and scrawnier, with a smaller head and a thinner, straighter neck. It also has narrower wings and more tapered primaries, which produce a quicker flight action. Of interest, Cormorant has 14 tail feathers, Shag 12: a surprisingly useful difference when identifying corpses (unless in tail moult). In breeding plumage, adult Shag shows a very distinctive forward-pointing crest on the forehead, while even non-breeding adults and first-winters may show ragged feathering on the forecrown (conversely, Cormorants often show ragged feathering on the *back* of the head).

Facial pattern *Adult summer* Shag lacks Cormorant's large bare 'gular patch' on the face and around the eye. Instead, adults in breeding condition have a relatively small area of bare skin confined to the chin, although this is often peppered with, and obscured by, minute black feathering. More distinctive is the swollen bright yellow gape line, which extends below and behind the eye. Unlike Cormorant, the rest of the face and throat is feathered, with the green eye isolated within the feathering. *Juvenile and non-breeding adult* The pattern is similar but the entire lower mandible is yellow or greenish-yellow, extending below and behind the eye to include the obvious gape line. Behind the bare chin, the feathered throat is noticeably white. The eye is completely enclosed by the facial feathering but, in juveniles, it is initially *yellowish-white* with a fine yellow orbital-ring.

Plumage *Adult and second-year* In favourable light, breeding adult Shag shows a subtle but obvious green sheen to the plumage (with narrow black feather fringes on the upperparts). Same-age Cormorants show a purple-blue gloss to the underparts and a bronze gloss to the upperparts, with broader and more obvious black feather fringing. Both species are duller in winter; Shags then gain a white throat but lose the forehead tuft (although some may regrow it by early December). *Juvenile/first-winter* Juvenile Shag is a much paler grey-brown than juvenile Cormorant, particularly on the underparts (it lacks Cormorant's bronze tint to the upperparts). It usually shows noticeable whitish fringes to the median and greater coverts, whitening with bleaching and wear so that, by their first summer, these feathers produce a conspicuous pale upperwing panel, obvious in flight. Young Shags have pale legs and feet (flesh-brown, yellow-brown or yellow; black on Cormorant).

Behaviour Although not diagnostic, behavioural differences are useful. Shags tend to fly low over the waves and avoid crossing land, whereas Cormorants often fly at considerable height (inland birds may soar high in the sky). Shags often feed in large, dense flocks whereas Cormorants are rather more solitary when fishing, although they too (perhaps mainly *sinensis*) sometimes feed in large flocks, with the birds at the back repeatedly flying to leap-frog those at the front. Both species jump clear of the water when diving, but this is particularly marked in Shag, which often enters the water almost vertically. Shags are often very tame, so an unusually approachable 'Cormorant' is always worth a second look. Both species roost communally (including by day).

Mediterranean Shag *Phalacrocorax aristotelis desmarestii*

Juveniles of the Mediterranean race are very pale, usually with extensively whitish underparts (with contrastingly dark thighs) and a noticeable pale area on the wing-coverts (which appears as a prominent pale panel in flight). The legs are obviously pale pink. Such birds are not infrequent in extreme SW England but, rather than originating in the Mediterranean, it is perhaps more likely that they are simply pale individuals thrown up by our own population. **References** Ekins (unpubl.), Garner (2008).

Red and Black Kites

Where and when Formerly confined to central Wales, as a result of reintroductions the Red Kite has increased spectacularly in recent years and is now a familiar sight in many parts of the country, even over some urban areas. Winter and spring dispersal produces records elsewhere including, on occasion, small flocks. Formerly an extreme rarity, Black Kite is now an annual visitor, currently averaging about 14 records per year (with a peak of 32 in 1994). Early birds appear in March, records increasing to a peak from April to June, with a few seen until late autumn (the nominate race has wintered once).

Red Kite *Milvus milvus*

Structure and flight A stunning raptor which, at distance, can look like a flying cross. It has long, rather narrow wings that are often angled back from the carpals. They are held flat or arched when soaring or gliding (inner wing slightly raised, outer wing slightly depressed, with the tips marginally upturned). Immediately separable from Common Buzzard *Buteo buteo* as the latter has shorter, more rounded wings and tail, and soars with its wings held in a V (although it frequently glides on flat wings). Red Kite's wingbeats are deep and elastic, producing a buoyant flight. The distinctive tail is long, deeply forked and often twisted in flight. When soaring, the fully spread tail looks much squarer, although it usually retains a noticeable notch. Note that the tail fork can be lost through heavy abrasion and in the late summer moult it can become very irregular in shape.

Plumage Rich orangey-brown, with a noticeably paler head. The following features are particularly obvious. **1** UNDERWING A large, square white patch across the inner primaries and the bases of the outer primaries. **2** TAIL A beautiful pale cinnamon-orange.

Juvenile Red Kite. Warm rufous wing panel and obviously forked rufous tail. ▶

Red Kites often gather in flocks at roost or feeding sites.

Juvenile Black Kite. Warm brown with paler coverts, dark mask, shallow tail fork. ▶

◀ **1st-summer male Marsh Harrier**

Adult Black Kite. ▶ More uniform than juvenile.

Adult Black Kite ▶

Marsh Harrier

Black Kite

Marsh Harrier may resemble Black Kite. Note 'V'-shaped wing posture of soaring Marsh Harrier. Kites have 'heavy' hands. Note tail shapes.

Marsh Harrier (right) similar to **Black Kite** (below) in gliding flight.

Juvenile Black Kite. Dark mask, warm brown body streaked paler, cream covert fringes.

Juvenile Red Kite. Warmer rufous than Black Kite, paler-headed; fresh scapulars and mantle brightly fringed.

3 WING-COVERT PANEL A striking pale panel across the median upperwing-coverts. *Juvenile* The general plumage tone is paler and buffier than the adult. Juveniles of both species can be aged by their immaculate plumage, which has better-defined pale fringing to the feathers of the back and upperwing-coverts. However, the most obvious difference from adults is that, on the upperwing in flight, juveniles in fresh plumage have very distinct whitish tips to the greater coverts (including the greater primary coverts) and also to the trailing edge of the wing. These form two narrow but noticeable parallel lines across the rear of the upperwing. Juveniles also show a buff tip to the tail. All these tips are gradually reduced by wear and juvenile plumage is gradually lost through a variable winter body moult, although some are still in complete juvenile plumage by the following spring.

Call A surprisingly thin, husky *weee-oooo wee-oo wee-oo* etc., very weak compared to Common Buzzard.

Black Kite *Milvus migrans*

Structure and flight It is essential that Black Kites are identified with care and only those seen well should be claimed. Red Kite should be used as the yardstick. Black Kite shares Red Kite's basic shape and jizz, being a long-winged, long-tailed raptor with a distinct tail fork. However, beware of Red Kites with a severely abraded or damaged tail, giving the impression of a shallow fork or even a square-ended tail (most likely in spring or summer). Like Red Kite, but unlike Common Buzzard and Marsh Harrier *Circus aeruginosus*, it flies and soars on flat wings that are often arched downwards from the carpal joint. It has deep and rather elastic wingbeats that produce a rather 'flappy' flight. Like Red Kite, the long tail is often used as a rudder, twisted and turned as the bird moves around. Black Kite is smaller and more compact than Red Kite, with shorter wings and a shorter tail (about 10% and 20% shorter, respectively). The tail lacks the deep and obvious fork of Red Kite, instead showing a shallower notch that is obvious only when the tail is closed (the tail fork may even be absent on heavily abraded birds). When soaring, the spread tail often loses the notch completely and instead shows a completely straight rear edge. This often causes confusion with other large raptors but it should be noted that, even when the tail is spread, it still shows sharply pointed corners, unlike the more rounded tail tips of Common Buzzard, Honey Buzzard *Pernis apivorus* and Marsh Harrier.

Plumage As its name suggests, the other main difference is its colour: whereas Red Kite is a richly coloured reddish or orangey-looking bird, Black Kite is a much darker chocolate-brown (sometimes tinged rufous below). This difference is most apparent with reference to the tail. On Red Kite, the upperside of the tail is strikingly cinnamon-orange (greyish on the underside) but on Black Kite it is dark chocolate-brown, concolorous with the rest of the plumage (sometimes with a slight cinnamon tinge). This difference is echoed by the rest of the plumage, Black Kite being a much more subdued, less colourful bird than Red Kite, lacking the latter's strong contrasts. Worn adults may be particularly dull. In particular, it has much less prominent pale patches on the bases of the under-primaries, where Red Kite shows conspicuous white 'windows'. The head too is duller, as is the pale panel across the upperwing-coverts, which is usually very obvious on Red Kite. *Juvenile* Ageing differences as Red Kite (see above) but juvenile Black Kite is distinctive in that it has extensive pale buff spotting and feather fringing on the back and upperwing-coverts, and obvious buff streaking on the underparts; it also shows a blackish mask through the eye. The under-primaries are barred black and white. This plumage is retained until the following summer, but wear and a variable body moult produce a more uniform appearance by then.

Call A drawn-out, rather tremulous *wee-o-o-o-o-o-o-o-o*.

The Marsh Harrier problem

Perhaps the commonest confusion arises with high-flying female, juvenile and immature Marsh Harriers. Like all harriers, when soaring or gliding Marsh holds its wings in a shallow V (which easily separates it from Black Kite) but, when directly overhead or in a steadily flapping flight, the wings can look flat (even at lower levels Marsh Harrier can, over short distances, look flat-winged). High-flying female and juvenile Marsh Harriers look uniformly dark, any yellow/golden on crown and throat being difficult to see. In addition, females and immatures can show a slightly paler patch at the base of the primaries while, from late spring onwards, first-summer males can show grey feathering which could perhaps be misinterpreted as the pale primary patch of a Black Kite.

Common and Honey Buzzards

Given its rather short-tailed/round-winged shape, Common Buzzard is a less likely confusion species, particularly since its V-shaped wings are usually so obvious. Another source of confusion is provided by Honey Buzzard, particularly darker individuals. Like Black Kite, Honey Buzzard has rather long, flexible wings that are held flat when soaring. Its tail is also longer than Common Buzzard's, but note that both its wings and its tail tend to look somewhat paddle-shaped. Like Black Kite, dark Honey Buzzards show pale areas on the inner primaries; indeed, the most common plumage of juvenile Honey Buzzard is all dark with pale inner primary windows, so be particularly careful in autumn.

Hybrids

There have been rare cases of hybridisation between Red Kite and Black Kite, and controversial birds have been seen in Britain. As with any rarity, make sure that a potential Black Kite does not show intermediate or anomalous characters.

Black-eared Kite *M. migrans lineatus*

There is a single British record (Lincolnshire and Norfolk, winter 2006/07 but subject to formal acceptance) of the e. Palearctic race *lineatus* or 'Black-eared Kite'. It differs from the nominate as follows: (1) wing-tip shows six deeply splayed primaries (rather than five) producing a squarer wing shape; (2) it has brighter and more distinct barring on the inner primaries; (3) the under-primary patches and the pale panel on the upperwing-coverts are whiter and more contrasting, almost suggesting Red Kite. Being a winter record, it occurred at a completely different time of year from our usual overshooting spring vagrants. Photographic evidence would be essential.

Reference Forsman (1999).

Hen, Montagu's, Pallid and Northern Harriers

Where and when Hen Harrier breeds mainly on upland moorland and in young conifer plantations in Scotland (particularly Orkney), Ireland, Wales and n. England; gamekeepers have exterminated it from many areas and it is almost extinct in England. More widespread in winter, but still scarce, frequenting all kinds of open country, particularly moorland and coastal marshes. It is most numerous on e. English coasts, but numbers of Continental immigrants are variable. Montagu's Harrier is a rare summer visitor, mainly from April to September, breeding mostly on heaths and arable land in s. and e. England (12–16 pairs in 2010). Pallid Harrier was formerly an extreme rarity, with just three records up to 1992. An influx of five in 1993 was then followed by a further 21 up to 2010. A remarkable influx then occurred in 2011, with 29 accepted, mostly juveniles in September–October. The upsurge is apparently related to a westward expansion of the species' European breeding range. Unlike Montagu's, records have also occurred in winter and early spring. There are three British and three Irish records (to 2012) of the North American Northern Harrier (late autumn and winter).

General features All have a long tail and relatively longish wings which they hold in a shallow V, most obviously when quartering low over the ground. Adult males are essentially grey and black, whereas females and juveniles are brown with prominent white uppertail-coverts (usually erroneously referred to as the 'rump'). Juveniles can be sexed by their eye colour (pale in males, dark in females) but this can be very difficult to determine in the field. Males acquire little if any adult-like plumage until a complete moult when one year old. They are then similar to adult males, but are generally browner and retain obvious traces of immaturity.

Hen Harrier *Circus cyaneus*

Structure Hen is larger and bulkier than Montagu's and about 30–40% heavier, with proportionately shorter, broader and more rounded wings which contribute to a heavier, less graceful flight. Particularly when high overhead, Hen may recall a giant *Accipiter*. At times, however, Hen Harrier's wing-tips can appear quite pointed.

Plumage *Adult male* A stunning bird: the soft pale grey head and upper breast contrast with the whiter underparts to produce a slightly hooded effect. The grey also contrasts smartly with the extensively black outer primaries and, most distinctively, a thick but variable blackish trailing edge to the underwing. It also has white uppertail-coverts and a fairly plain tail. *Adult female* Brown above with prominent white uppertail-coverts and noticeably streaked below on a buff or whitish background (streaking strongest across the breast). The underwings are heavily barred across the primaries and secondaries. It has white surrounding the eye but the face is otherwise relatively plain and owl-like, with a narrow whitish border. *Juvenile* Compared to adult female, juvenile shows neat and immaculate plumage with a narrow whitish trailing edge to the wing and buff tips and fringes to the dark brown upperwing-coverts (the latter forming a paler panel). Buffer below than the female and variably streaked over the breast, belly and flanks; some are quite rusty below (Stoddart 2012). The under-secondaries are darker than adult female and more diffusely barred. Unlike Montagu's and Pallid Harriers, deep orange individuals are rare (such birds are discussed under Northern Harrier).

Adult female Hen Harrier

Juvenile female Hen Harrier. Warm fresh plumage, underparts without club-like spots typical of adult female.

Adult female Hen Harrier. Wings broad and heavy, some wing barring. Five-fingered wing-tip with four long primaries.

Adult female Hen Harrier. Clearer banded secondaries, boldly streaked body.

Juvenile female Hen Harrier. Plain remiges, pale wing-coverts. Large white rump.

Adult female Pallid Harrier. Clean white collar, face mask reaches bill. Tail longer than wing-tip.

Juvenile Pallid Harrier. Note face pattern: clear buff 'ruff', dark 'boa'. Face mask reaches bill, thick weak supercilium.

Adult female Pallid Harrier. Plain upperwing, clear collar, usually dark breasted. Lacks barring to axillaries.

Pale stripe to secondaries reduced to spots near body. Compare with female Montagu's Harrier.

Juvenile Montagu's Harrier (left) lacks strong head pattern of **juvenile Pallid Harrier** (right). Compare underwing patterns. Dark 'boa' on neck-sides.

Juvenile female Montagu's Harrier. Wide supercilium, whiter around eye, restricted dark mask.

Adult female Montagu's Harrier. Wide pale bar on secondaries, coverts barred.

Adult female Montagu's Harrier. Note dark bar on upperwing.

Adult female Montagu's Harrier. Pale faced, eye crescent joins wide supercilium.

110

Two 1st-summer female Pallid Harriers. Very Similar to Montagu's, compare underwing and axillary pattern.

1st-summer female harriers show contrast between faded coverts and darker secondaries. 1st-summer female Montagu's Harrier lacks adult's upperwing bar.

Two 1st-summer female Montagu's Harriers. Note underwing pattern: usually some barred axillaries and rusty outer-tail feathers.

1st-summer female Pallid Harrier. Some moult into adult plumage. Note faded worn wing-coverts.

1st-summer female Montagu's Harrier. Head moulted to adult, but old wing-coverts faded in advanced birds.

Fading head pattern and collar become more Pallid-like.

1st-summer male Montagu's Harrier. The acquisition of adult-type plumage variable in all species. Both male and female may show a faded collar like Pallid; note more white around eye.

1st-summer female Pallid Harrier. Most stay in rather faded juvenile plumage.

Central tail feathers new, grey on males.

1st-summer male Montagu's Harrier in more advanced plumage.

1st-summer male Pallid Harrier. Underwing and tail patterns are the best features, just as for juveniles. Typically less advanced into male plumage.

Two 1st-summer male Montagu's Harriers. Note tail, underwing pattern and barred axillaries. Face pattern useful.

1st-summer male Hen Harrier. Note broad, five-fingered wing, remiges still from juvenile with diffuse pale secondary bars. Obvious white uppertail coverts.

Adult male Pallid Harrier. Little black in wing-tip, pale head.

2nd-summer male Hen Harrier. Pale, faded unmoulted outer primaries may give a Pallid-like wing pattern.

2nd-summer male Montagu's Harrier. Barred axillaries, upperwing bar.

Two 2nd-summer male Pallid Harriers. Darker to head, note wing-tip fingers and plainer underwing.

111

Montagu's Harrier *Circus pygargus*

Structure Montagu's is slimmer and more graceful than Hen Harrier with a long, narrow tail and long, narrow wings with a distinctly tapered wing-tip (only three or four prominently fingered primaries, compared with five on Hen). Whereas Hen may suggest a giant *Accipiter*, Montagu's may suggest an oversized falcon. The long, supple wings of Montagu's produce a slow, easy flight action which may recall that of Common Tern *Sterna hirundo*. **Adult male** Montagu's can be separated by the following features. **1** GENERAL APPEARANCE It has a darker grey head, mantle and leading wing-coverts, all of which contrast with paler grey greater coverts and secondaries to produce a two-toned area of grey on the upperwing. Below, the darker head and upper breast contrast with the paler belly and underwings; the belly and flanks are lightly streaked with rufous. **2** UPPER SECONDARIES There is a black bar across the base of the upper secondaries. **3** UNDERWINGS Two black bars cross the bases of the under-secondaries (with a dark grey trailing edge) and variable brown flecking on the underwing-coverts. **4** TAIL Barred sides to the tail (best seen from below or when spread). **Adult female** Plumage very similar to Hen Harrier and best identified by shape (see above). At close range, the most useful difference is facial pattern: whereas Hen shows a relatively plain, owl-like face, Montagu's has a more contrasting face with a broad whitish ring around the eye, contrasting with a thick, dark brown crescent-shaped cheek patch. On the upperwing, there is a distinct dark band across the secondaries, immediately behind the greater coverts. The underwing-coverts and axillaries are distinctly barred. **Juvenile** Immaculate in fresh plumage, with a narrow whitish trailing edge to the wing and obvious paler feather tips and fringes to the dark chocolate-brown upperparts. The most striking differences from juvenile Hen Harrier are as follows. **1** UNDERPARTS Juvenile Hen usually has a slight but distinct ginger tone to the underparts, with *heavy streaking on the breast and flanks*. Juvenile Montagu's is very variable in plumage tone. Some are pale orangey-buff on the underparts but darker individuals have the breast, belly and leading underwing-coverts intensely coloured, varying from dark buff to deep rufous-orange, *with little obvious streaking* (but more than on Pallid). The under-secondaries are dark. **2** HEAD PATTERN Hen shows white above and below the eye but has a rather plain, owl-like face, which is surrounded by a narrow pale collar. Montagu's shows a more heavily patterned face with a broad and obvious broken white or buff ring around the eye (forming a short supercilium and a thick sub-ocular crescent) and this contrasts markedly with a broad, dark chocolate crescent-shaped patch that curves under the eye and onto the ear-coverts. Below this is a variable narrow buff collar. Note that juvenile Montagu's lacks Pallid's thick brown 'boa' on the neck-sides (see below). **First-year male** A variable post-juvenile body moult in their first year produces a mixture of juvenile and adult feathering (thus Montagu's tend to be more advanced than equivalent-aged Hen Harriers) **Melanism** Both sexes and also juveniles have a rare but very distinctive melanistic form, males being sooty-grey above and blackish below; some have a contrasting silvery area on the bases of the primaries.

Pallid Harrier *Circus macrourus*

Occurrence patterns Given the recent upsurge in records, Montagu's can no longer be considered the 'default' small harrier and it therefore follows that any small, slim-winged harrier should be very carefully scrutinised and, if possible, photographed. Since Montagu's Harriers occur mainly from May to September, any small harrier seen in early spring (March to April), late autumn (October to November) or winter is much more likely to be a Pallid.

Structure Very similar to Montagu's but with same-sex birds in direct comparison, Pallid appears slightly smaller, with narrower, more pointed wings and shallower, less elastic wingbeats.

Plumage *Adult male* Entire plumage strikingly pale whitish-grey, the *only black* being a narrow but prominent black wedge on the four longest primaries. The tail is only slightly barred on the outer feathers (best seen when spread). *Adult female* Very difficult to separate from Montagu's, but the following subtle differences are most significant, although photographic evidence would be beneficial to confirm them. **1** HEAD PATTERN Pallid may show a stronger narrow pale collar below the ear-coverts (in front of the darker neck-sides). **2** SECONDARIES Pallid has dark secondaries, both above and below, on the underwing contrasting with the paler under-primaries (forming a two-toned appearance: pale outer wing/dark inner wing). Also, the under-secondaries usually show one or two pale bands that taper and dissipate towards the body (secondaries usually pale and more obviously barred on Montagu's, but can be dark). **3** UPPERWING Usually lacks Montagu's dark bar across the base of the secondaries. **4** UNDER-PRIMARIES Weaker dark trailing edge (on Montagu's, the entire rear of the wing shows a dark border). **5** OUTER UNDER-PRIMARIES The dark bars are more irregular and the pale bars wider (usually more evenly barred on Montagu's). **6** BOOMERANG There is usually a pale crescent (or 'boomerang') on the base of the under-primaries, immediately behind the carpal. *Juvenile* Similar to juvenile Montagu's (note that the under-secondaries also appear blackish at any distance, but barring is visible at close range). The following are the main features. **1** UNDERPARTS Vary from orangey-buff to a deep orange-brown with little or no streaking. **2** HEAD PATTERN The most obvious difference is that Pallid has *a very contrasting and striking head pattern* with a buff or whitish eye-ring and a prominent buff or whitish collar, sandwiched between the dark brown ear-covert crescent and *a thick, dark brown half-collar or 'boa' on the neck-sides.* Montagu's has a much weaker pale collar and lacks the obvious dark 'boa'. Other features are much more difficult to evaluate in the field (again, photographic evidence would be beneficial). **3** UNDER-PRIMARIES Like adult females, the under-primaries have a weaker, more blurred dark trailing edge than Montagu's (the latter's primary tips are more solidly dark). The under-primaries are also contrastingly pale but the barring is somewhat irregular compared to Montagu's. **4** BOOMERANG The pale bases of the under-primaries often form a pale crescent or 'boomerang' around the dark carpal. *First- and second-year males* Plumage sequences are similar to Montagu's (see p. 112) but, on average, immature Pallid is less advanced than similarly aged Montagu's (Forsman 1999).

Hybrids As Pallid Harrier apparently spreads westwards, hybridisation with Hen Harrier has been recorded. An extremely thorny subject, but something to bear in mind if faced with a particularly difficult individual.

Northern Harrier *Circus hudsonius*

All but one of the British and Irish records have related to juveniles.

Plumage *Adult male* Distinctive, differing from nominate Hen Harrier as follows. **1** UNDER-WINGS White, contrasting strongly with a more obvious broad jet-black trailing edge (strongest on secondaries). **2** OUTER-PRIMARIES Five black outer primaries (Hen shows six). **3** UPPERPARTS Darker grey, with more brown feathering admixed. **4** HOOD AND BELLY Grey hood contrasting with a white belly spotted orangey-brown. *Juvenile* Plumage may suggest a juvenile Pallid or Montagu's but it has the structure of a Hen Harrier. The main differences from Hen are as follows. **1** BOA/HOOD 'Classic' juveniles have a uniform or heavily streaked dark brown neck-band (or 'boa') which almost joins across the foreneck. When viewed side-on, this gives the impression of *a dark hood*. **2** UNDERPARTS The hood contrasts strongly with very obviously *orange-toned underparts and leading underwing-coverts*, which appear either uniform or only lightly streaked. **3** UNDER-SECONDARIES Tends to show a darker patch at the base of the under-secondaries, although some Hen Harriers also show this. **4** UPPERPARTS Distinctly darker brown above with a large and contrasting white uppertail-covert patch. It may show warm rusty tones on the wing-coverts and tail. Identification is, however, complicated by the occasional occurrence of juvenile nominate Hen Harriers with rufous underparts. Consequently, observers of a potential Northern Harrier should try to obtain good-quality photographs and seek expert advice. For more detailed discussions, see Grant (1983) and Martin (2008).

References Grant (1983), Forsman (1999), Holling *et al.* (2012), Martin (2008), Stoddart (2012).

Goshawk and Sparrowhawk

Where and when Sparrowhawk *Accipiter nisus* is common throughout Britain and Ireland, but is less abundant in e. England. Extinct in Britain until the 1950s, Goshawk *A. gentilis* has become re-established, mainly through falconers' escapes and re-introductions, and currently has a population of *c.* 450 pairs. It occurs in extensive tracts of coniferous woodland, particularly in Scotland, n. England, on the English/Welsh border and in the New Forest, and is now common in a few areas.

General approach Goshawk is a notorious 'beginner's bird' and many birders pass through a phase of misidentifying Sparrowhawks as Goshawks. Its identification requires extreme caution and it is strongly recommended that observers gain experience of the species by visiting one of the well-publicised Goshawk viewpoints established in several areas (details online).

Size Larger than Sparrowhawk (female Goshawk averages about eight times heavier than male Sparrowhawk) but it can be extremely difficult to accurately judge the size of a lone bird, particularly against the sky or low over distant woodland. Despite statements to the contrary, Goshawks do not look 'buzzard sized', but intermediate between Common Buzzard *Buteo buteo* and Sparrowhawk. In direct comparison they are about three-quarters the size of a Buzzard and maybe twice the size of a Sparrowhawk, depending on the bird's sex. Female Goshawks are *c.* 10% larger than males, but this difference is not usually obvious unless seen

Peregrine

Goshawk

Common Buzzard

Sparrowhawk

Soaring raptors showing relative shapes and sizes.

'Slow-flapping' display flight of female Sparrowhawk very different from normal flight, undertail-coverts spread.

◄ Adult female Sparrowhawk. Slight build, strong barring, long slender legs and square-ended tail.

◄ Adult Goshawk. Tapered 'hand' with subtly bulging secondaries and rounded tail-tip. Entire underparts much whiter than most Sparrowhawks.

Adult female Sparrowhawk. ► Short, rounded wings (all strongly barred) short head, long tail which is square-ended.

Adult female Goshawk. Fine barring renders adult Goshawk very white below. Note strong legs and feet and rounded tail-tip.

Juvenile Goshawk. ◄ Tail may appear almost diamond-shaped.

▲ Juvenile Goshawk. Protruding head and tapered wings, rounded tail-tip.

▲ Juvenile Sparrowhawk. Short head, rounded wings and long, square-ended tail. Underparts barred.

Adult male Goshawk. ◄ Undertail-coverts often obvious from behind on relaxed perched birds.

Juvenile Goshawk (left). Note head, bill and feet proportions. Cream underparts with streaking, not barring as in juvenile Sparrowhawk. **Sparrowhawk** (above) has a small head and bill, and long, slender legs.

115

together. Goshawks tend to look 'big' rather than 'huge' but distant or lone birds high in the sky may not look particularly large at all.

Flight One of the reasons why they look big is because of their flight. Whereas Sparrowhawk has an energetic 'flap-flap-glide' flight with quick, shallow wingbeats, Goshawk has slow, deliberate, deep and rather elastic wingbeats. This creates a 'flappy' action that is somewhat crow-like (a useful analogy to remember). As well as soaring, they may hang in the wind, barely flapping, sometimes for long periods.

Display Goshawks and Sparrowhawks have similar displays. Goshawk has a protracted display period from late November onwards, reaching a peak in February/March, whereas Sparrowhawk's display is confined to a more discrete period in early spring (late February to April). In addition, Goshawks display much more frequently, more habitually and more persistently. This normally consists of slow, lazy but exaggerated flapping in which the wings are raised and lowered well above and below the horizontal, recalling a displaying harrier *Circus* or a European Nightjar *Caprimulgus europaeus* (they may also slow their wingbeats in normal flight). Much more spectacular is the 'sky dance', which consists of a series of 'switch-backs' that involve a fast, deep plunge (with slow-motion flapping) before shooting vertically upwards like a bullet, with the wings held tight against the body. Although both species do this, its larger size renders Goshawk's display by far the more impressive.

Shape Goshawk has a distinctive shape of its own: it is not simply a 'big Sparrowhawk'. **1** WINGS Whereas Sparrowhawk has short, evenly rounded wings, Goshawk's look distinctly paddle-shaped, particularly from a distance or when soaring. This is because the wings are proportionately longer and more tapered, and the trailing edge shows slightly bulging secondaries (not always noticeable). However, wing shape varies depending on what the bird is doing. In a descending glide or the initial stages of a stoop, the primaries may be swept back and pointed, prompting confusion with Peregrine *Falco peregrinus*. **2** TAIL Another useful feature is that the tip of the tail is rounded (noticeably square-ended on Sparrowhawk). However, the obviousness of this varies. When closed, the tail can look quite broad and tapered, with obviously rounded corners if looked for. However, when soaring it can appear strikingly paddle-shaped or, when more fully spread, diamond-shaped. Because of their long, rather tapered wings and rounded tail tip, their shape can be oddly reminiscent of Raven *Corvus corax*. **3** HEAD Also significant is a noticeably protruding head (apparent at most angles) creating a pointed head profile. This, combined with the longer, more tapered wings and longer tail, gives Goshawk an overall 'rakish' appearance. A very useful analogy is that Goshawk resembles a 'flying crucifix' whereas, because of its smaller, less protruding head and more compact shape, Sparrowhawk resembles a 'flying T' (i.e. † compared with T). **4** UNDERTAIL-COVERTS In display the undertail-coverts are often fluffed out sideways (see below) often giving a distinct 'stepped effect' to the base of the tail. The spread undertail-coverts can make the tail look quite short, but note that this may also apply to Sparrowhawk.

Plumage *Adult* **1** UNDERPARTS An important feature is the colour of the underparts. Because the barring is fine and the background colour white, adults usually look uniformly whitish below, obvious even at long range (when the underwings may 'flash' white in flapping flight). In good light, their underparts and underwings may look strikingly white, especially when viewed against the trees. Although variable, Sparrowhawks look duller below as a consequence of having broader barring and slightly buffier background tones (orange barring and shading on adult males). **2 HEAD** In close views, Goshawks often look obviously capped or hooded owing to their dark head and ear-coverts, which sharply contrast with the whitish underparts. This is usually most obvious on adult males, but some females have a very similar pattern (immature males and many adult females show a more diffuse head pattern). A white supercilium may be noticeable at close range, but not at a distance. **3 UNDERTAIL-COVERTS** Despite statements to the contrary, in normal flight the white undertail-coverts are not obvious as they are sleeked down and do not contrast with the rest of the white or whitish underparts. However, when displaying, they are fluffed out to the side and even wrap around the sides of the rump; this means that they are most obvious not from below but from above or from the side, when they contrast with the blue-grey or grey-brown upperparts. They can even give the impression of a white rump or white sides to the rump (recalling a flying auk). The undertail-coverts may be much more obvious when the bird is perched, especially from behind when they may stick out either side of the tail base. *Juvenile* In a reasonable view, easily separated from adult by yellowish-buff underparts, with messy vertical *streaking* (not barring, as on adults and all ages of Sparrowhawk). However, the streaking is surprisingly hard to see at any distance so juveniles may simply appear rather uniformly buff below. They also have pale fringing to the upperpart feathers, rendering them browner above than adults. Juvenile plumage is retained through the first year of life and they may look rather worn by spring.

At rest Unlike Sparrowhawks, Goshawks often perch conspicuously in the tops of pine trees, their white underparts making them visible over considerable distances. Sparrowhawks are much more furtive and invariably perch within the canopy.

Calls Both species are largely silent but in the breeding season males utter an accelerating slow, rhythmic *ki-ki-ki-ki-ki-ki-ki-ki*; juveniles give a persistent begging *peee-u peee-u peee-u* that sounds plaintive and mournful. Inevitably, the calls of Goshawk are louder and stronger than Sparrowhawk's, sometimes much harsher or with a ringing quality.

Other pitfalls High-flying female or immature Hen Harrier *Circus cyaneus* can appear quite *Accipiter*-like. Note the harrier's slimmer proportions, rounded, well-fingered wing-tips, square tail and heavily barred primaries and secondaries (see p. 109). Prolonged views should reveal Hen Harrier's V-shaped wings (flat on Goshawk) and white uppertail-coverts. Goshawk is not confusable with Common Buzzard but is slightly more similar to Honey Buzzard *Pernis apivorus* (see p. 119): note differences in shape, plumage and flight.

Common, Rough-legged and Honey Buzzards

Where and when Common Buzzard is now by far the commonest bird of prey in n. and w. Britain, and is continuing to increase and spread in eastern areas. Rough-legged Buzzard is a rare winter visitor, mainly to coastal e. England (sometimes e. Scotland). It currently averages *c.* 50 records a year but there are periodic influxes related mainly to the lemming cycle (exceptionally 319 in 1994/95). October records from western areas (e.g. Scilly in 1984 and 2001) are thought likely to relate to the North American race *sanctijohannis* (Flood *et al.* 2007). Honey Buzzard is a very rare summer visitor to selected woodlands in s. England, Wales and Scotland. It is also a rare late spring and autumn migrant averaging *c.* 160 records a year, but a huge influx in September–October 2000 brought over 2,200 into the country.

Common Buzzard *Buteo buteo*

Structure and flight A rather stocky, evenly proportioned, broad-winged raptor, typically seen soaring over hillsides and woodland, often calling. Soars with wings held in a shallow V, but may glide on flat wings.

Plumage *Adult* Extremely variable but most are dark brown, often showing white on the throat and belly and/or a broad but diffuse whitish crescent across the lower breast, most obvious at rest. The bases of the under-primaries and secondaries are paler and greyer (barred darker) and there is a dark trailing edge to the underwing (better defined on male). The paler under-primaries can almost 'flash' as the bird flaps. Some are very pale, with a mainly white head, underparts and underwings (with dark crescent-shaped carpal patches) and a pale rump and tail. Such birds can easily provoke confusion with Rough-legged Buzzard (see below) or even Osprey *Pandion haliaetus* (note Osprey's thick black eye-stripe and long, narrow, almost gull-like wings, which are distinctly down-kinked, never raised in gliding flight). *Juvenile* Similar to adult but in late summer they look immaculate at a time when adults are rather scruffy through moult and wear. They can also show paler upperwing-coverts, a rather diffuse trailing edge to the underwing and a narrower tail-band; the eye is pale (dark on adult). Juvenile plumage (and pale eye) is retained throughout winter but gradually wears.

Calls Adults have a well-known mournful mewing call. Begging juveniles are extremely vocal, giving a loud, mournful and rather irritating *plee-u plee-u*, commonly heard in late summer and early autumn (may persist into New Year).

Rough-legged Buzzard *Buteo lagopus*

The English east coast in October/November is the most likely place to see this species and winterers may linger into April or May. Elsewhere or at other times of year, identification requires extreme caution; note that, in western areas, pale Common Buzzards are far more likely. Genuine Rough-legged Buzzards are very striking – if not, think again.

Structure and flight A typical view is of a large, pale buzzard persistently hovering over coastal fields and marshes (with ponderous wingbeats) or hanging motionless, the tail twisted and turned like a kite's. Common Buzzard also hovers, but less persistently. It is similar in shape to Common Buzzard but slightly larger, longer-winged and sturdier (but the head may look rather rounded at rest). It soars and glides with the inner wing slightly raised and there is a noticeable kink between the inner and the flat outer wing.

Plumage *Juvenile* Most of our Rough-legged Buzzards are juveniles. In flight they are striking, standing out through a combination of the following contrasting characters. **1** TAIL From above, a gleaming white base to the tail and white uppertail-coverts contrast with a thick black terminal band (narrower, diffuse and less distinct from below). **2** HEAD AND BREAST Predominantly creamy-white (may look very white-headed) contrasting with an obvious brownish-black belly. **3** CARPAL PATCH A large black carpal patch on the underwing. **4** UNDER-PRIMARIES Thick black tips to the outer under-primaries, but the tips to the inner primaries and secondaries are narrow and dusky. **5** UPPER-PRIMARIES A variable pale or white patch at the base of the upper-primaries. *Ageing* The best ageing character is the tail-band: on juveniles it is wider, duller and less clear-cut, whereas on adults it is narrow, black and sharply defined (both above and below) often with two or three narrower black bands towards the base (especially on males). Adults have dark eyes, juveniles pale. *Adult male* Compared with juveniles, adult males average darker (but lack the warmer tones of Common Buzzard). They have a dark head and breast (forming a hood) whilst the black belly patch is finely barred. They also have more heavily patterned underwing-coverts and perhaps more heavily barred under-secondaries and inner primaries. *Adult female* More like juvenile, having a paler hood than the male, less patterned underwings and a more solidly dark belly; unlike juveniles, they show a distinct black trailing edge to the underwing. The upperwings of adults are darker than juvenile's with a less conspicuous pale patch at the base of the primaries. Note in particular that adult Rough-legged (especially females) often has a whitish U-shaped area on the lower breast, between the dark upper breast and the black belly (although many Common Buzzards also show this).

Pale Common Buzzards Some pale Common Buzzards can look similar to Rough-legged (see above) so the objective evaluation of size, structure, behaviour, date and locality are essential. Remember that pale Common Buzzards rarely show (a) a clear-cut black-and-white tail (any white is *usually* restricted to the uppertail-coverts) or (b) a pale patch at the base of the upper-primaries. Their carpal patches tend to be crescent-shaped (round or oblong on Rough-legged). Also, unlike Rough-legged, pale Commons often show creamy-white upperwing-coverts and usually lack a black lower-belly patch.

Honey Buzzard *Pernis apivorus*

Structure and flight Similar in size to Common Buzzard but noticeably slimmer and less heavy. It floats around like a sailplane, often twisting the tail like a kite. Clearly establish the following features. **1** WING POSTURE Honey soars (and glides) on *flat* wings, with the tips often slightly *drooped* (never in Common Buzzard's shallow V, but note that Common Buzzard sometimes glides on flat wings). **2** HEAD SHAPE A *small* pointed head protrudes prominently (strongly recalling Common Cuckoo *Cuculus canorus*); note also that the head is narrow, not broad. **3** WING SHAPE In flapping flight the wings are proportionately longer and narrower than Common Buzzard's, the primaries often looking straighter and narrower. However, when soaring, the wings appear distinctly broad and rounded (or 'paddle-shaped') with very rounded tips and rather bulging secondaries that have a 'pinched-in' effect where they meet the body. **4** WINGBEATS Slow, deep and rather elastic, with the wings noticeably bowed on the down-stroke, recalling a kite. **5** TAIL The tail looks proportionately long and narrow, and rather paddle-shaped when closed; when soaring it appears very full with a

Female Marsh Harrier. Slim bodied with relatively long tail.

Juvenile Honey Buzzard. Small head, long tail.

Juvenile Rough-legged Buzzard. Larger and longer-winged than Common

Rough-legged Buzzard. Raised 'arm', flatter 'hand'. ▼

Common Buzzard ▶

▲ Wings broadest where secondaries and primaries meet.

◀ **Common Buzzard.** Stocky, shortish tail.

◀ **Honey Buzzard.** Flat 'arm', drooped 'hand'.

◀ **Dark juvenile Honey Buzzard.** Small protruding head, long tail.

Adult Common Buzzard. Typically, pale primary bases, extensive barring and dark trailing edge to wing, pale crescent across breast.

◀ **Common Buzzard**, pale type. Has dark primaries and pale wing-coverts.

'Nipped in' base to tail, rounded tip.

Marsh Harrier. All-dark variant. Unbarred primaries and tail.

Common Buzzard, very ▶ pale type. Note underbody pattern compared to Rough-legged Buzzard. Upperparts often extensively white too. ▼

Juvenile Rough-legged Buzzard. White bases to primaries and tail, ▶ black belly patches.

▲ ◀ **Adult male Rough-legged Buzzard.** Several distinct tail-bars and dark carpal patch.

Juvenile Rough-legged Buzzard. Pale headed, black belly.

◀ **Adult female Rough-legged Buzzard.** Single dark tail-bar. ▶

Adults have strong black trailing edge to underwing.

Adult male Honey Buzzard. Note yellow eye, tiny bill and dull cere. Long tail and weak feet. Typical males have greyish face and secondaries with broad black tips.

Adult male Honey Buzzard. Note tail pattern and sharp black trailing edge to wing.

Adult male Honey Buzzard. Barred type.

Adult male Honey Buzzard. Rufous type.

Adult male Honey Buzzard. Dark type. Tail pattern distinctive.

Adult male Honey Buzzard. Pale type. All males show sharp black trailing edge to wing and wide pale area to secondary bases.

Adult female Honey Buzzard. Note wing and tail patterns.

Adult female Honey Buzzard. Compare tail and underwing with male.

Adult female Honey Buzzard. Typically browner than males.

Juvenile Honey Buzzard

Juvenile Honey Buzzard. Note long tail, dark eye and yellow cere.

Juvenile Honey Buzzard, dark type. Dark secondaries evenly barred, fewer (but broader) bars than on Common Buzzard.

Juvenile Honey Buzzard of rufous type. May resemble Common Buzzard, but note long tail, barred body sides and axillaries, and extensive black to the fingers.

Juvenile Honey Buzzard. Pale type.

121

rounded tip. **6** The overall effect sometimes recalls a huge *Accipiter*, particularly when viewed side-on. Note that the juvenile is slightly shorter-winged and shorter-tailed than the adult.

Plumage Underparts extremely variable and complicated, from dark and rather uniform through medium and rufous to very striking individuals with a white head, underparts and underwing-coverts; see Forsman (1999) for more detail. Note that all show a large oval or rectangular black carpal patch on the underwing (as opposed to the more crescent-shaped patch on pale Common Buzzards). *Adult* On darker birds, there is a marked contrast between the dark body/underwing-coverts and paler primaries and secondaries. The following should be carefully noted. **1** UNDERPARTS BARRING Most show obvious barring on the breast, belly and underwings. **2** WING-TIPS Black extends around the tips and then along the tips of the secondaries to form a black trailing edge to the underwing. **3** PRIMARY AND SECONDARY BARRING Black bars across the bases. **4** TAIL A black band across the tip with one or two narrow, less obvious bands towards the base. *Sexing* **1** MALES Distinctly grey on the head and upperparts, and barred on the upperwing. The black trailing edge to the underwing is broad and well defined, but the under-primaries and secondaries are barred only across the bases, so the distal part of the feathers (behind the dark wing-tips and trailing edge) is unbarred. The tail usually shows only one narrow bar across the base. They also tend to be barred on the whitish belly. **2** FEMALES Browner above, lacking prominent barring. The under-primaries and secondaries are more evenly barred, with a narrower and less obvious trailing edge. Unlike males, they may show translucent inner primaries when backlit. The tail usually shows *two* narrow bars across the base (not one). They tend to be blotchy on the underparts, hence more uniform at a distance. *Juvenile* **1** GENERAL APPEARANCE Most are a uniform warm dark brown with noticeably paler under-primaries. As a consequence of their dark plumage, many juveniles are surprisingly similar to Common Buzzard, but are much plainer (the yellow cere being prominent as a result). However, like adults, the head and body plumage varies considerably and intermediate and very white individuals also occur. **2** PRIMARY AND SECONDARY BARRING Three or four rather faint bars across the primaries and heavier bars across the secondaries. **3** TRAILING EDGE Also a *diffuse* dark trailing edge to the underwing (note that juvenile's under-secondaries appear dark in all morphs). **4** TAIL A less clear-cut terminal band to the tail and the dark bars across the base are also less noticeable. The upper tail is plainer than the adults' and the feather tips are more tapered. **5** UPPERTAIL-COVERTS Many show contrasting whitish uppertail-coverts. **6** UPPERWING PANEL Many show a pale brown panel on the upperwing-coverts. **7** BARE PARTS At close range, they have a dark eye and a yellow cere (yellow eye and blue-grey cere on adult). Dark juveniles could also be confused with a high-flying juvenile Marsh Harrier, but that species glides on raised wings. The harrier's tail is longer and slimmer in travelling flight and it usually lacks a pale under-primary patch.

Display Adults have a unique display that facilitates even long-range recognition: they raise their wings vertically over their back and then quiver them in a most peculiar manner. This occurs after an upwards swoop or in a series of descending swoops.

Calls Adults and begging juveniles have a disyllabic or slightly trisyllabic call, rather penetrating and more mournful than Common Buzzard.

References Flood *et al.* (2007), Forsman (1999).

Golden and White-tailed Eagles

Where and when Golden Eagle breeds only in mountainous areas of n. and w. Scotland (including the Hebrides), with a population of nearly 450 pairs. It has bred in the English Lake District, but not in recent years. It is rarely seen away from the Scottish breeding range, but dispersing immatures occasionally reach as far south as n. England. Having become extinct in Britain in about 1916, White-tailed Eagle has been re-established in w. Scotland with an increasing population (57 breeding pairs in 2011) mainly in the Hebrides (a re-introduction project is also underway in Ireland). Dispersing Continental migrants sometimes occur in winter in e. England, but their appearances remain erratic. Such birds frequent coastal marshes and farmland.

Moults Although it moults annually, Golden Eagle does not moult all of the body plumage or the flight feathers in any given moult, so they usually show different ages of feathers. White-tailed has a complete annual moult, but does not moult all of the flight feathers (Forsman 1999, which see for further details of moult and ageing).

Golden Eagle *Aquila chrysaetos*

Size and structure A huge raptor, with a wingspan *c.* 70% greater than Common Buzzard *Buteo buteo*. Despite this enormous difference, bear in mind that it can be surprisingly difficult to judge the size of high-flying raptors. With Golden Eagles, this problem may be exacerbated by the huge scale of the landscapes that they inhabit. Consequently, structure is especially important in their identification, particularly in their separation from White-tailed Eagle. Golden has very long, broad, well-fingered wings. They are narrower at the base, being 'pinched in' where the rear edge of the wing meets the body, producing a curved trailing edge to the inner wing (most obvious on juveniles). The tail is long (similar in length to wing width) and broad and square in shape. The head and neck protrude noticeably (an obvious difference from Common Buzzard). The wings are held either flat or in a distinct but shallow 'V', the latter a distinctive difference from White-tailed Eagle. The wings may, however, be arched, with a drooping 'hand' (Forsman 1999). In flapping flight, it has deep, slow wingbeats, followed by a prolonged glide (with primaries swept slightly back, the profile of the leading edge appearing rather S-shaped). When soaring, the wings are pushed slightly forward.

Plumage *Adult* Uniformly dark brown but, in a reasonable view, the crown and nape appear paler, the golden colour often clearly visible at closer range. Other plumage features are subtle: on the underwing, the basal primaries and secondaries are grey (lightly barred darker), forming a contrast with the brown underwing-coverts. The tail may also appear distinctly two-toned, the base being similarly grey with a broad dark tip (but some adults have an all-dark tail). When visible, the upperwing shows an irregular and variable pale panel across the coverts. Yellow legs may also be noticeable in a good view, particularly in favourable light. Unlike adult White-tailed, the bill is dark at all ages (with a yellow cere and gape-line). *Juvenile* The most distinctive plumage. Initially, juveniles are immaculate, very dark and uniformly brown, with a pale blond crown and nape (whitish to yellowish-golden). Most distinctive is the strikingly white tail with a broad and prominent black terminal band.

In flight, they also have a large white patch at the base of the under-primaries (extending onto a few outer secondaries). They may also show a smaller white patch on the bases of the upper-primaries, but juveniles lack the adult's pale panel on the upperwing-coverts. *Second-year* Similar to juvenile but plumage more worn and so distinctly paler; also shows a faded panel on the upperwing-coverts. *Older immatures* Golden Eagles take about five years to reach maturity, the areas of white in the wings and tail gradually diminishing over that period (rate varies individually). However, white in the base of the tail may be retained for five or six years, distinguishing subadults from full adults.

White-tailed Eagle *Haliaeetus albicilla*

Size and structure Appears huge, particularly if seen in the context of the English country-side (in flight, aptly likened to a 'barn door'). Long, broad wings, with slightly narrower 'hands', rounded wing-tips and fingered primaries. In flight, easily separated from Golden Eagle by a combination of (1) its front-heavy appearance, produced by a long, thick neck and a noticeably heavy bill, and (2) its short, slightly wedge-shaped tail (because of their white tail, adults can appear almost tail-less when viewed against the sky). It flies on flat or slightly bowed wings, with the primaries gently curved up at the ends (when soaring it can raise the wings into a very slight V). Juveniles show a distinctly uneven, 'serrated' trailing edge to the wing (straighter in adults). In flapping flight, it has deep wingbeats and a lazy, lolloping flight action, vaguely recalling Grey Heron *Ardea cinerea*. With its large bill, short tail and bulky body, it appears almost vulture-like on the ground.

Plumage *Adult* Brown, but distinctly paler yellowish-brown or whitish-brown on the head, neck and upperwing-coverts (a mixture of pale and dark brown often gives a rather 'moth-eaten' appearance). Older birds may look very pale-headed. A large, deep and strongly decurved yellow bill should be obvious; legs also yellow (unfeathered, unlike Golden Eagle). Tail white, very striking and obvious against a dark background but can often 'disappear' when viewed against the sky. *Juvenile* Very dark brown with feathers that are regularly patterned and of uniform age. The head is dark brown, but the neck and underparts are thickly streaked creamy-buff or golden-buff; more heavily patterned on the mantle, scapulars and wings (buff feathers with a brown teardrop or arrowhead pattern). The underwings are darker with a distinctive large, contrasting, messy pale patch on the axillaries and inner wing-coverts (retained almost into adulthood; Forsman 1999); also variable lines of white streaks on the underwing-coverts. The tail feathers have dark tips and outer webs but pale inner webs; the tail therefore appears darker when closed, but predominantly pale buffy-white when spread (although some show less white). Bill dark (with pale lores often obvious) and legs yellow. Like Golden Eagle, it takes about five or six years to reach adulthood. *Second-year* Exact pattern variable but often uniquely pale, with a dark head and a patchy mixture of a whitish-streaked upper breast and a large pale area on the lower breast, which in turn contrasts with the dark thighs. The back, scapulars and upperwing-coverts are often heavily patterned with whitish; in flight, the latter sometimes appear quite plain and sandy at a distance, contrasting strongly with the dark brown primaries and secondaries. The bill may begin to turn pale yellow. *Older immatures* Third-year plumage is usually somewhat intermediate between immature and adult-type. Thereafter, the plumage gradually becomes more adult-like. For more details, see Forsman (1999).

Adult Golden Eagle. Pale nape and patchy plumage.

Adult Golden Eagle. Wings held forward and raised in soaring flight. Adult dark below, large tail.

Subadult Golden Eagle.

Juvenile Golden Eagles. White bases to primaries and tail above and below.

Adult Golden Eagle. Strong upperwing-covert panel.

Juvenile Golden Eagle. Uniformly dark plumage, golden nape, white tail base.

White-tailed Eagle. Soars on flat wings.

Juvenile White-tailed Eagle. Huge, parallel-edged wings, short, wedge-shaped tail. Centres of tail feathers white.

Juvenile White-tailed Eagle. ▶ Dark droplet pattern of wing-coverts diagnostic.

2nd-year White-tailed Eagle. Often pale, and moult increases pale marks in wings, axillaries and lower body. Dark head and breast.

3rd-year White-tailed Eagle. Tail now mostly white. ▲

125

Behaviour Often feeds over water, plucking fish and birds from the surface (sometimes being attracted to offal from fishing and 'eagle-watching' boats). Also feeds on carrion.

Reference Forsman (1999).

Falcons: Peregrine, Merlin, Hobby and Red-footed Falcon

General identification problems Given a good view, the separation of Peregrine, Merlin and Hobby is not difficult but, when seen poorly, all three can be confused. Remember that falcons seen very briefly are a real problem, as are birds high in the sky, when size cannot be accurately judged. Be prepared to leave such birds unidentified.

Peregrine *Falco peregrinus*

Where and when Peregrine breeds mainly on cliffs, in both coastal and inland locations, as well as on buildings in towns and cities. It has increased spectacularly in recent years, although it remains commonest in the north and west. Outside the breeding season, many disperse to coastal areas, particularly estuaries, or to inland sites where there is an abundance of suitable prey.

Flight, structure and behaviour A large, spectacular falcon. Shape distinctive: heavy, thickset and deep-chested, with broad-based pointed wings (although the extreme tips can appear rather blunt, particularly when soaring); it has a medium-length tail. Viewed from the side, the body looks rather cigar-shaped. It is likely to be seen high in the air, hanging in the wind or circling upwards before stooping at great speed at waders, pigeons, thrushes or other suitable prey. Compared to Merlin and Hobby, it is rather stiff-winged with shallow wingbeats in which only the tips are obviously moved (flight can recall Fulmar *Fulmarus glacialis*). When soaring and gliding, the wings are often held slightly bowed below the horizontal.

Plumage *Adult* Blue-grey above, with a variably barred, darker tail. It has a prominently white breast but the rest of the underparts look pale whitish-grey at a distance, an effect produced by delicate black barring on a white background. Thick black moustachial stripes stand out strongly against the white face and throat. *Juvenile* Browner above and darker below, the underparts heavily streaked brown; dark moustaches stand out strongly against a creamy face. It has a slightly paler forehead and sometimes a pale supercilium, which curves back towards the nape. Cere blue (yellow on adults).

Calls May be quite vocal in breeding areas, adults having a variety of calls including a persistent anxious slow, gruff *kyaa kyaa kyaa kyaa kyaa* ... or a loud and emphatic *kee kee kee kee kee*.... Also a weak *chjik*. In winter, a gruff, slightly muffled barking *achik achik* or *archk archk* may be heard.

Pitfalls Besides confusion with both Merlin and Hobby, the identification of Peregrine and other large falcons is complicated by the not infrequent occurrence of falconers' escapes, such as Lanners *F. biarmicus*, Sakers *F. cherrug*, Laggars *F. jugger* and even hybrids. When faced with an odd falcon, in particular with a possible dark-plumaged Gyrfalcon *F. rusticolus*, be sure to eliminate these various escapes. A complicated subject: see Forsman (1999).

Tundra Peregrine

In recent years there have been a number of reports of juvenile Peregrines resembling the Arctic races *calidus* (n. Europe) or *tundrius* (Canada and Greenland). Being a long-distance migrant, the latter is probably a likely vagrant to Britain. Such birds may appear slightly slimmer and longer-winged than nominate *peregrinus* and show extensive white on the cheeks, a prominent supercilium that curves down the sides of the nape, narrower moustaches and narrow brown underpart streaking on a creamy background. Such birds may suggest juvenile Lanner. The problem, however, is that nominate *peregrinus* may also produce juveniles that closely resemble these races, so the positive identification of such birds is fraught with difficulty.

Merlin *Falco columbarius*

Where and when Merlin breeds on moorland, mainly in Scotland, N. England and Wales (and locally in Ireland) but it is more widespread in winter (mostly early September to early April) when there is a more general dispersal and some immigration. It is typically found on coastal and estuarine marshes, but also very sparsely inland, generally in open country.

Flight, structure and behaviour A small falcon, the male being only about the size of a Mistle Thrush *Turdus viscivorus*, but the female is nearly as big as a Kestrel *F. tinnunculus*. Compact with sharply pointed but relatively short, angled, swept-back wings and a medium-length tail. This distinctive shape readily separates it from both Hobby and Kestrel. However, when circling, the wings can look quite rounded at the tips. It is most likely to be seen flying low over moorland or coastal marshes with a fast, dashing flight in which the sharply pointed, angled-back wings produce a rather 'flicking' flight action. It will sometimes rise to some height and fly at speed towards unsuspecting prey, with the wings intermittently closed-in tight against the body, producing a flight action strongly reminiscent of a fast-flying Mistle Thrush (but note that Sparrowhawk *Accipiter nisus* also does this). When chasing birds at close range, it can be extremely persistent, following every twist and turn of its quarry. It frequently uses prominent perches, such as posts and stones, when the wing-tips fall well short of the tail-tip (unlike Hobby, which has the wings similar in length to the tail).

Plumage Note that all ages show weak moustachial stripes and a dark area on the ear-coverts, so they appear markedly plain-faced compared to Hobby and Peregrine. *Adult male* Adult males are in the minority and note that they are actually quite rare in southern areas in autumn and winter. Blue-grey above, with a broad, dark subterminal tail-band (some also show narrower dark barring towards the base). The underparts vary from beige to dark orange, with broad brown streaking. *Adult female* Similar to juvenile (see below) but tends to be greyer above with pale feather fringes producing a somewhat barred appearance. The outer tail feathers (visible from below) are less regularly barred than juvenile. *Juvenile* Most autumn Merlins seen in lowland Britain are juveniles. They appear neat, with dark brown upperparts and whitish-buff underparts heavily lined with thick chocolate streaks (maybe some barring on the flanks). From above, the primaries, secondaries and tail are heavily barred with buff (unlike Sparrowhawk); the tail shows a noticeable whitish tip. In fresh plumage, the brown upperparts are tinged rufous (a result of rufous feather tips and fringes).

Falcons in soaring flight. Peregrine appears heavy and compact. Merlin has a similar shape but is small and slight. Hobby appears longer-winged and shorted-tailed whilst Kestrel can appear long-tailed.

Juvenile Peregrine. Thick set, blunt wing-tips.

Female Merlin

Juvenile Hobby. Dark, catches insects in flight.

Juvenile Merlin. Although similar in shape, Merlins are much smaller than Peregrine.

Juvenile Peregrine is heavy-bodied and compact. Wing-width equals tail-length.

Female Kestrel. Slim-winged, long tail with obvious terminal band.

Juvenile Merlin. Appears slight-bodied and large-headed.

Juvenile Peregrine. Dark hood, streaked underparts.

Merlin in hunting flight. Low and rapid interspersed with short glides, wings held close to body.

Juvenile Merlin. Small and big-headed. Neatly barred tail.

Juvenile Sparrowhawk. Note yellow eye, barred underparts and long, slender legs.

'Merlins' seen hunting along country lanes in front of the car are invariably Sparrowhawks.

128

Calls On the breeding grounds, its calls include a shrill, chattering *kik-ik-ik-ik* and an anxious *kee-kee-kee...* recalling Kestrel.

Pitfalls Confusable with both Hobby and Peregrine but note that, by October, when most migrant Merlins appear in lowland Britain, Hobbies are rare. A notorious beginner's bird and claims of Merlins in atypical situations, particularly inland, frequently involve male Sparrowhawks, which occasionally bunch their primaries to produce a pointed wing shape. When hunting, Sparrowhawks may behave like Merlins, flying at speed close to the ground and partially closing their wings in pursuit of prey (note that 'Merlins' flying low along the road in front of the car are invariably male Sparrowhawks). Because of these problems, special care is needed when views are brief and a Merlin should not be claimed unless a clear and unambiguous view is obtained of its wing-shape. Note also that, whereas Merlins tend to land on exposed perches, Sparrowhawks usually land in trees. In summer, a less likely confusion species is Common Cuckoo *Cuculus canorus*, which also has sharply pointed wings but is easily separated by its long, graduated tail, pointed head/bill profile and slower, unhurried flight with stiffer, shallower wingbeats.

Hobby *Falco subbuteo*

Where and when A summer visitor from late April to early October, breeding mainly in s. England, with a small number in Wales and a few pairs in Scotland. Like Peregrine, it has increased significantly in recent decades. The largest populations occur in s. England, particularly on heathland, but it also breeds widely in agricultural areas. Large spring concentrations hawk dragonflies over selected southern wetlands.

Flight, structure and behaviour Arguably the most beautiful of our commoner falcons, with a distinctive silhouette: slim, with very long, narrow, pointed, scythe-shaped wings and a shortish to medium-length tail. Its shape can suggest a gigantic Common Swift *Apus apus* or, when hawking insects low over the water, a Black Tern *Chlidonias niger*. Far more conspicuous in their aerial feeding behaviour than other falcons, often feeding in loose parties, floating around like sailplanes over heaths, downs, lakes and marshes, twisting and turning as they snatch dragonflies and other insects in their talons, before transferring them to the bill. They also pursue birds and, like Peregrine, spend some time circling to a great height before stooping into flocks of swifts and hirundines, their favoured prey. Note that House Martins *Delichon urbicum* give an anguished shrill *shrip shrip* bird-of-prey alarm. It is worth learning this call as it is often the first indication of the presence of a Hobby. Also, feeding Common Swifts (and associated hirundines) will often, in unison, suddenly and silently fly fast and direct in the same direction; when this happens, look *behind* the moving flock as there may well be a Hobby in pursuit. Although they perch on fence posts and bare branches, they usually land within the tree canopy (unlike Merlin). Like Red-footed Falcon, they will sometimes feed from fence posts in a shrike-like manner.

Plumage *Adult* Grey above with a heavily streaked white breast and belly; distinctive reddish thighs and vent may 'flash' at a distance. Black moustachial stripes stand out prominently against the white face; at close range, it also shows a narrow, short white supercilium. *Juvenile* Pale brown feather fringes on the upperparts produce a slightly browner appearance than the adult, the rump being browner still; the underparts are buff (heavily streaked dark) and note that the vent and undertail-coverts are also buff, an obvious difference from the adult. Juveniles also show a pale trailing edge to the wing and, more obviously, a whitish tip to the tail. *First-summer* Some are similar to adults but many are browner on the upper-parts, a result of variable amounts of retained, faded, juvenile feathering; also, the vent and undertail-coverts are a pale washed-out orange.

Calls Often quite vocal and calls are useful in locating Hobbies hidden in trees. The commonest call is a distinctive *kee kee kee kee kee...* distinctly lower, deeper and slightly huskier than Kestrel. It also gives a shrill *kerr-it-it* in aggressive encounters. Juveniles may give a muffled dry, throaty *ick ick.*

Pitfalls High-flying Hobbies can be confused with Peregrine but note the latter's greater bulk and broader wings. High-flying Merlins are often misidentified as Hobbies, but note the shape differences outlined above. Kestrel will, on occasion, also hawk flying insects just like a Hobby, so be cautious when identifying distant individuals.

Red-footed Falcon *Falco vespertinus*

Where and when A rare spring and summer vagrant (very rare in autumn), currently averaging *c.* 14 records per year although as many as 125 were seen following persistent easterly winds in the spring of 1992. Most occur in s. and e. England, in similar habitats to Hobby, but it has occurred throughout Britain; very rare in Ireland.

Behaviour Most likely to be found hunting from bushes, telephone wires or fence posts in a rather shrike-like manner, often perching with partially drooped wings, rather like a Cuckoo. Alternatively, they may hunt from the ground, flapping short distances or awkwardly leaping and hopping after prey. They also hawk flying insects like a Hobby. Sometimes remarkably tame.

Flight, structure and shape Similar in shape to Hobby, for which it could be mistaken in a brief view, but the wings are not quite as long, while the tail is longer and more obviously rounded when spread; thus its shape is slightly more Kestrel-like. Flight similar to Hobby but more leisurely, slightly flappier, with perceptibly deeper wingbeats. Unlike Hobby, it persistently hovers, but with noticeably deeper wingbeats than Kestrel.

Plumage Great care is required when identifying poorly seen Red-footed Falcons (particularly 'fly-overs'), records of which are always carefully scrutinised by records committees. Note in particular that Hobby is superficially similar in shape, and in certain lights it can look uniformly dark below; therefore, it is essential to obtain a good, prolonged look at a potential Red-foot. Four basic plumage types occur here. *Adult male* Note that this is one of the rarest plumage types to occur in Britain (*c.* 15% of all records). Uniformly blackish-grey but, importantly, *on the upperwing the primaries and secondaries are always conspicuously pale silvery-grey.* Rufous vent and undertail-coverts, bright orange feet, cere and eye-ring. *First-summer male* This age occurs much more frequently than adult male. It has a variable body moult in its first-winter but the juvenile flight feathers and tail are retained as are, usually, most of the juvenile greater coverts. It therefore lacks the strong silvery-grey colour of the adult male's upper-primaries and secondaries. It shows obvious traces of immaturity as follows. **1** An off-white throat and orange coloration on the sides of the neck and upper breast (exact colour individually variable). **2** A strongly-barred juvenile tail. **3** Juvenile barring on the greater coverts and tertials. **4** Heavily barred juvenile under-

Hobby in typical hunting flight.

1st-summer male Red-footed Falcon. Appears dark below, recalling Hobby.

Adult Hobby. Dark, long-winged with shortish tail. White on head often obvious.

Adult female Red-footed Falcon. Orange underparts.

Juvenile Hobby. Browner above. Note head pattern. Juvenile lacks russet 'trousers'. Tail is plain above with whitish tip.

Adult male Red-footed Falcon. Pale wing-tips diagnostic.

Juvenile Hobby. Scythe-shaped wings, dense patterning below.

2nd-autumn Red-footed Falcon. Contrast in old and new wing feathers.

1st-summer male Red-footed Falcon. Dark above and below with pale 'face' and barred tail.

Juvenile Red-footed Falcon. Thin breast streaking.

Melanistic Kestrel. Wing-tips fall well short of tail-tip. Lacks red 'trousers'. Note yellow feet.

Juvenile Red-footed Falcon. Breast streaks thinner than on Hobby. Tail barred above.

Juvenile Hobby. Finely fringed pale above, long wings.

1st-summer female Red-footed Falcon. 'Washed out' underparts, mixed age feathers in wing-coverts.

Adult female Red-footed Falcon

131

wings. **5** Orange-yellow legs, cere and eye-ring. Since most first-summer males have a paler throat/upper breast, this age/sex is that most likely to be confused with Hobby. *Adult female* Perhaps more likely to be passed off as a Kestrel, but the underparts and crown are orange-buff, sometimes strikingly dark. It has a prominent black 'highwayman's mask' around the eye (extending into a short moustache) and a partial white collar that extends around the neck-sides towards the nape. The mantle and wings are blue-grey, barred dark grey. Both the tail and the under-primaries/secondaries are barred, with noticeable dark tips to the under-primaries. The underwing-coverts are conspicuously orange-buff (although this colour may be restricted to the leading coverts). Eye-ring, cere and legs orange or orange-yellow. *First-summer female* Far more likely to be seen than adult female (see 'first-summer male' above for an outline of first-winter moult). Separable as follows: **1** Streaked on the crown and nape (plain on adult). **2** Larger dark face mask. **3** Streaked below (little or no streaking on adult). **4** Noticeably browner above, with plain outer greater coverts (barred on adult); the rump and barred tail may look quite pale when spread (almost sandy). **5** Underwings entirely barred. Beware of confusion with browner first-summer Hobby. *Juvenile* This age and plumage is extremely rare in Britain. Surprisingly similar to juvenile Hobby, but buff and rufous feather fringes make it scalier above. It has an obviously barred tail (barring is confined mainly to the inner webs on Hobby, so is not readily apparent when viewed from above). It also has a paler forehead and forecrown with a dark 'highwayman's mask' and a short moustache; also a more extensive white collar around the neck-sides. Streaking on the underparts is not usually as heavy as on juvenile Hobby.

Pitfalls Besides Hobby (outlined above) another pitfall with males is melanistic Kestrel. The bird portrayed on p. 131 is loosely based on one such individual seen near Cardiff in July 1986. It had a blackish head and underparts and very dark upperparts, but it was easily identified at rest by the wing/tail ratio (on Kestrel the wing-tips fall 1–2cm short of the tail-tip, on Red-foot they project *c.* 0.5cm *beyond*); in addition, it showed typical Kestrel bare-part colours, it lacked the rufous vent and undertail-coverts of Red-foot, and showed traces of barring right across the upperparts.

Amur Falcon *Falco amurensis*

An extreme vagrant with just one British record (Yorkshire, September–October 2008). However, possibly overlooked. Very similar to Red-footed Falcon but adults are easily separated by their unbarred gleaming white underwing-coverts. First-years do not show white until it gradually appears during the autumn moult into second-winter plumage. Any Red-foot showing obvious white on the axillaries and underwing-coverts should be photographed and expert assistance obtained.

Reference Forsman (1999).

Ringed, Little Ringed and Kentish Plovers

Where and when Ringed Plover is a widespread breeding and wintering bird (mainly coastal) with numbers greatly inflated in spring and autumn by passage migrants (mainly Greenland breeders); also occurs inland on migration, sometimes in reasonable numbers if water levels are low. Little Ringed Plover is a relatively recent colonist (first bred in 1938) but is now a widespread summer visitor, with more than 900 pairs in a 2007 survey. Although associated with English gravel pits, it has now spread on to rivers in Wales and Scotland. It occurs more widely on migration, from March to May and June to September. A former breeder, Kentish Plover is now a rare migrant, currently averaging 25 records a year, mainly on south and east coasts; it is very rare elsewhere and is unlikely to be seen inland.

Ringed *Charadrius hiaticula* and Little Ringed Plovers *C. dubius*

General features Little Ringed Plover is smaller and slimmer than the dumpy Ringed Plover, with a distinctly more attenuated and tapered rear end. These differences are very obvious when the two species are together. In flight, the best distinguishing feature is Little Ringed's lack of a wing-bar: Ringed has a conspicuous broad white bar, whereas Little Ringed shows at best only tiny pale tips to the greater coverts, which are very difficult to detect in flight.
Calls Totally different: Ringed gives a soft, mellow *poo-ip*, with an upward inflection, whereas Little Ringed gives a thin, abrupt, rather whistling *tee-u*, inflected downwards.
Songs Both have a slow, bat-like display flight. Ringed Plover's song is a musical *tuluwee tuluwee tuluwee...* (snatches sometimes given in winter). Little Ringed gives a fast, pulsating *pre-pre-pre-pre...* and also a high, rolling *arrreeoo...arrreeoo...*
Plumage and bare parts *Adults* At rest, Little Ringed has dull horn or flesh-coloured legs and a fine black bill. In contrast, summer Ringed Plover has bright orange legs and a stubby black-tipped orange bill; winter bare-part colours are much duller. Further differences include Little Ringed's obvious yellow orbital-ring (conspicuous in breeding condition) and much narrower breast-band. Female Little Ringed often has extensive brown feathering on the otherwise black ear-coverts and are generally duller and less smart than the males. Winter adults of both species have the blacks duller and browner. *Juveniles* Juvenile Ringed Plovers may appear obviously darker and more 'chocolate' looking than accompanying adults, with yellowish-green, not orange legs. When identifying juvenile Little Ringed at rest, concentrate on the head: unlike Ringed, the forehead is buff, often golden-buff, merging *gradually* with the rest of the crown; the buff supercilium is faint or virtually lacking, and the ear-coverts are more or less concolorous with the crown. All this combines to produce a distinctive *hooded effect*, which Ringed Plover lacks; there is also a fine, inconspicuous pale orbital-ring. Juvenile Ringed Plover has a clear-cut white forehead and supercilium, and darker ear-coverts, producing a far more contrasting pattern. Like adults, juvenile Little Ringed has a narrower breast-band, while the paler, sandier upperparts may be surprisingly obvious when the two species are together. Very fresh juveniles have golden fringing to the upperparts, which quickly fades to buff.

Kentish Plover *Charadrius alexandrinus*

Size, structure and behaviour Intermediate in size between Ringed and Little Ringed, but with rather a front-heavy 'chick-like' shape (much less attenuated at the rear than either

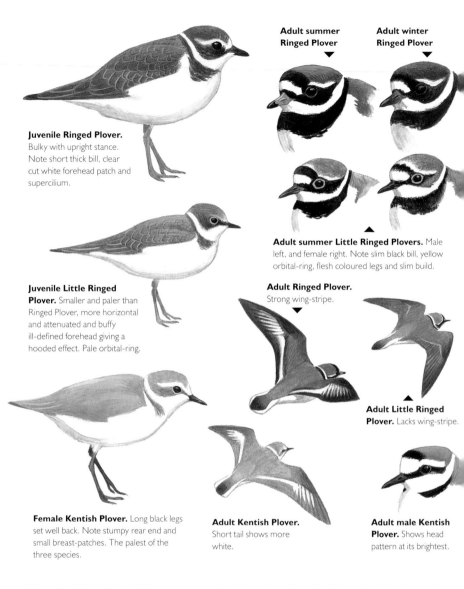

Juvenile Ringed Plover. Bulky with upright stance. Note short thick bill, clear cut white forehead patch and supercilium.

Adult summer Ringed Plover

Adult winter Ringed Plover

Juvenile Little Ringed Plover. Smaller and paler than Ringed Plover, more horizontal and attenuated and buffy ill-defined forehead giving a hooded effect. Pale orbital-ring.

Adult summer Little Ringed Plovers. Male left, and female right. Note slim black bill, yellow orbital-ring, flesh coloured legs and slim build.

Adult Ringed Plover. Strong wing-stripe.

Adult Little Ringed Plover. Lacks wing-stripe.

Female Kentish Plover. Long black legs set well back. Note stumpy rear end and small breast-patches. The palest of the three species.

Adult Kentish Plover. Short tail shows more white.

Adult male Kentish Plover. Shows head pattern at its brightest.

Ringed or Little Ringed). It may appear rather long-legged, particularly when the plumage is sleeked down in hot weather. Tends to be rather active, often running along the beach like a Sanderling *Calidris alba*.

Call Migrants give a rather soft, Sanderling-like *tip* or *fwit*, occasionally sounding rather more metallic; sometimes a faintly disyllabic *ki-kip*.

Plumage Individuals of all ages are easily identified by their pale sandy plumage (much paler than Ringed Plover), blackish or dark grey legs, fine black bill and narrow patches confined to the breast sides. Females and juveniles are pale sandy-brown and white, juveniles having pale feather fringes and a less well-defined head pattern; on both, the black eye and black legs stand out from the pale plumage. By late summer, males attain female-like winter plumage, but close scrutiny usually reveals slightly darker breast patches and eye-stripe. Males acquire breeding plumage early, and by the New Year they are rather dandy with neat black breast patches, a thick black eye-stripe and a black forecrown bar. The rear crown is variable, but most males show at least some bright cinnamon-orange coloration and some have the whole rear crown this colour and are very striking. In flight, Kentish shows a narrow white wing-bar (narrower than Ringed) and white sides to the tail.

Semipalmated Plover *Charadrius semipalmatus*

A very rare North American vagrant (three records to 2012) but probably overlooked. Very similar to Ringed Plover but most likely to be located by its distinctive call: a markedly disyllabic upslurred *ch-wee*, reminiscent of Spotted Redshank *Tringa erythropus*. Detailed examination of the bill, head and feet would then be required (and photographs essential).

Large plovers: Grey, European Golden, American Golden and Pacific Golden Plovers, and Dotterel

Grey Plover *Pluvialis squatarola*

Where and when A common coastal winter visitor and passage migrant, some remaining throughout the summer; small numbers occur inland on passage and in winter.

Structure A large, bulky, rather hunched, long-legged plover with a large black eye and a hefty, thick, blunt bill; the latter is particularly useful when separating distant individuals from European Golden Plover.

Flight identification and call A large, grey plover with a prominent white wing-bar that contrasts with the black primaries and primary coverts. Easily separated from Golden by its square white rump and prominent black axillaries, which contrast strongly with the white underwing-coverts. The call is diagnostic: an evocative clear, mournful whistle: *wee-oo-eeeee*.

Plumage *General* Compared to Golden, Grey Plover is a pale, colourless, grey-and-white wader, but note that juveniles are distinctly buffer. If in doubt, wait until it flies. *Adult summer* Easily identified by blackish upperparts, spangled with white (lacking the yellow tones of Golden), a striking white forehead and a supercilium that runs down the sides of the neck and bulges on the breast-sides. The underparts are black, but with a white rear belly and undertail-coverts. In full plumage, females are slightly browner below and show white feathering within the black. First-summers remain in their winter quarters and do not acquire full breeding plumage; they may become very bleached and worn. *Juvenile* Distinctive, but also potentially confusing. Fresh plumage is immaculate: grey-brown above, neatly and beautifully spangled with pale yellow. The underparts may also look smoothly and uniformly pale yellowish-buff, but closer views reveal delicate breast streaking. The pale yellow tones may provoke confusion with Golden Plover or, especially, American Golden. Compared to the latter, Grey has a browner crown, a less distinct supercilium and an obviously larger bill. Note, however, that the yellow tones soon fade to whitish as autumn progresses. *Winter* Essentially a pale grey plover with diffuse whitish upperparts spotting and fringing, and a mottled grey upper breast. It has a subdued supercilium and a darker patch through the eye.

European Golden Plover *Pluvialis apricaria*

Where and when A familiar farmland species in winter (abundant in some areas) and also breeds on moorland in upland areas of n. and w. Britain, as well as in Ireland. It occurs on passage in a wide variety of habitats, and in winter will feed and roost in more saline environments.

Structure A medium-sized plover, similar in shape to Grey, but slightly smaller and stockier with a narrower, weaker bill.

Flight identification and calls Distinctly yellowish-brown in overall plumage tone. Looks rather uniform above, lacking Grey's white rump. The wing-bar is less conspicuous, confined mainly to the bases of the primaries. The underwing-coverts are silvery-white, lacking Grey Plover's prominent black axillaries. It forms large and impressive flocks, often with Lapwings *Vanellus vanellus*. Golden Plovers tend to fly at some height, bunching together and flying fast and direct, their underwings flashing white from a distance. From below, the wings look long, narrow and pointed. Easily identified by call: a soft, rather mournful *too-lee* or *tloo*, unobtrusive yet distinctive, and peculiarly evocative of frosty winter mornings. The calls vary both in length and composition, and large flocks are noisy and conversational. On the breeding grounds, the male sings a beautiful, mournful *poo-wee-oo* (rising on the middle syllable).

Plumage Easily identified by its obviously brown-and-yellow plumage tones. *Adult summer* Varies according to range (with much individual variation). Northern individuals acquire more extensive summer plumage and are completely black from the face to the belly, with a broad white supercilium extending down the neck-sides, broadening on the breast-sides before continuing down the flanks to the vent/undertail-coverts. Southern birds are greyer and more mottled on the face and throat, with black only in the centre of the belly; consequently, the white on the sides of the neck and breast is more extensive. The upperparts are strongly spangled with yellow. *Winter and juvenile* Blackish upperparts are liberally spangled dark yellow, while the dingy underparts have a streaked/mottled yellow-buff breast and flanks, and a whiter belly. Tends to look rather plain-faced with a large black eye; the supercilium is ill-defined and subdued (although stronger on some) and there is a dark patch on the rear of the ear-coverts. Adult winter and juvenile plumages are very similar, but autumn juvenile is neater, slightly more streaked on the breast and slightly greyer on the belly.

American Golden *Pluvialis dominica* and Pacific Golden Plovers *P. fulva*

General These two species were formerly lumped together as 'Lesser Golden Plover' and the old name provides the first clue to their identification. Both are distinctly smaller and slighter than European Golden. The key feature on which to concentrate is the underwing.

Whereas European Golden has pure white underwing-coverts and axillaries, both American and Pacific are grey in this area. When separating these two vagrants from each other, their occurrence patterns are also relevant. Whereas American tends to occur in the same kind of dry habitats as European Golden, Pacific is usually encountered in the same type of saline environments as Grey Plover. In addition, whereas most Americans occur in autumn (mainly juveniles) most records of Pacific relate to summer-plumaged or moulting adults in late summer (July/August).

American Golden Plover

Where and when An annual vagrant, currently averaging 16 records a year. Most are juveniles in northern and western areas in September/October, but adults and first-summers also occur in spring and late summer, often on the east coast. Very rare in winter.

Structure In direct comparison, American is smaller, slimmer and proportionately taller looking than European (*c.* 80% its weight). When active, it often appears very slim, sleek and attenuated, with a small head, long neck and slim, rather pear-shaped body and long legs; consequently, it usually looks more upright and rather gangly. European usually looks quite fat and rounded in comparison but, when resting, the two species can appear more similar. The most obvious structural difference is that American has long primaries that extend in a scissor-like manner well beyond the tail. Equally important is the relative length of the exposed primaries in relation to the overlying tertials. On American, the primary projection is approximately equal in length to the overlying tertials, whereas on European it is about one-quarter the length (about three-quarters on juvenile Grey Plover, usually shorter on adults). Also, the tips of the tertials on American fall well short of the tail-tip, whereas on European they fall closer to the tip itself (but note that some Europeans show shorter tertials). American's long-winged appearance is related to the fact that many undertake a huge trans-oceanic migration over the w. Atlantic to wintering grounds in s. South America.

Behaviour Unlike most Europeans, often very tame. When feeding, it pivots forward at nearly 45 degrees, whereas European tips its body much less obviously, simply reaching down with its head and bill.

Flight identification Similar to European Golden but, in direct comparison, obviously smaller and slighter with shorter, narrower wings. Most important is a *dull and dusky underwing*: the lesser and median coverts and axillaries are quite dark grey (silvery-white on Golden) with the greater coverts and remiges paler grey. The wing-bar is also pale grey (white on European).

Calls See below.

Plumage At all ages it is greyer, less yellow than European. *Adult summer* Note that *American Golden usually retains much of its summer plumage until arrival on its winter quarters*, so autumn adults are often conspicuous amongst winter adult or juvenile Europeans; consequently, any 'Golden Plover' in autumn showing obvious traces of summer plumage is always worth a second look. In summer plumage, adult American has more extensive black on the underparts than European, males being solidly black *right down to the undertail-coverts*. A smartly contrasting area of white extends from the supercilium, down the neck-sides, bulging prominently onto the sides of the breast. Females, however, may be less solidly black below, while moulting birds usually display a patchy mixture of black and white. Like juveniles,

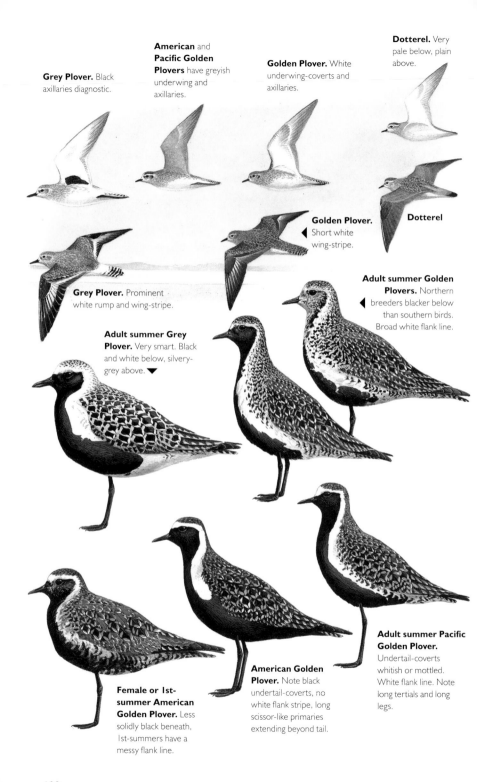

Grey Plover. Black axillaries diagnostic.

American and **Pacific Golden Plovers** have greyish underwing and axillaries.

Golden Plover. White underwing-coverts and axillaries.

Dotterel. Very pale below, plain above.

Dotterel

Golden Plover. Short white wing-stripe.

Grey Plover. Prominent white rump and wing-stripe.

Adult summer Golden Plovers. Northern breeders blacker below than southern birds. Broad white flank line.

Adult summer Grey Plover. Very smart. Black and white below, silvery-grey above. ▼

Female or 1st-summer American Golden Plover. Less solidly black beneath, 1st-summers have a messy flank line.

American Golden Plover. Note black undertail-coverts, no white flank stripe, long scissor-like primaries extending beyond tail.

Adult summer Pacific Golden Plover. Undertail-coverts whitish or mottled. White flank line. Note long tertials and long legs.

Juvenile American Golden Plover.
Strong white supercilium, dark cap, plainer breast, much greyer than European Golden Plover – lacks yellow tones.

Juvenile European Golden Plover. Yellowy tint to head and breast. Large dark eye. Yellow-spangled upperparts. Wing-tips equal tail length.

Note scissor-like primaries extend well beyond tail, tips of tertials fall short of tail-tip.

Juvenile Pacific Golden Plover. Yellow plumage tones and obvious supercilium. Note longer tertials equal to tail length.

Juvenile Dotterel. Prominent white supercilium, buff below, scalloped above, yellow legs.

Both American and Pacific Golden Plovers are smaller, slighter and longer-legged than Golden Plover.

Juvenile Grey Plover. Spangled above with pale buff, often suggesting Golden Plover.

Adult winter Grey Plover. Plumage essentially grey. Note heavy bill.

summer Americans are greyer and less colourful above, although closer views reveal fine but extensive pale yellow notching across the back and scapulars. Late autumn adults are more faded, such birds often looking worn and very 'monochrome' (sometimes lacking all yellow tones). They then show a striking thick white supercilium and a solid black cap, delicate spangling and fringing across the upperparts and patchy black underparts. *Juvenile* Once they lose their summer plumage, European Golden Plovers are not easy to age, winter adults and juveniles both appearing rather coarsely patterned and distinctly yellowish. In comparison, juvenile Americans appear cold, grey and colourless. Their upperparts are delicately spangled with buff or white, but they lack the *strong* yellow tones of European (but may appear yellower on the rump); their buff underparts are paler and plainer, immaculately streaked and mottled with grey (lacking European's stronger breast-band). Most distinctive is a prominent whitish supercilium that contrasts with a dark crown, the latter producing a capped effect. Front-on, they can look strikingly white-faced (obvious also in flight). *Winter* Remember that, unlike European, late autumn individuals may retain strong traces of summer plumage. In full winter plumage, it has an obvious white supercilium and capped appearance, but the upperparts are plainer and greyer with the spotting more diffuse (many feathers are fringed rather than notched). The underparts are whiter, diffusely mottled and shaded with pale grey. Its plumage is less neat than juvenile and plainer when worn. *First-summer* Acquires very little summer plumage until late April (O'Brien *et al.* 2006), appearing 'winter-like' before then: dull, colourless, worn and often rather messy across the upperparts. Consequently, spring vagrants show variable but often only limited amounts of black feathering on the underparts, the white supercilum and general greyness then being the most obvious differences from European. Some remain on their wintering grounds and acquire very little summer plumage.

Pitfalls Double check that any potential juvenile American is not a juvenile Grey Plover (see p. 135). Note also that very grey Golden Plovers occasionally occur, so it is essential to check not only the plumage features, but also the structural differences and the underwing.

Pacific Golden Plover

Where and when A very rare vagrant, currently averaging three records a year. Most are adults in July and August, mainly in the Northern Isles and on the east coast. Records have, however, occurred in most months, but it is very rare in winter. As with all vagrant e. Palearctic waders, juveniles are peculiarly rare. Most occur in saline habitats, whereas American tends to occur in fields and freshwater environments.

Structure Breeding as it does in Siberia, between European and American Golden Plovers, Pacific is intermediate in appearance. It is useful to think of it as being structurally similar to American but with plumage more similar to European. By weight it is the smallest of the three golden plovers. Being *c.* 60% the size of European, it looks noticeably smaller, slighter and proportionately longer-necked and longer-legged; it can look quite thin and lanky, particularly in hot weather with the plumage sleeked down. Note that photographs often show Pacific with very long legs (with long, exposed tibia) but many of these are taken in the Far East where hot weather induces feather-sleeking; vagrants at cooler latitudes appear more fluffed-out and, consequently, shorter-legged. Compared with American, it tends to look longer-legged, rounder-bodied and perhaps longer-necked, making it look somewhat front-heavy when feeding; it also has a longer bill. Its primaries do not project beyond the tail in the same scissor-like manner as American. It is much more similar to European in this respect, the *exposed primaries being only about one-quarter to half the length of the overlying tertials* (equal on American). Also, whereas the tertial tips fall well short of the tail-tip on American, they fall much closer to it on Pacific.

Flight identification Appears similar to European Golden, but smaller and slighter. Most importantly, the underwing-coverts and axillaries are dusky-grey, like American (white on Golden). The feet project beyond the tail to produce a more attenuated rear end, more so than both American and European.

Calls See below.

Plumage Its plumage is yellower than American, being similar to European. *Adult summer* Pacific is similar to 'Northern' European Golden Plover, being yellowish above (although perhaps more coarsely patterned). Compared to American, the white supercilium, neck and breast stripe bulge less prominently into the breast-sides; also, the white extends narrowly right down the flanks, where it becomes rather messy due to the intrusion of a certain amount of black feathering (note, however, that some female Americans and also moulting late summer males may be similar in pattern). Although variable, like American the black extends from the belly onto the vent and undertail-coverts but only rarely are the undertail-coverts completely black. *Juvenile* In overall appearance, juvenile Pacific suggests a small, slim, long-legged, gangly, front-heavy European Golden Plover with grey underwings and axillaries. Its plumage is quite different from juvenile American, being yellower and more coarsely patterned (therefore, much more similar to European). Pacific also differs from American by its more subdued facial pattern, showing a less obvious supercilium and capped effect (face and supercilium yellower in tone). *Adult winter* Unlike European, many retain traces of summer plumage into early winter. The upperparts are brown with old feathers showing whitish feather notching and fringing. Any new feathers, particularly on the scapulars, have noticeably *yellow* notching or fringing. The supercilium is more prominent than European and, unlike American, both it and the face may show a distinct yellow tone. *First-summer* As American Golden, there is a variable body moult in their first spring. Some acquire summer plumage by early May, but others acquire only partial summer plumage, while others retain winter-type plumage, the latter usually remaining behind on the wintering grounds (O'Brien *et al.* 2006).

Hybrids Presumed hybrids between European Golden and Pacific Golden have occurred. One seen in Somerset superficially resembled Pacific but had white axillaries and underwing-coverts (Vinicombe 1988). It is, therefore, essential to check for such anomalies when faced with a potential Pacific Golden Plover.

Calls Although American and Pacific may sound noticeably different from European, separating them from each other is much more difficult, particularly since they have a variety of calls with which most European observers are unfamiliar. Although similar, it is their quality and delivery that are different. American sounds slower and squeakier, Pacific quicker, more cheerful and more urgent. The latter's 'Spotted Redshank *Tringa erythropus* call' is perhaps the most distinctive. *American Golden* Gives a rather slow, upslurred *tu-wee* (sometimes down-slurred), strikingly shrill and squeaky compared to European Golden, being higher-pitched, less mournful and less haunting; it can also give a 'thicker' more urgent version of this: *che-wit*. Also a more mournful *t-wee-oo* (emphasis on the middle syllable) which may recall a squeaky swing. ***Pacific Golden*** Gives a mellow but quicker, more cheerful, more urgent *tu-weet* or *chu-wee*, again with the second syllable upslurred, or a *tu-wee-u* (the second syllable upslurred and the third simply an insignificant lower-pitched add-on). It also gives a quicker, more energetic *tu-ick* or *chu-it* that sounds like a quick Spotted Redshank *Tringa erythropus*, but fuller and mellower.

Dotterel *Charadrius morinellus*

Where and when Breeds almost exclusively in Scotland (500–750 pairs during a 1999 survey) but occurs more widely on migration, often on hilltops and moorland, or in coastal fields; in some areas, small flocks may appear annually in the same favoured fields.

General features Smaller and stockier than the previous four species and easily separated, particularly in summer plumage. Unlike the three 'golden plovers', its legs are pale: yellow or yellowish-brown. It is often very tame, but larger flocks may be more timid. Often hunkers low to the ground when 'spooked'.

Flight identification Resembles a small Golden Plover and looks similarly narrow-winged. High overhead, it may not be readily identifiable if its smaller size is not apparent. The upper-wings and rump are plain, but the tail is noticeably blacker, with white right around the edge. The leading underwing-coverts are white, the greater underwing-coverts and remiges pale grey.

Calls The flight call is a low, rolling, downward-inflected guttural purring *prrrrr*, rather odd but distinctive (a quiet version is sometimes given on the ground). The contact call in flight is a subtle, rather abrupt, soft, plaintive, slightly mournful *pyoo*, vaguely resembling an abrupt European Golden Plover.

Plumage As their sexual roles are reversed, females are brighter than males. ***Adult summer*** Easily identified. At all ages has stunning, broad white supercilia, starting above the eyes and meeting in a V on the back of the head. In summer, these are offset by a dark brown crown and a dark line through the eye. The throat is white but the upper breast is grey, separated from the orange underparts by a narrow white band across the breast. The belly is blackish on females, duller on males. The upperparts are brown, with buff feather fringes. Females are very bright, but males are duller both on the crown and below, and some may be difficult to sex. ***Juvenile*** The vast majority of autumn migrants in Britain are in juvenile plumage. Immaculate and rather buff-looking. Prominent creamy-buff or golden-buff supercilium, strongest from the eye back, contrasting with a blackish crown. The underparts are golden-buff, with fine dark grey streaking and mottling across the upper breast, a narrow pale band on the lower breast, which in turn is bordered below by a dark band of diffuse grey mottling. Blackish-brown upperparts are prominently spangled whitish. ***Winter*** Similar to juvenile but duller and browner, and the grey-brown upperparts are fairly plain, showing chestnut or buff feather *fringes*, not notches.

References O'Brien *et al.* (2006), Pym (1982), Vinicombe (1988).

Little and Temminck's Stints and Sanderling

Where and when Little Stint is a rather scarce passage migrant, mainly from late April to early June and from July to October. Numbers of autumn juveniles vary annually, with late August to early October being the peak time; a few winter, mainly in s. England. Temminck's Stint is a much rarer migrant, currently averaging 120 a year, with a peak of 309 in 2004. It occurs mainly in May and August/September, mostly in e. England; it occasionally breeds in Scotland and very rarely winters. Sanderling is a winter visitor to sandy coastlines, but larger numbers of Greenland migrants pass through western areas mainly in May and July–September; small numbers occur inland on migration (and occasionally in winter, particularly during freezing weather).

Little Stint *Calidris minuta*

General features A tiny wader, whose small size should always be apparent, even if no other species are present for comparison. It has quick, energetic, rather jerky feeding actions, but at other times can be slower and even plover-like. It is often prone to fast, Sanderling-like runs. A dumpier, longer-legged and more upright bird than Temminck's. The primaries are usually rather long, extending well beyond the tertials and just beyond the tail, forming quite a pointed rear end (which is often angled upwards when feeding, emphasising this pointed impression). Note, however, that on adults in particular, the primary projection is variable, with some showing a much shorter projection. The short bill easily separates it from Dunlin *C. alpina* and the black legs are an important difference from Temminck's. It looks very small in flight and is easy to pick out among Dunlin flocks.

Call An unobtrusive yet distinctive *tip* or *tip tip tip*.

Plumage *Adult summer* A bright and well-patterned stint compared with Temminck's. In spring, their overall plumage tone is very variable, mainly because fresh migrants have delicate whitish fringes to their new summer-plumaged head, breast and upperpart feathers; these produce a pale whitish-grey or 'frosted' look. However, these fringes soon wear away to reveal a much more colourful, rich buff or chestnut appearance, the strong chestnut tones being an important distinction from Temminck's (and also the rare Semipalmated Sandpiper *C. pusilla*). The breast-band is strong but also variable, buff to chestnut, delicately mottled blackish and usually fading to white in the middle. Pale edgings to the mantle usually form a V when viewed from the rear, but this is rarely as obvious as on autumn juveniles. It has a variable supercilium and an important feature of Little Stint in all plumages is that it subtly forks on the crown, just before the eye (most other stints lack this faint 'upper supercilium'). Behind the eye is a variable buff-brown to chestnut ear-covert patch (with a narrow white eye-ring). Note also that, *at a distance*, summer-plumaged Little Stints can look rather uniform, thereby prompting confusion with Temminck's. Returning autumn migrants in July and August are much more worn, having lost all traces of their grey spring 'frosting'. Again, they are very variable, many showing a messy mixture of black, buff and chestnut on their upperparts as well as a fairly obvious mantle V. However, heavily worn individuals look much blacker above, the more colourful buff and chestnut tones having largely worn away and/or faded to whitish. Some can, therefore, look quite black, buff and white in overall plumage tone. The acquisition of their first grey winter feathers can reinforce this impression of drabness. ***Winter*** The upperparts become uniformly pale grey but each feather shows a noticeable black line down the centre and this often extends either side of the shaft to form a narrow, pointed black wedge; on better-marked individuals, this produces a distinctly patterned impression, even at a distance. The breast-sides are also grey, mottled darker. It has an indistinct supercilium (with a split supercilium effect visible in closer views). Some, presumably first-winters, may also retain a hint of a paler mantle V. Some acquire winter plumage by late summer, and the presence of such grey individuals among the more familiar juveniles often perplexes the inexperienced. ***Juvenile*** Most autumn Little Stints, from mid August onwards, are juveniles. They are immaculately patterned with golden, chestnut and buffish feather fringes (strongly rufous-fringed on the upper scapulars at least). Most distinctive is the characteristic white V down the sides of the mantle, formed by pale feather edgings. The crown and ear-coverts are also rich buff, with a white 'split supercilium'. The underparts appear very white, with shading and diffuse streaking confined to a patch on the breast-sides. By late autumn, juveniles fade considerably and look much greyer, particularly when their first grey winter feathers appear; also, the white mantle Vs gradually become much less obvious. Note that a minority begin body moult by early September, some virtually losing their white mantle Vs and becoming very drab on the mantle. The overall appearance of such birds may suggest Semipalmated Sandpiper *C. pusilla* (see p. 147).

Temminck's Stint *Calidris temminckii*

General features An unobtrusive, secretive stint that is easily overlooked. Occurs almost exclusively in freshwater habitats. Tends to be rather solitary, creeping around on flexed legs in a slow, furtive, mouse-like manner (although it can be more energetic). The body looks long, low and horizontal and, unlike Little Stint, the tail projects beyond the primaries to produce a rather attenuated rear end. Compared to Little Stint, the plumage always looks dull and plain, and at all ages its general appearance recalls a diminutive Common Sandpiper *Actitis hypoleucos*. The head and upperparts are dull brownish-grey with an obvious and well-defined breast-band. The head is relatively plain but shows traces of a faint supercilium and a distinct but narrow pale eye-ring. *The key feature* is leg colour, which varies from yellowish through greenish-yellow and olive-green to brown (always black on Little Stint, but beware of the effects of mud staining). The flight is fast and twisting on swept-back wings and a flicking action, perhaps recalling a hirundine at a distance; it frequently towers when flushed. Pure white sides to the tail can be obvious on take-off or landing, but frustratingly difficult to see in normal flight. These are a feature not shared by any other *Calidris*.

Call Completely different from Little Stint: a soft, fast, mouse-like trill *si-si-si-si-si* that can be surprisingly loud at close range, perhaps sounding more like *prrrrooooip*. Note that Little Stint sometimes strings several tip calls together, but the sound is harder, drier and quite different.

Plumage *Adult summer* Pale brownish-grey above, close views revealing black feathering on the scapulars, often forming a band. The black feathers are edged with buff or chestnut. ***Winter*** Similar to summer, but plainer and greyer, lacking the black scapular feathering and obvious pale fringes to the wing-coverts. ***Juvenile*** Similar to winter but, at close range, the

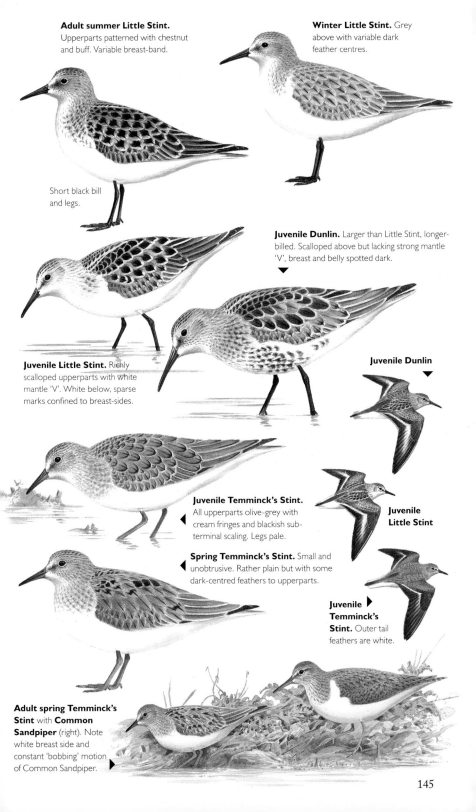

Adult summer Little Stint. Upperparts patterned with chestnut and buff. Variable breast-band.

Winter Little Stint. Grey above with variable dark feather centres.

Short black bill and legs.

Juvenile Dunlin. Larger than Little Stint, longer-billed. Scalloped above but lacking strong mantle 'V', breast and belly spotted dark.

Juvenile Little Stint. Richly scalloped upperparts with white mantle 'V'. White below, sparse marks confined to breast-sides.

Juvenile Dunlin

Juvenile Temminck's Stint. All upperparts olive-grey with cream fringes and blackish sub-terminal scaling. Legs pale.

Spring Temminck's Stint. Small and unobtrusive. Rather plain but with some dark-centred feathers to upperparts.

Juvenile Little Stint

Juvenile Temminck's Stint. Outer tail feathers are white.

Adult spring Temminck's Stint with **Common Sandpiper** (right). Note white breast side and constant 'bobbing' motion of Common Sandpiper.

145

upperparts are neatly fringed black and buff, producing fine, delicate and subtle scaling. It may show small black centres to the rear scapulars, forming a slight dark V when viewed from behind.

Sanderling *Calidris alba* (illustrations on pp. 152 and 161)

General features Although quite different from the previous two stints, Sanderling does, nevertheless, recall an oversized stint. This is mainly because, unlike Dunlin, it has a relatively short black bill. The traditional image of Sanderling is of a very pale hyperactive 'clockwork toy' chasing the waves up and down the beach. In such circumstances, it is easily identified but, in its less familiar summer and juvenile plumages, the species can prove far more confusing. This is particularly the case if seen out of context at, for example, an inland reservoir. In such 'atypical' situations it may be far more lethargic (although still prone to running off at speed). Easily identified in flight by its very broad white wing-bar (much broader than Dunlin). Winter Sanderling looks very pale in flight. Unlike other waders, it lacks a hindclaw (close-range views are required to establish this).

Call A hard, dry, monosyllabic *kik* or *pit*.

Plumage *Winter* In its most familiar plumage, it appears rather white-headed with very pale grey upperparts and contrasting black bill and legs. It often shows an obvious blackish area at the bend of the wing, but this may be concealed by adjacent overlapping breast feathering. *Adult summer* A body moult into summer plumage starts on the wintering grounds, so spring migrants, which are commonest in May, are in very fresh summer plumage. Two distinct types occur in spring. Many appear 'cold and frosty' in tone, all the small feathers of the head, breast and upperparts having black centres with broad white fringes, producing profuse but delicate black-and-white mottling. The scapulars and wing-coverts have black centres but they too have white fringes. Some retain a few winter feathers mixed-in. There may be a slight chestnut tone to both the ear-coverts and scapulars. As the spring advances, the white fringes wear away to reveal rich chestnut on the feather bases, so later birds are often strikingly chestnut-brown on the head, upperparts and breast-band, appearing rather stint-like as a consequence. There is, however, considerable individual variation and some look distinctly two-toned: predominantly grey on the head and mantle, but chestnut on the scapulars and wing-coverts. Returning adults reappear in July–August, by which time many have become severely worn, faded and rather messy, with winter plumage often starting to appear. Such birds may again appear rather 'frosty' and colourless, lacking the strong chestnut tones of late spring. *Juvenile* Immaculate and quite unlike both adult summer and winter. The entire mantle and scapulars are black, with heavy white spotting that produces a beautifully spangled appearance to almost the entire upperparts. The crown is also black (spangled white) but the face and underparts are predominantly pure white, lacking summer adult's breast-band; instead, a small area of shading and streaking is restricted to the breast-sides. This plumage is lost in the autumn body moult.

Reference The identification of stints, including the rarer species, is very thoroughly dealt with by Grant & Jonsson (1984).

The rare stints: Semipalmated, Western and Least Sandpipers, and Red-necked and Long-toed Stints

Semipalmated and Western Sandpipers

Where and when Semipalmated Sandpiper is an annual vagrant, currently averaging three or four records a year. Most are storm-driven juveniles in September–October, mainly in western areas, with a few summer adults in July–August (sometimes on the east coast) and also the occasional spring bird. Western Sandpiper is a great rarity (eight records to 2011). Again, there have been juveniles in September–October, a few adults or first-summers in late summer and a single spring record. Both species have also wintered.

Semipalmated Sandpiper *Calidris pusilla*

Juvenile The first step is to separate 'Semi-p' from Little Stint *C. minuta*. The most obvious initial difference is that juvenile Semipalmated is much duller, colder, greyer and more uniform, appearing grey-buff or grey-brown above, lacking both the rich rusty tones and the white mantle Vs of juvenile Little Stint. The following are the main features to check. **1 UPPER-PARTS** Juvenile Little Stint has well-patterned upperparts, the most distinctive feature being the characteristic white mantle Vs (two white lines down the sides of the mantle). Two less well-defined white lines are also present on the tips of the second row of scapulars. The rest of the upperparts are black and brown, the individual feathers showing a mixture of white and rich buff or strong rusty fringes. To all intents and purposes, Semipalmated lacks white mantle Vs (it may show just a faint pale line down the sides of the mantle), while the regularly patterned upperparts are duller and browner, a consequence of paler and colder-toned feather fringing. The overall impression is much more uniform than Little Stint; their upperparts are in fact reminiscent of juvenile Curlew Sandpiper *C. ferruginea*, with similar neat scalloping. One specific feature shown by Semipalmated is the presence of black anchor-shaped markings on the lower scapulars and greater coverts, which Little Stint lacks (it has heavier, more diamond-shaped marks). These take the form of a thick black shaft streak at the base of the feather, which broadens into a black anchor shape towards the tip (although there is some variation in the precise shape). ***Early-moulting juvenile Little Stints*** A significant pitfall to bear in mind is that some juvenile Little Stints commence their post-juvenile body moult by early September, with advanced individuals acquiring first-winter mantle feathers as early as mid September. Such birds appear very drab on the upperparts and lack white mantle Vs, so they bear a strong resemblance to juvenile Semipalmated. If in doubt, such birds appear to be more lined on the mantle than 'Semi-p' (rather than scalloped) and may show a *hint* of a paler V; check also structural differences, particularly the primary projection (longer on Little Stint) and foot-webbing. By October, such drab greyish birds inevitably become much more frequent, so particular caution needs to be exercised in late autumn. **2 HEAD PATTERN** Semipalmated has a uniformly streaked forehead and crown, producing something of a capped effect (it lacks the obvious 'split supercilium' of Little Stint). Below this, a white supercilium extends back from the bill, narrowing over the eye and then flaring at the rear. A dark line extends from the bill back across the lores, through the eye and fans out behind the eye to form an ear-covert patch which is often clear-cut and well defined. Another important point is that the eye shows a narrow but quite noticeable white eye-ring, most of which is contained entirely within the ear-covert patch. This head pattern gives Semipalmated a characteristic facial expression that is very distinctive once learnt and is subtly but distinctly different from that of Little Stint. The latter has a fairly white forehead, a 'split supercilum' (the supercilium splits so that there is a narrow 'upper supercilium' forking up into the sides of the crown) and a much weaker and more diffuse area behind the eye that fails to form a discrete and solid ear-covert patch as it does on Semipalmated. **3 BILL** Little Stint's bill tapers to a relatively fine tip, whereas Semipalmated's is slightly thicker at the base and is relatively broad right down to the rather blunt-ended tip (close views may reveal slight lateral broadening at the tip). This creates a characteristic 'tubular' impression. Many Semipalmated have relatively short and rather thick-based bills, but note that some eastern females are much longer-billed (see below). **4 UNDERPARTS** Little Stint has an area of rather diffuse streaking on the breast-sides, whereas Semipalmated *tends* to show a patch of more neatly defined, crisper streaking. **5 PRIMARY PROJECTION** Little Stint has a long primary projection which, surprisingly, is about half to two-thirds the length of the overlying tertials. Semiplamated's primary projection is much shorter, perhaps approaching one-third of tertial length at a maximum (but some larger females can apparently show a longer projection). Most have the primaries projecting only slightly beyond the tertial tips and some have their primaries virtually covered by them. Consequently, Little Stint looks significantly longer-winged (often with quite a pointed rear end at a distance). **6 CALL** Little Stint gives a quiet, insignificant *tip* or *tip tip*. Semipalmated has a soft, thin, slightly more rolling call, variously transcribed as *cherk, chlip, chip* or *prip*, distinctly softer and less penetrating than that of Little Stint. It is sometimes repeated in quick succession. **7 SIZE AND STRUCTURE** Semipalmated is slightly larger, bulkier, more 'hump-backed' and longer-legged than Little Stint, but these differences are minor and they are, of course, much less apparent in the absence of a direct comparison. **8 PALMATIONS** As its name suggests, Semipalmated shows palmations (tiny webbing) between its toes, which Little Stint lacks (this is strongest between the middle and outer toe). In the field, the angle between the toes therefore looks rounded, whereas it appears pointed on Little Stint. It must be stressed that close views are required to see this, preferably with the bird front-on and on dry, unvegetated ground. Beware of the effects of mud between the toes.

Adult Adults do not normally occur here in late autumn; instead the few that turn up are much more likely to be encountered in July and August. Adult Semipalmateds at this time resemble juveniles in that they are dull, cold and rather colourless, but they appear greyish, rather than brownish – in fact they lack any brown tones to their plumage. In consequence, they often give the impression of being in winter plumage rather than summer, but careful scrutiny reveals that their head, breast and upperparts are in fact heavily patterned. Unlike adult Little Stint, which is quite a rich rufous-brown in summer, late summer adult Semipalmated is essentially a cold, greyish-toned bird, because the black upperparts feathering has pale buff or white fringing. The breast is finely streaked black on a cold whitish background, petering out towards the centre, and the head pattern is similar to that of juvenile, having a uniformly streaked crown (with no split supercilium), a well-defined ear-covert patch, a narrow pale eye-ring and a whitish supercilium. Like the juvenile, it does not usually commence its moult into winter plumage until arriving on the wintering grounds, although a few grey feathers may start to appear in the upper scapulars by mid August (these have a fine black 'hair-line' central shaft-streak, rather than Little Stint's thicker, more diffuse black central streak).

Western Sandpiper *Calidris mauri*

Juvenile 1 STRUCTURE Although Western is traditionally viewed as a confusion species with Semipalmated Sandpiper, it shares many of its characters with Dunlin. In fact a useful *aide-mémoire* is to think of Western not as a 'long-billed stint', but as a 'miniature Dunlin'. It has a similar shape to Dunlin, being slightly longer-legged, more upright, larger-headed and flatter-backed than Semipalmated, but the most obvious similarity is its bill, which often appears slightly downcurved. Although quite thick at the base, it is quite long and *tapers to a fine tip*. This is fundamental to its separation from Semipalmated which, as stated above, has a rather thick, blunt-ended, 'tubular' bill. It must be stressed, however, that the bill is longest on females and shortest on males, so that some male Westerns show shorter bills than some female Semipalmateds (this is where the *shape* of the bill becomes important). Like Curlew Sandpiper *C. ferruginea*, Western also has the habit of wading into deep water and immersing its head below the surface to feed. It is thus less typically 'stint-like' than Semipalmated. **2** UPPERPARTS Much more colourful than Semipalmated, with two lines of *bright rufous-fringed* upper scapulars that contrast with the duller, greyer lower scapulars and wing-coverts. Its back and mantle feathers may also be rufous-fringed and it may show quite noticeable narrow white mantle Vs. The more rufous appearance to the crown, mantle and scapulars is best appreciated when the bird is front-on. Like Semipalmated, the lower scapulars and greater coverts are greyer and they too show dark anchor-shaped marks behind the tip, but the bases of these feathers are slightly paler and the black marks are more pointed in shape – appearing as 'arrowheads' rather than 'anchors' – and they are less well defined. Like Little Stint, however, the plumage tends to fade somewhat late in autumn, prior to the post-juvenile moult. **3** HEAD PATTERN Like Semipalmated, it is has a uniform crown – although this tends to be fairly rufous – a whitish supercilium that curves up behind the eye, a well-defined ear-covert patch and a narrow white eye-ring, but it tends to differ in that the supercilium is often broader and the ear-covert patch paler and more diffuse. **4** UNDERPARTS Whiter than Semipalmated, no doubt because of the smaller, less well-defined area of more random streaking on the breast-sides. **5** MOULT A very useful short cut in their separation is the timing of their moult. Semipalmated migrates whilst still in juvenile plumage and most do not start to moult until they arrive on their wintering grounds (although the odd grey winter scapular or wing-covert can occasionally be seen by late autumn). Western, however, is like Dunlin in that it often – but not always – commences its post-juvenile moult whilst on migration (Grant & Jonsson 1984). This means that many autumn migrants show plain grey winter feathering (with a narrow black shaft-streak) mixed in with the rufous-fringed juvenile upper scapulars. This area of its plumage may also appear somewhat messy, as a consequence of it having already lost some of its juvenile feathering. **6** CALL Also reminiscent of Dunlin: a thin, shrill, high-pitched *chreeep*, *treet* or short *jeet*. Also a *teet teet treep* when flushed. **7** PALMATIONS Like Semipalmated, Western also has palmations between its toes.

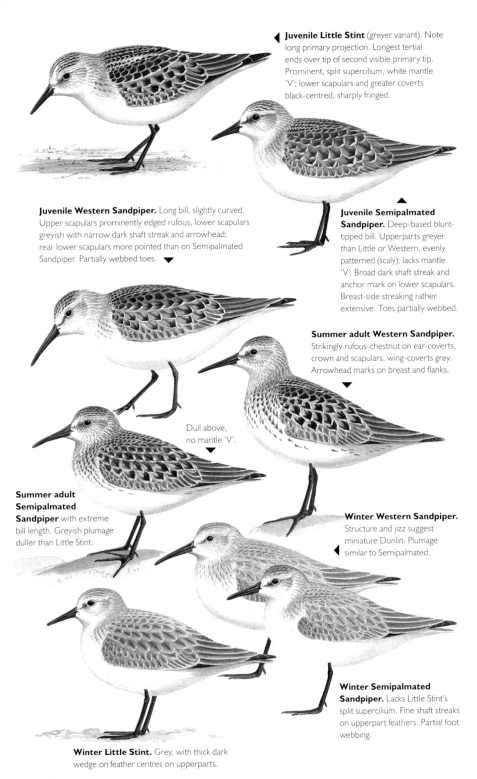

Juvenile Little Stint (greyer variant). Note long primary projection. Longest tertial ends over tip of second visible primary tip. Prominent, split supercilium, white mantle 'V'; lower scapulars and greater coverts black-centred, sharply fringed.

Juvenile Western Sandpiper. Long bill, slightly curved. Upper scapulars prominently edged rufous, lower scapulars greyish with narrow dark shaft streak and arrowhead; rear lower scapulars more pointed than on Semipalmated Sandpiper. Partially webbed toes.

Juvenile Semipalmated Sandpiper. Deep-based blunt-tipped bill. Upperparts greyer than Little or Western, evenly patterned (scaly); lacks mantle 'V'. Broad dark shaft streak and anchor mark on lower scapulars. Breast-side streaking rather extensive. Toes partially webbed.

Summer adult Western Sandpiper. Strikingly rufous-chestnut on ear-coverts, crown and scapulars, wing-coverts grey. Arrowhead marks on breast and flanks.

Dull above, no mantle 'V'.

Summer adult Semipalmated Sandpiper with extreme bill length. Greyish plumage duller than Little Stint.

Winter Western Sandpiper. Structure and jizz suggest miniature Dunlin. Plumage similar to Semipalmated.

Winter Semipalmated Sandpiper. Lacks Little Stint's split supercilium. Fine shaft streaks on upperpart feathers. Partial foot webbing.

Winter Little Stint. Grey, with thick dark wedge on feather centres on upperparts.

150

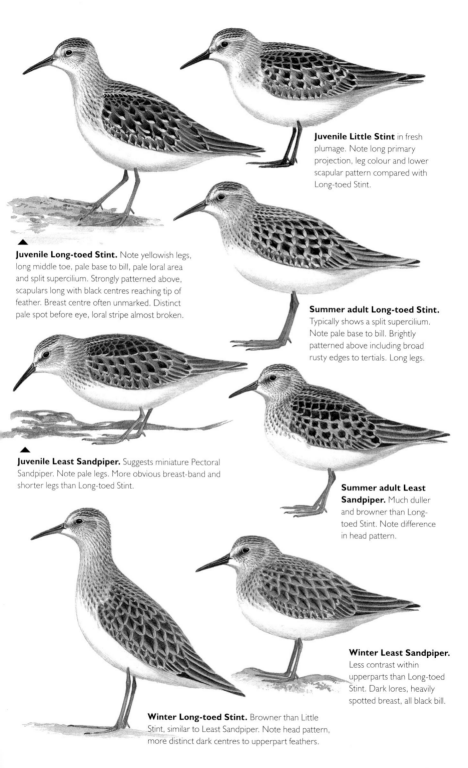

Juvenile Little Stint in fresh plumage. Note long primary projection, leg colour and lower scapular pattern compared with Long-toed Stint.

Juvenile Long-toed Stint. Note yellowish legs, long middle toe, pale base to bill, pale loral area and split supercilium. Strongly patterned above, scapulars long with black centres reaching tip of feather. Breast centre often unmarked. Distinct pale spot before eye, loral stripe almost broken.

Summer adult Long-toed Stint. Typically shows a split supercilium. Note pale base to bill. Brightly patterned above including broad rusty edges to tertials. Long legs.

Juvenile Least Sandpiper. Suggests miniature Pectoral Sandpiper. Note pale legs. More obvious breast-band and shorter legs than Long-toed Stint.

Summer adult Least Sandpiper. Much duller and browner than Long-toed Stint. Note difference in head pattern.

Winter Least Sandpiper. Less contrast within upperparts than Long-toed Stint. Dark lores, heavily spotted breast, all black bill.

Winter Long-toed Stint. Browner than Little Stint, similar to Least Sandpiper. Note head pattern, more distinct dark centres to upperpart feathers.

Juvenile Little Stint. Note dark central crown, split supercilium and white mantle 'V', also lower scapular pattern and black centred tertials. Buffy or orangey breast patch. A few individuals are much greyer and less strongly patterned.

Summer adult Red-necked Stint. Plain brick-red face and throat, wing-coverts largely grey.

Juvenile Red-necked Stint. Note exact pattern of lower scapulars and grey-centred tertials. Breast streaking greyish, faint but extensive. Head pattern relatively plain.

Summer adult Sanderling can resemble summer Red-necked Stint but note larger size. Pale area beneath eye, streaked ear-coverts, distinct pale centres to scapulars, greater coverts and patterned tertials. Sanderling lacks a hind toe.

Winter Red-necked Stint. Often appears more elongated than shown here.

Summer adult Little Stint. Note white throat speckled face and cream mantle 'V', warm non-contrasting wing coverts and tertials (greyer and contrasting on Red-necked Stint).

Winter Red-necked Stint. Dark of lores often runs through eye and across ear-coverts.

1st-winter Little Stint. Dark of lores separated from eye, usually more prominent breast patch and darker feather centres to upperparts, but often impossible to separate from Red-necked Stint on plumage.

Adult In marked contrast to Semipalmated, summer-plumaged Western is very distinctive and quite a striking bird. It is strongly rufous on the crown, ear-coverts, mantle and scapulars, while the breast is heavily streaked with black, with small black chevrons extending in a characteristic manner well down the flanks. In consequence, such birds are not particularly difficult to identify. Like adult Dunlin, its moult into winter plumage is often carried out whilst on autumn migration, so not only should July and August birds show evidence of upperparts moult but, more importantly, also the presence of plain grey winter feathering in the back and scapulars. *First-summer* It has been thought any 'winter-plumaged' Semipalmated/Western seen in Britain in summer is more likely to be a Western. This is because most such birds seen in North America are indeed Westerns. However, first-years of both species undergo a partial moult in spring, replacing variable numbers of first-winter head and body feathers. Some acquire nearly full breeding plumage and head north to their breeding grounds, whereas others remain on or near their winter quarters and either retain or moult into a winter-type plumage. About two-thirds of first-summer Semipalmateds remain behind (Chandler & Marchant 2001, O'Brien *et al.* 2006). Westerns winter further north than Semipalmateds and it may be that, in spring, some are prone to moving relatively short distances northwards into temperate latitudes, whereas non-breeding first-summer 'Semi-ps' do not. Although this may well be the case, it remains to be confirmed.

Red-necked Stint *Calidris ruficollis*

Where and when The E. Palearctic Red-necked Stint has occurred on just seven occasions (to 2010). All except one (a late August juvenile) have been worn adults from mid July to late September.

Structure Appears slightly shorter-legged than Little Stint with a shorter, stubbier bill (more like a short-billed Semipalmated Sandpiper). Also tends to look rather long-bodied and 'horizontal'.

Plumage *Summer adult* In early spring, the fresh summer plumage is very grey, but broad grey feather fringes wear off to reveal a distinctive brick-red face and throat, fading to reddish-orange in late summer (when variable in extent but often strongest on the throat). Beneath the red is a variable but distinctive band of thick brown streaking, extending from the nape right around the upper breast. Brown across the lores, with a variable but subdued whitish supercilium (broadening behind the eye) and a narrow white eye-ring (like Semipalmated). The back and scapulars are a mixture of black, grey and white, as well as some brick-red, strongest in early summer, with variable white mantle Vs. Like Semipalmated, the primary projection is often short (but seemingly variable). Note that richly coloured summer-plumaged chestnut Sanderling *C. alba* have been confused with Red-necked Stint (Sanderling is much larger with a broad white wing-bar and lacks a hind toe in close views). *Juvenile* Superficially similar to juvenile Little Stint (but bill shorter) with head pattern more reminiscent of Semipalmated (lacks a split supercilium and has a dark ear-covert patch with a narrow white eye-ring enclosed within the grey). Overall, its plumage looks much greyer than juvenile Little Stint and this is most likely to draw attention to a vagrant: the head, breast patches and centres to the upperwing-coverts and tertials are distinctly *pale grey* (fringed white with a black shaft-streak). The back and scapulars, however, show contrasting bright orangey feather fringes, strongest and most consistent on the upper two rows of scapulars, but variable in extent and fading as autumn progresses. A weak white mantle V may be apparent, and white tips to many of the lower scapulars. The primary projection is longer than most adults, and is quite similar to juvenile Little Stint.

Call Quite distinctive: a soft, rolling, rather musical *pleep* or *p-r-leep*.

Least Sandpiper *Calidris minutilla*

Where and when A very rare vagrant (36 records to 2011). Records fall into two categories: summer-plumaged adults in July/August and juveniles in September/October, with two spring (May) and one winter record.

Structure As its name suggests, it is the smallest stint, with a short, squat body and rather short legs; note that its toes are distinctly long and 'spidery'. The bill is rather heavy based but tapers to a fine point. An important difference from Little Stint is that it shows only a very short primary projection, reaching just beyond the tertials.

Plumage and leg colour As a useful *aide-mémoire*, it may be helpful to think of Least as resembling a miniature Pectoral Sandpiper *C. melanotos* (even its calls are vaguely similar, albeit much higher-pitched). The key difference from Little Stint is leg colour: dull green or greenish-yellow (but beware of mud making the legs look dark). *Juvenile* Fresh plumaged individuals are richly coloured and suggest Little Stint in an initial view. Close views should reveal a distinct capped effect (may show a *faint* lateral crown-stripe) a dark line across the lores and a browner ear-covert patch (plus a narrow white eye-ring). The upperparts feathers are neatly fringed with rich buff, those of the scapulars being more richly fringed with chestnut (lower scapulars with white tips). It shows a slight mantle V, but this is narrower and weaker than on Little Stint. Most importantly, it has a complete breast-band of fine brown streaking (which may be weaker and narrower in the middle). Note that the background colour of the band is white, particularly in the middle, so the streaking may be surprisingly difficult to detect at a distance (when it may appear as side patches). Some individuals commence their post-juvenile moult by September, acquiring grey back and scapular feathers (with black shaft-streaks). *Adult summer* In many ways similar to juvenile, but early spring birds show greyish feather fringing that gradually wears off. Plumage then appears very dark above, with ill-defined whitish mantle and scapular Vs, the scapulars showing a mixture of white and chestnut feather fringes (wing-coverts browner). It too shows a distinct cap and a brown face patch. Returning autumn adults may appear very dark and worn. The breast-band and leg colour remain the key features. *Winter* Upperparts, head and breast-band brownish-grey, with thick black shaft-streaks to the upperparts and black streaking on the crown and breast, creating a more heavily patterned appearance than Little Stint.

Calls Readily separated from Little Stint: a soft, trilling *treep*, *prreep*, *tr-rrr*, or *s-r-eep*, longer when excited: *s-s-s-s-sip*.

Long-toed Stint *Calidris subminuta*

Where and when A major rarity with just two accepted British records (1970 and 1982). The first was in June, the other a juvenile in August/September. There was another in Ireland in June 1996.

Structure In many ways the Asian equivalent of Least Sandpiper (it also has pale legs), but appears more upright, with longer legs and neck, producing a more 'scrawny' appearance. At a distance, its shape may suggest a miniature Wood Sandpiper *Tringa glareola*, an impression heightened by its whitish supercilum, capped appearance and breast-band (see below). It can also look quite square-headed, flat-backed and pot-bellied, with a rather truncated rear end (like Least, it has little or no primary projection beyond the tertials). The legs appear about equal to the body depth, with a lot of tibia showing above the 'knee'. Because of its long legs, it often 'tilts forwards' when feeding. The 'jacana-like' toes are strikingly long and well splayed. The bill appears relatively straight, being slightly less tapered than Least's.

Plumage *Juvenile* A much more patterned and more richly coloured bird than juvenile Least. It has a dark cap, which is lined with chestnut (including a fairly obvious narrow 'split supercilium'). The dark of the crown curves forward to connect with the lores, producing a 'reversed J' mark. It has a prominent whitish supercilium, a dark line across the lores, a narrow white eye-ring and a grey or brown fan-shaped ear-covert patch. At certain angles, the facial pattern can suggest juvenile Sharp-tailed Sandpiper *C. acuminata* or even a Dotterel *Charadrius morinellus*. The upperparts are well patterned with a prominent whitish mantle V, chestnut lines on the mantle and rich chestnut fringes to the upper scapulars and tertials. The lower scapulars and wing-coverts are duller, fringed with a mixture of pale buff and white. Like Least, it has a complete breast-band, but this is coarsely streaked (on a white background) and fades in the middle. The base of the bill is faintly horn or olive-coloured and the legs are bright lime-green. *Adult summer* Pattern similar to juvenile, including the 'split supercilium', the 'reversed J' mark on the face and complete breast-band. In fresh summer plumage it has prominent chestnut or orangey fringes to the upperparts, this rich coloration giving it a 'foxy appearance'. However, plumage tones vary individually, with worn autumn individuals appearing much duller and browner. *Winter* Like Least, grey-brown above, strongly patterned with streaking on the head, neck and breast, and obvious dark centres to the upperparts feathers.

Calls Very similar to Curlew Sandpiper *C. ferruginea*, but much softer: a low and liquid *chirrup* or *chree*. Also, a more whistling upslurred *poweep*.

References Chandler & Marchant (2001), Cramp & Simmons (1983), Grant & Jonsson (1984), O'Brien *et al.* (2006).

White-rumped and Baird's Sandpipers

Where and when White-rumped Sandpiper is an annual migrant that occurs in two distinct waves. Firstly, worn or transitional adults appear in July and August, almost exclusively on the east coast (one theory is that they arrive across the Arctic from Canada). The second wave relates to late autumn juveniles, mainly in west coast locations, from late September to early November. The species currently averages 19 records a year, with a peak of 39 in 2005. Baird's is essentially a storm-driven vagrant, mainly juveniles in western areas from late August to October. It is extremely rare both in spring and late summer (and has wintered once). It currently averages seven records a year, with a peak of 12 in 2005.

White-rumped is mainly a saltwater species whereas Baird's tends to prefer freshwater habitats (often being found in grassy environments, such as airfields) although it too may be found with coastal waders.

General These two long-distance migrants are treated together as they are very similar in size and shape. Both are small, intermediate between Dunlin *Calidris alpina* and Little Stint *C. minuta*, and both are very long-winged, a result of their very long 'scissor-like' primaries. Three visible primaries extend well beyond the tail, producing a strikingly attenuated look to the rear end (primaries roughly equal in length to overlying tertials). When viewed front-on, Baird's has a peculiarly broad, flat-backed appearance, almost as if it has been trodden on.

White-rumped Sandpiper *Calidris fuscicollis*

Plumage *Summer adult* In late summer, predominantly greyish, well streaked on the breast, extending down the flanks; chestnut is often present in any unmoulted summer scapulars. Many late summer adults, however, are already acquiring winter plumage on the back and scapulars, showing a mixture of plain grey feathers (with a black shaft-streak) and worn and messy white-fringed blackish or brownish summer feathers. The most obvious plumage feature is a well-defined whitish supercilium that tends to curve down and then up behind the eye. It also usually shows a fleshy or orangey base to the lower mandible, which Baird's always lacks. If in doubt, wait until it flies; then, the key feature is the obvious, curved, white uppertail-covert patch, immediately in front of a plain, dark grey tail. ***Juvenile*** Also rather greyish on the head, neck and breast (the latter delicately streaked, forming a pectoral band), the head also showing a fairly prominent white supercilium. The upperparts are blackish, with all of the feathers neatly fringed white (but less scaly than Baird's) and some rufous on the upper scapulars (and perhaps also on the crown and some wing feathers).

Call Markedly different from Baird's: a quiet, unobtrusive, thin, mouse-like *jit* or *j-jit*, or a more rapid *si-si-sit*.

Baird's Sandpiper *Calidris bairdii*

Plumage *Summer* Breeding plumage similar to juvenile (which see). Rather buff-coloured except that the upperparts patterning is more diffuse and less regular than juvenile. Late summer adults are likely to be worn and messy compared to pristine juveniles. ***Winter*** Breast and upperparts retain golden-buff tones but duller with less contrasting brownish feather fringes. ***Juvenile*** Neat and delicately patterned. Obviously buff-toned compared to White-rumped, but also relatively featureless. Rather plain-faced, the supercilium varying from buff and inconspicuous to whitish and fairly noticeable. At close range, it shows an obvious narrow whitish eye-ring. The most distinctive feature is a well-defined buff pectoral band that is profusely but delicately streaked with brown. The overall impression may therefore suggest a miniature, long-winged Pectoral Sandpiper *C. melanotos*, but it lacks that species' white V's on the mantle and scapulars. However, the upperparts are distinctive, being prominently scalloped whitish (strongest on the back and scapulars). Some have narrower, buffier and more subdued scalloping (and the scalloping is always less obvious at a distance). In flight, it lacks White-rumped's white uppertail-coverts (dark instead). It also

Juvenile Dunlin. Note short primary extension in relation to tail compared with Baird's and White-rumped Sandpipers.

Juvenile White-rumped Sandpiper. Long-bodied with long primary extension. Rusty upper scapulars, two pale lines down back, dark crown and pale supercilium, grey nape. ▼

Pale-based lower mandible. ▶ Breast streaks extend to flanks.

Adult summer Baird's Sandpiper. Bill straighter and finer than White-rumped Sandpiper; bill all black. Buffish head and breast, flanks unmarked, dark rump. ▼

▲ **Juvenile Baird's Sandpiper.** Buffy, very neatly fringed above (scaly), supercilium indistinct, white spot above lores.

Adult summer White-rumped Sandpiper. Distinct supercilium, rusty tones to head and upperparts. ▼

Juvenile Curlew ▶ Sandpiper

Heavily streaked over entire breast, with markings continuing down flanks.

When feeding, both Baird's and White-rumped Sandpipers look very elongated with typical horizontal posture.

Juvenile White-rumped Sandpiper. Darker than Curlew Sandpiper, short bill, no legs beyond tail. Distinctive call.

Winter White-rumped Sandpiper. Drab brownish-grey upperparts. Breast streaked greyish, a few streaks extending to flanks. Pale supercilium.

Winter Baird's Sandpiper. Golden-buff tones to breast and upperparts. Paler supercilium. Flanks unmarked.

shows a variable wing-bar, often broad and obvious across the bases of the primaries but sometimes inconspicuous (narrow white tips to greater coverts). It may look long-winged in flight (although it can be surprisingly difficult to pick out from accompanying waders). Baird's tends to be an active feeder, always on the move and often running at speed but, at other times, it may feed more furtively on flexed legs.

Call A thin, high-pitched *kreep* or *prrrrt*, vaguely recalling a thin, high-pitched Pectoral Sandpiper.

Confusion with Little Stint Baird's may be confused with Little Stint, mainly because that species also has quite long primaries. However, juvenile Little Stints show stronger facial patterning, a noticeable white mantle V and the underpart streaking is usually confined to the breast-sides. Summer adults, however, may be buffier with a better-defined breast-band, but are more coarsely patterned. If in doubt, Little Stint's primaries are usually only half to two-thirds the tertial length (or less). Little Stint's call is also different: a quiet, unobtrusive *tip* or *tip tip tip*.

Dunlin, Curlew Sandpiper, Broad-billed Sandpiper and Knot

Where and when Dunlin is our commonest small shorebird, often abundant on estuaries in winter and large numbers of migrants pass through in spring and autumn, when it may also be numerous inland (usually when water levels are low). Small numbers breed in upland areas, mainly in n. England and Scotland. Curlew Sandpiper is a migrant from Siberia. Adults pass through mainly in eastern areas in late spring and early autumn, and larger but variable numbers of juveniles are more widespread from August to October; very rare in winter. Broad-billed Sandpiper is a very rare migrant, mainly in spring (May/June) and very occasionally in autumn (July to September) mainly to English east coast (averages about four a year). Canadian and Greenland-breeding Knot winter on selected estuaries, principally in NW England and the Wash, where they may occur in huge concentrations; more widespread on migration when small numbers turn up inland, mostly juveniles in August/September.

Dunlin *Calidris alpina*

General features This abundant small wader should act as the yardstick when identifying all similar species. A small, hunched, rather dumpy bird with a medium to long, gently decurved bill. Bill length varies: the migrant races *schinzii* (SE Greenland, Iceland, Britain and s. Scandinavia) and the much rarer *arctica* (NE Greenland) – both of which winter mainly in w. and NW Africa – have shorter bills than *alpina* (n. Scandinavia and NW Russia) which is common in winter. They feed in large flocks on mudflats, not bunching as tightly as Knot, picking and probing with the bill held downwards.

Flight identification Forms tight flocks like Starlings *Sturnus vulgaris*, flashing grey and

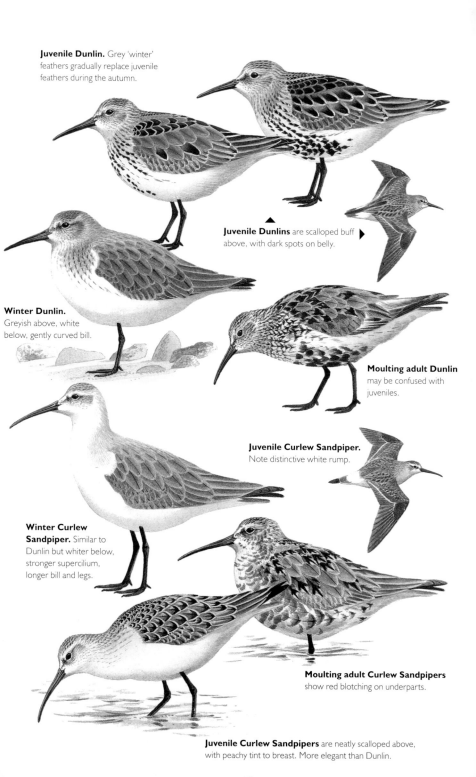

Juvenile Dunlin. Grey 'winter' feathers gradually replace juvenile feathers during the autumn.

Juvenile Dunlins are scalloped buff above, with dark spots on belly. ▶

Winter Dunlin. Greyish above, white below, gently curved bill.

Moulting adult Dunlin may be confused with juveniles.

Juvenile Curlew Sandpiper. Note distinctive white rump.

Winter Curlew Sandpiper. Similar to Dunlin but whiter below, stronger supercilium, longer bill and legs.

Moulting adult Curlew Sandpipers show red blotching on underparts.

Juvenile Curlew Sandpipers are neatly scalloped above, with peachy tint to breast. More elegant than Dunlin.

159

white as they twist and turn over the mudflats. Closer views reveal a rather plain pattern, with a narrow white wing-bar and a dark line down the centre of the rump.

Call A distinctive, drawn-out, slightly rasping *treeeeep*.

Song Similar in quality to the call: a fast, dry and almost rasping *treee-treee-treee...*, developing into a drawn-out rapid, pulsating, almost buzzing *tree-ee-ee-ee-ee-ee-ee....*

Plumage Three distinct plumages. *Adult summer* Easily identified by the strong buff or chestnut-patterned upperparts and large black belly patch; upperparts wear darker by late summer. Continental *alpina* averages more chestnut than the buffier *schinzii* and *arctica*. *Winter* Rather nondescript, with a grey head and upperparts and a strong grey suffusion to the breast-sides (and light streaking across the centre); slight supercilium. *Juvenile* Rather like a dull version of adult summer. The upperparts feathering is broadly edged chestnut and buff, and the buffish breast is lightly streaked brown. Most importantly, the belly is noticeably streaked blackish, usually forming a messy patch that mirrors the black belly patch of breeding adult. Juveniles generally look neat compared to contemporary summer adults, which often look patchy and scruffy when moulting into winter plumage. Note that, unlike most small waders, juvenile Dunlins commence their moult into winter plumage in late summer, so many start to show grey patches on the scapulars whilst on migration, these increasing in area as autumn progresses.

Curlew Sandpiper *Calidris ferruginea*

General features Slightly larger, taller, 'leggier' and more elegant than Dunlin; some larger individuals can be very obviously taller (females average larger than males). Legs and bill longer, but differences in bill length and curvature subtle and should not be too heavily relied upon for identification. Juvenile has a longer primary projection than most Dunlin (less apparent on adults). Often feeds in deep water, immersing the bill and head below the surface (Dunlin also feeds like this, but less persistently).

Flight identification Easily separated from Dunlin by its striking *white rump*, contrasting with the grey tail.

Call Flight calls similar in quality to Dunlin, but slightly softer and markedly disyllabic: a soft, rolling *shirr-up*, very distinctive once learnt.

Plumage *Adult summer* Easily identified by reddish underparts. Size and structural differences separate it from summer Knot. Fresh summer plumage has white tips to the red feathers, so spring birds may show a strong grey cast (strongest on females) which gradually wears off as summer progresses. *Winter* Full winter plumage is uncommon in this country, as most spring and autumn adults show traces of summer plumage (noticeable red blotching on underparts in autumn). Winter plumage is very similar to Dunlin, Curlew Sandpipers being surprisingly difficult to pick out at any distance; the most obvious differences are a slightly stronger, thin white supercilium, extending well behind eye, and a slightly whiter breast (although this is variable on Dunlin). Perhaps best located by the subtle structural differences outlined above. *Juvenile* From mid August onwards, the vast majority of migrants are juveniles. For separation from Dunlin, concentrate on the underparts. Juvenile Curlew Sandpiper always looks neat and immaculate, with smooth, clean, white underparts with a noticeable soft peachy tint to the breast (gradually fades as autumn progresses). Any streaking on the breast is fine and inconspicuous. This is in marked contrast to accompanying juvenile

Dark above, scalloped paler. Note flecking on flanks and double 'split' supercilium.

Winter Broad-billed Sandpiper. Grey above, retains 'split' supercilium.

Adult spring Broad-billed Sandpiper

Adult summer Broad-billed Sandpiper. Worn birds are very dark.

Juvenile Broad-billed Sandpiper. Neat with poor breast-band, plain flanks.

Juvenile Broad-billed Sandpiper in flight is rather dark.

Juvenile Dunlin. Poorly developed mantle 'braces', weak head pattern, dark blotches on belly.

Juvenile Dunlin

Juvenile Sanderling. White below, spangled above.

Winter Sanderling. Note broad white wing-bar.

Adult summer Sanderling

Winter Sanderling. Very pale, dark 'shoulders', shortish bill and black legs. Runs rapidly along tideline.

Adult summer Sanderling. Brighter individuals patterned above with chestnut. Strong pectoral band.

Juvenile Knot. Scalloped above.

Adult winter Knot. Pale grey above, whiter below. Appears stocky with shortish dark bill and short greenish legs.

Winter Knot in flight. Grey plumage, white wing-bar and pale grey rump and tail.

Dunlins, which show noticeable blackish belly streaking, markings never shown by juvenile Curlew Sandpipers. Whereas juvenile Dunlin has thick and coarse upperparts patterning (often with patches of grey first-winter feathering on the scapulars), the upperparts of Curlew Sandpiper are finely and neatly scalloped, the rather washed-out pale greyish-brown feathers having a narrow black subterminal line and narrow buff fringes. It also has a better-defined supercilium than Dunlin and often a slight capped appearance. Perhaps more likely to be confused with winter adult Dunlin, but the latter is uniformly pale grey above, with grey shading on the breast-sides.

Broad-billed Sandpiper *Calidris falcinellus*

General features Structurally similar to Dunlin, but noticeably smaller and slightly shorter-legged. Noticeably thicker bill and the tip often looks distinctly down-kinked. Despite statements to contrary, not particularly lethargic.

Flight identification Summer adults and juveniles look very dark and blackish, and show an obvious breast-band and a narrow white wing-bar.

Call A distinctive dry, hard trilling *pprrrrrk*.

Plumage *Adult summer* Most of those occurring here are summer-plumaged adults, which are easily identified. At a distance, they look very dark – blackish – with a dark, buffy-brown breast-band and messy dark mottling on the flanks. At closer range, the upperparts feathers are neatly fringed whitish, very obvious when fresh; two sets of mantle and scapular lines recall Common Snipe *Gallinago gallinago*. Also, there is some chestnut fringing, especially on the upper scapulars. Most distinctive is the striped head pattern, which strongly recalls that of Jack Snipe *Lymnocryptes minimus*: a thick white supercilium forks before the eye, producing a very distinctive 'split supercilium'. The prominence of this is highlighted by the very dark crown and dark eye-stripe, which broadens into a thick patch on the ear-coverts. (Note that the 'upper supercilium' but may be difficult to detect on distant worn birds in late summer.) Some fresh spring migrants show frosty grey feathering about the head and breast (that gradually wears off). Late summer and autumn adults lose much of the pale upperparts fringing through wear and can appear rather uniformly blackish above. *Winter* Full winter plumage is unlikely to be seen in this country. The predominantly blackish summer feathering is replaced with grey, but the characteristic head pattern is retained (although the split supercilium is subdued and not obvious). Darker leading lesser coverts form a dark patch at the bend of the wing. Winter adults can be surprisingly difficult to pick out from Dunlin. *Juvenile* Very dark like summer adult, but fresh and immaculate at a time when adults are scruffy and worn. The breast-sides are finely and evenly streaked, and the upperparts show well-defined white feather fringes and mantle and scapular stripes. When very fresh, they may show a faint buff wash on the breast.

Knot *Calidris canutus*

General features A rather bland-looking bird that inexperienced observers may struggle to identify. At rest, best identified by size and structure: a rather bulky, medium-sized wader (much larger than Dunlin) with a short neck and long body. Note in particular that the bill is relatively short, as are the legs (greenish or greyish, not black). Unlike Dunlin, it has an attenuated rear end with long primaries that project well beyond the tertials (exposed primary length is approximately equal to tertial length). A rather slow, methodical feeder. At its main wintering sites it forms huge, tightly packed flocks that carpet the ground when roosting.

Flight identification Easily distinguished by the combination of obvious thick white wing-bar and pale, greyish-white rump and tail (rump and uppertail-coverts finely barred grey and white at close range).

Call An unremarkable unobtrusive, soft, quiet *oo–ik* or *oo–ik–ik,* unlikely to attract attention.

Plumage *Adult summer* Easily identified by brick-red underparts (females tend to be paler than males and both sexes may fade by late summer). Moulting adults are patchy. Structure and bill length easily separate it from Curlew Sandpiper. *Winter* A very pale grey bird, plain above (with faint pale feather fringes) and a lightly mottled grey breast; grey chevrons extend onto the flanks. Lacks a strong facial pattern (weak whitish supercilium). *Juvenile* Similar to winter, but upperparts feathers neatly and finely scalloped, each feather having a narrow dark subterminal line and a narrow whitish fringe. The breast and flanks are finely streaked but the belly has a distinctive soft peachy tint, which gradually fades as autumn progresses. A noticeable white supercilium curves down and then up behind the eye. Plumage usually looks smooth and immaculate.

Ruff, Buff-breasted and Pectoral Sandpipers

Where and when Ruff is a very rare and erratic breeding bird but is far more numerous as a migrant, mainly in freshwater habitats; variable numbers winter, mostly in s. Britain, tending to occur in fields with Lapwings *Vanellus vanellus* and Golden Plovers *Pluvialis apricaria.* Buff-breasted Sandpiper is a North American visitor, mainly to w. Britain and Ireland in September (records in spring and late summer, often on the English east coast, may relate to birds that arrived in previous years). Numbers vary, but it currently averages 25 a year; small parties sometimes occur (a remarkable arrival occurred in 2011, with provisional estimates of 75 in Britain and 90 in Ireland, including a flock of 28 at Tacumshin, Co. Wexford). Pectoral Sandpiper is another North American species, currently averaging 110 records a year, with a peak of 192 in 2003. Most occur in September/October with records in spring and late summer, often in e. England, thought to relate to birds that crossed the Atlantic in previous years or possibly to individuals from e. Siberia, where it also breeds. It has very rarely wintered. A pair almost certainly bred in Scotland in 2004.

Ruff *Calidris pugnax*

General features Apart from the male's remarkable display plumage, in many ways a rather nondescript wader, variable in plumage, bare-parts colour and size; consequently, it is often confusing to inexperienced birders. Males and females differ markedly in size, males being about the size of Common Redshank *Tringa totanus,* females (also known as Reeves) *c.* 25% smaller. Rather a gangly looking bird: short-billed, small-headed, long-necked and bulky bodied (sometimes looking humpbacked) and often showing a bulging 'Adam's apple' at the front of the neck. It strides around in a purposeful manner, picking at the mud, but will also wade, immersing both its bill and its head below the surface. Distant individuals can be confused with Redshank, which has a similar walk, but the latter has a longer, straighter bill and darker, more uniform plumage. Adult Redshank has red legs, but late summer juvenile's are orange, a colour also shown by adult Ruffs. If in doubt, wait until the bird flies: Redshank's white secondaries and rump should be obvious.

Plumage Four basic plumage types occur. *Juvenile* Most August to October migrants are juveniles, which are easily separated from adults. The underparts are a distinctive orangey-buff (white on the rear belly and ventral area). The upperparts are neatly patterned with buff feather fringes that produce a distinctive scalloped appearance. The bill is dark and the legs dull grey-green. Juvenile plumage is gradually lost in late autumn/early winter body moult. *Winter* Acquired by adults from late summer onwards, by juveniles from late autumn onwards. More distinctive, being essentially pale grey above (with noticeable pale grey or whitish feather fringes) and white below (with subdued grey mottling on the breast-sides and flanks). It has a white eye-ring and many show a noticeable white patch at the base of the bill (generally strongest on adult males). Some, nearly always males, have variable patches of white on the nape, sides of the neck and/or breast, and some are completely white-headed and very striking. The bill usually has a pink or orange base (especially males) and the legs are bright pink or orange; first-winters, however, retain a black bill and dull, greenish legs. *Adult summer* Males readily identified by their remarkable ornamental head plumes, which vary from white, black and white, through black and brown to ginger; some plumes are heavily barred. This plumage is, however, rarely seen away from the Continental breeding grounds, although passage males often show obvious traces of it, as well as variable black markings on the upperparts. Confusingly, some breeding males are female-like. Adult females are also variable: essentially brown, variably and irregularly patterned with coarse black markings above, and more delicately patterned on the head and breast. The bill is dark, often with a pink base, while the legs vary from pinky-orange to green (the latter no doubt mainly first-summers).

Flight identification A noticeable whitish wing-bar becomes broader and more diffuse across the bases of the primaries; distinctive white V-shaped patches on the lateral upper-tail-coverts (some virtually lack the dark central dividing bar, creating a white crescent-shaped 'rump patch'). Long-winged, with an easy, languid flight action.

Voice Oddly silent, but rarely gives a low, hoarse grunt.

Buff-breasted Sandpiper *Calidris subruficollis*

Size and shape Superficially similar to a small juvenile female Ruff, but easily identified. Male slightly larger than female but, whereas Ruff is a medium-sized wader, Buff-breast is basically a *small* wader, only slightly taller than a Dunlin *C. alpina*. Although its shape resembles Ruff, it is less gangly, being more horizontal, proportionately longer-bodied and squarer-headed. Unlike Ruff, juveniles show two or three long primaries projecting noticeably beyond the tertials, producing an attenuated rear end (although adults have longer tertials).

Plumage *Juvenile* The entire underparts are uniformly pale buff, paler and less orange than juvenile Ruff. The vast majority of September/October vagrants are juveniles, which appear immaculate; a few of these may be much whiter on the belly and a small minority show abrupt demarcation between the buff and the white. They also show delicate black spotting on the breast-sides. A large dark eye and pale eye-ring stand out on the bland, plain-looking face,

Adult winter male Ruff. Grey-brown above, white below. Often very pale or white on head. Orange or pink legs.

Summer female Ruffs show considerable variation in plumage.

Leg colour varies from orange (adult) to green (1st-summer).

Winter female Ruff. Smaller than male, typically darker to head. ▼

Juvenile Ruff. Long-necked. White wing-bars and thin dark stripe down rump. Flight languid.

Juvenile Ruff. Golden plumage, scalloped feathers dark-centred. Considerable variation in size between the sexes. Legs greenish.

Juvenile Buff-breasted Sandpiper. Smaller than Ruff with finer scalloping. Legs yellow. ▼

Juvenile Buff-breasted Sandpiper. Plover-like in flight, poor wing-bar. Lacks white rump sides.

Juvenile Pectoral Sandpiper. Weaker wing-bars and wider dark rump stripe than Ruff.

Adult Pectoral Sandpiper. Coarse streaking to head and breast.

Juvenile Pectoral Sandpiper. White mantle 'V', a weak split supercilium and neat pectoral band clearly demarked from white belly. Note long primary projection. Legs yellowish.

while the streaked crown can give a slight capped effect. Like Ruff, the upperparts feathers are fringed buff, but the edgings are narrower so the upperparts appear much less coarsely patterned. Legs pale yellow (greenish on juvenile Ruff). *Summer adult* Broader, less well-defined buff fringes to the upperparts feathers, lacking juvenile's dark sub-terminal crescents. Autumn adults may begin to show varying amounts of winter upperparts feathering, which has broad buff fringes.

Habitat and behaviour Frequents short-grass habitats, such as golf courses and airfields, but will also associate with other small waders in more typical freshwater environments. An active feeder, walking quickly and daintily on slightly flexed legs, picking every two or three steps, but actions rather erratic, with frequent changes of direction (it lacks Ruff's smoother, more confident walk and slower, more deliberate picking action). Constantly bobs its head whilst walking. Unlike Ruff, does not wade into water to feed. Often very tame, sometimes crouching low and freezing when approached.

Flight identification Always looks small and rather plover-like, lacking Ruff's easy, languid flight action. The upperparts pattern is completely different from Ruff: it lacks a prominent wing-stripe and prominent white patches on the sides of uppertail-coverts. Instead, it shows a faint and diffuse pale bar across the base of the upper primaries, and the rump and upper-tail-coverts are plain. The underwings are dull silvery-white, with a thick dark crescent across the under-primary coverts. Unlike Ruff, the feet do not project beyond the tail, so it lacks an attenuated rear end. It may suggest a miniature Golden Plover *Pluvialis apricaria* in flight.

Voice Like Ruff, strangely silent, but occasionally utters an insignificant, quiet, soft, downward *cheu*.

Pectoral Sandpiper *Calidris melanotos*

Size and shape A medium-sized wader with a mid-length, slightly decurved bill and medium-length legs. Although superficially similar to Ruff, it is smaller, shorter-legged and quite different in shape: less upright and more horizontal with a rather long, pear-shaped body and an attenuated rear end. It has a very long primary projection, the exposed primaries being about equal in length to the overlying tertials (the tertials virtually cover the primaries on Ruff). Perhaps 10–15% larger than Dunlin although, like Ruff, size varies sexually, some small females being much closer in size to Dunlin.

Plumage *Juvenile* Despite an unremarkable face pattern (it has an inconspicuous narrow whitish split supercilium) easily identified by its underparts pattern: a clear-cut, well-defined, neatly and finely streaked buff breast-band (lower border pointed in centre) clearly demarcated from the white belly. Like juvenile Ruff, it has neat, pale feather fringing on the upperparts, but the edgings to the mantle feathers produce a noticeable white back V, with a second, less well-defined line formed by white tips to the second row of scapulars. In fresh plumage, the crown and upper scapular fringes are tinged rufous. The legs are usually yellowish, but sometimes lime-green; the basal half of the bill is dull orange. *Adult summer* Summer adults and juveniles are similar, but adults are less neatly patterned and slightly messier looking, with coarser head streaking, less obvious mantle Vs and plainer, greyer-fringed median coverts that contrast more with the pale-edged scapulars (late summer adults are also more worn). Compared to females, summer males are rather mottled on the breast. *Adult winter* Winter plumage is very rare in this country as the post-breeding moult takes place mainly in the winter quarters. However, winter individuals are plainer above, with broad and rather diffuse brown feather fringes to predominantly blackish-centred mantle and scapular feathers; also, they lack the white mantle and scapular Vs.

Flight identification Resembles Ruff in having two white patches on the sides of the upper-tail-coverts, but is fairly plain-winged, lacking an obvious wing-bar. Flight more Dunlin-like, lacking Ruff's languid action.

Call Calls frequently in flight: a distinctive low, soft, rolling, *chrrrp*, *trrrp* or *k-r-r-r*, perhaps recalling a low, subdued Curlew Sandpiper *C. ferruginea*.

Behaviour Less mobile than Ruff with two quite different feeding styles. Solitary birds are often very slow, furtive and surprisingly inconspicuous, creeping around on flexed legs with the head down and a constant rapid vertical picking action. However, amongst other waders, such as Dunlin, it usually feeds just like them in an active and mobile fashion.

Sharp-tailed Sandpiper *Calidris acuminata*

Pectoral is superficially similar to Sharp-tailed Sandpiper, which is a very rare vagrant, mainly adults in late summer (breeds in e. Siberia). Summer adult shows a capped effect, a white eye-ring and extensively patterned underparts with messy streaking on the breast and chevrons on the flanks. Juvenile has a rusty cap, a stronger face pattern (white supercilium and dark ear-covert patch) and plainer underparts, with a strong orangey wash across the breast.

Common, Jack and Great Snipes

Where and when Common Snipe is a widespread but declining breeding species, now mainly found in upland areas of n. Britain (and Ireland). Much commoner and more widespread in autumn and winter, frequenting all kinds of wet and damp freshwater habitats, its numbers augmented by Icelandic and Continental immigrants. Jack Snipe is a scarce and rather local winter visitor, mainly from late September to April. Great Snipe is a very rare vagrant, currently averaging two or three records per year, the majority in the Northern Isles with a few on the east coast (extremely rare elsewhere). Most occur in autumn, from late August to October, with a peak in September. There are also a few May records, including a male displaying in Norfolk in May 2011.

Common *Gallinago gallinago* and Jack Snipes *Lymnocryptes minimus*

Habitat Jack Snipe is a peculiarly secretive bird, rarely straying from well-vegetated soggy places. Edges of reedbeds, water meadows, overgrown ditches and coastal *Spartina* marshes are favoured habitats. Unlike Common Snipe, it usually avoids open mud at all costs and does not wade into open water to feed. Generally speaking, the only chance of seeing it on the ground is from the seclusion of a well-sited hide.

Flight identification Unlike Common Snipe, Jack Snipe waits until almost underfoot

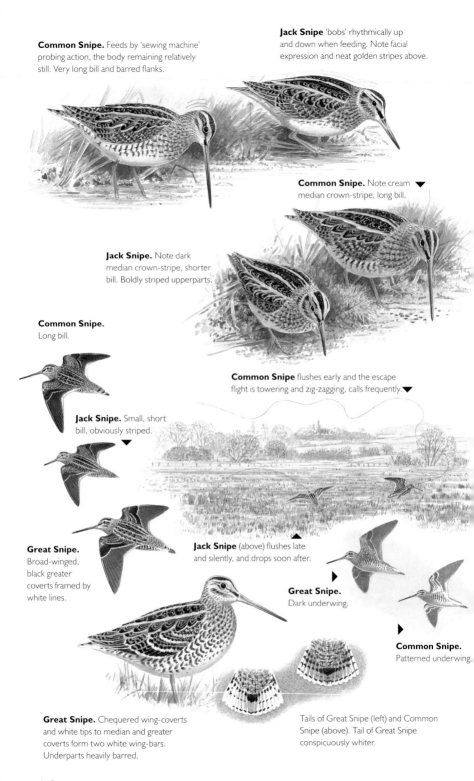

Common Snipe. Feeds by 'sewing machine' probing action, the body remaining relatively still. Very long bill and barred flanks.

Jack Snipe 'bobs' rhythmically up and down when feeding. Note facial expression and neat golden stripes above.

Common Snipe. Note cream ▼ median crown-stripe, long bill.

Jack Snipe. Note dark median crown-stripe, shorter bill. Boldly striped upperparts.

Common Snipe. Long bill.

Common Snipe flushes early and the escape flight is towering and zig-zagging, calls frequently.▼

Jack Snipe. Small, short bill, obviously striped. ▼

Great Snipe. Broad-winged, black greater coverts framed by white lines.

Jack Snipe (above) flushes late and silently, and drops soon after. ▲

Great Snipe. Dark underwing. ▶

Common Snipe. Patterned underwing. ▶

Great Snipe. Chequered wing-coverts and white tips to median and greater coverts form two white wing-bars. Underparts heavily barred.

Tails of Great Snipe (left) and Common Snipe (above). Tail of Great Snipe conspicuously whiter.

before taking flight, often startling the observer as it does so. Once airborne, it tends to fly low and straight, often rising rather half-heartedly and quickly dropping back into cover on half-closed wings (although, like Common, it may fly some distance). It can sometimes be repeatedly flushed. In contrast, Common Snipe's panicking escape flight is towering and zigzagging, often disappearing high into the distance before resettling. Whereas Common Snipe takes to the air with a characteristic squelching call, Jack Snipe is usually silent, uttering at most a low, barely audible, rather desultory nasal call, similar to Common's but weaker. Beware, however, of occasional tight-sitting Common which may fly off without calling or zigzagging. In flight, Jack Snipe is easily separated by its small size, short bill, prominent golden back and scapular stripes and shorter, more rounded wings.

Behaviour On the ground, Jack Snipe is decidedly crake-like in behaviour, feeding furtively on the edge of reeds or amongst low vegetation. It uses its short bill to pick at the surface of the mud, probing less than Common Snipe, and it has a peculiar and very distinctive habit of *rhythmically bobbing its body up and down*, like some sort of bizarre clockwork toy. Feeding Common Snipe, on the other hand, has a more vigorous 'sewing machine' probing action which enables distant recognition simply by the rhythm of its head movements. Note, however, that Common Snipe will also occasionally bob, but not as persistently as Jack Snipe.

Plumage The head pattern of Jack Snipe is diagnostic: its dark crown lacks the buff central crown-stripe of Common; instead it has a narrow buff lateral crown-stripe (or 'upper supercilium') which produces a distinctive head pattern. For the more rarity minded, this pattern is very similar to that of the vagrant Broad-billed Sandpiper *Calidris falcinellus* (p. 162). In addition, the dark eye-stripe curves strongly around the lower edge of the ear-coverts. Golden mantle and scapular stripes stand out much more strongly than Common's, while the underparts lack Common's flank barring. The bill is short (about equal to the head length) and has a pale base. A squat body and a rather truncated rear end somehow complement Jack Snipe's furtive feeding behaviour.

Faeroeensis **Common Snipe** Common Snipe breeding in Orkney, Shetland, the Faeroe Islands and Iceland belong to the race *faeroeensis*. These birds winter on mainland Britain, particularly in western areas, and in Ireland. They are much more colour-saturated than nominate *gallinago*, appearing distinctly browner and buffier in the field (some are rufous or even orangey). However, outside the breeding range it will be difficult to assign any individual to this race with absolute certainty.

Great Snipe *Gallinago media*

In flight Most likely to be seen when flushed, flying away low and direct for a relatively short distance before dropping back into cover. Appears large, bulky and heavy (perhaps two-thirds heavier than Common) with rather rounded wing-tips, all of which may give it something of the feel of a Woodcock *Scolopax rusticola*. Most distinctive are noticeable white tips to the median and greater coverts, extending onto the tips of the median and greater primary coverts, either side of a dark mid-wing panel. Unlike Common Snipe, the white trailing edge is inconspicuous. It has prominent pure white corners to the tail (less obvious on juveniles) best seen when spread on take-off and landing. Also significant in flight are the heavily barred underparts and dark underwings.

On the ground Crouches low and keeps to cover. When visible, it appears rounded and pot-bellied with a slightly shorter and thicker bill than Common. The mantle and scapular stripes are prominent, although the latter tend to be less well defined than Common. The wing-coverts are chequered with white (most obviously on adults) and the white tips to the median and greater coverts form two white wing-bars. More significant is that the underparts are heavily and extensively barred, with thick chevrons covering the flanks, leaving only a restricted area of white in the centre of the belly. The thicker legs are greyer than Common Snipe's. It bobs up and down when feeding although its movements are less exaggerated than Jack Snipe.

Call and song Rarely calls when flushed, although it may give a low, throaty *ugh*. The remarkable undulating song is a combination of high-pitched whistling combined with a strange noise likened to the sound of rapidly falling ice crystals.

Godwits

Where and when Two races of Black-tailed Godwit occur in Britain and Ireland: nominate *limosa* breeds in small numbers, mainly in e. England; it is rare on passage and winters in W. Africa. Alarming declines have become apparent over its Eurasian breeding range. The Icelandic race *islandica* breeds in small numbers in n. Scotland, but is common on passage and in winter. It has seen a huge population increase since the 1980s and is now beginning to approach Bar-tailed Godwit in abundance. The latter breeds from Arctic Scandinavia east to Siberia (some summer here) and it is widespread around the coast in winter (but declining). As well as estuaries, it also frequents open shorelines, more so than Black-tailed. A large passage up the English Channel in April and May involves birds that winter in W. Africa. Both species occur inland, mainly on autumn passage, Black-tailed being much the commoner.

Bar-tailed Godwit *Limosa lapponica*

Distant godwits can be difficult to identify, but they are easy at close range, especially in direct comparison. Firstly, concentrate on structure.

Structure and bill colour Rather squat-looking and less gangly than Black-tailed, averaging slightly smaller, shorter-necked and shorter-legged (especially shorter tibia). The bill is slightly but noticeably upcurved and generally slightly shorter than Black-tailed's, but female Bar-taileds have obviously longer bills than males, with little overlap in measurements. Except in breeding condition, the bill shows a prominent bright pale pink base (Black-tailed also shows pink in winter, yellow or orange in summer).

Plumage *Winter* Far more variegated above than Black-tailed, the feathers being brownish-grey with whitish fringes, producing a pattern more reminiscent of Curlew *Numenius arquata* and quite unlike the plain brownish-grey of winter Black-tailed. The underparts are distinctly whiter than Black-tailed, but lightly streaked with brown across the breast. The head is also whiter and more strongly patterned, with an obvious long white supercilium that extends noticeably beyond the eye (unlike Black-tailed). *Juvenile* Buffer, neater, more regularly patterned and more contrasting than winter adult. Like adult, it also shows

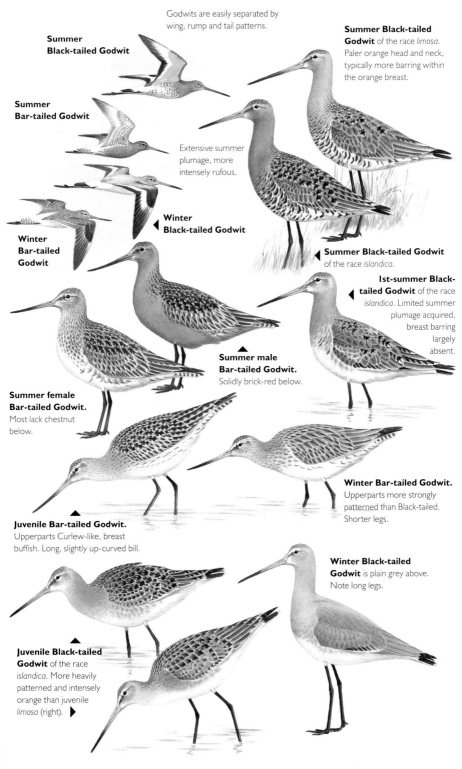

Godwits are easily separated by wing, rump and tail patterns.

Summer Black-tailed Godwit

Summer Bar-tailed Godwit

Extensive summer plumage, more intensely rufous.

Winter Black-tailed Godwit

Winter Bar-tailed Godwit

Summer Black-tailed Godwit of the race *limosa*. Paler orange head and neck, typically more barring within the orange breast.

Summer Black-tailed Godwit of the race *islandica*.

1st-summer Black-tailed Godwit of the race *islandica*. Limited summer plumage acquired, breast barring largely absent.

Summer male Bar-tailed Godwit. Solidly brick-red below.

Summer female Bar-tailed Godwit. Most lack chestnut below.

Winter Bar-tailed Godwit. Upperparts more strongly patterned than Black-tailed. Shorter legs.

Juvenile Bar-tailed Godwit. Upperparts Curlew-like, breast buffish. Long, slightly up-curved bill.

Winter Black-tailed Godwit is plain grey above. Note long legs.

Juvenile Black-tailed Godwit of the race *islandica*. More heavily patterned and intensely orange than juvenile *limosa* (right).

171

a Curlew-like pattern to the upperparts, quite unlike the tortoiseshell scalloping of juvenile Black-tailed. Each feather has a dark brown centre with a broad buff fringe, but many feathers also have pale notching or spotting; note especially the obvious notching on the tertials (thin whitish fringe on adult). Unlike Black-tailed and winter Bar-tailed, the underparts are evenly suffused with a soft orangey-buff (wearing paler) with delicate brown streaking across the upper breast. This plumage is gradually lost during an autumn body moult. *Adult summer* Obvious sexual dimorphism. Males are uniformly orange-chestnut over the entire underparts, lacking both a white belly and barring. Females usually gain some chestnut or cinnamon but lack the colour intensity of males; some retain winter-like plumage throughout the summer. The entire upperparts are browner than Black-tailed, strongly patterned with chestnut, brown and black. On breeding males, the supercilium is more or less concolorous with the rest of the head. *First-summer* Some acquire traces of adult-like summer plumage but others apparently moult directly from first-winter to second-winter (*BWP*); certainly, many winter plumaged-type birds in late spring have already acquired fresh non-breeding type plumage, with thick buff or whitish fringes to the upperpart feathers. *Spring moult* Note that, in spring, British wintering birds depart in February and March to moulting grounds on the Wadden Sea. They do not usually acquire summer plumage prior to their departure. In contrast, the W. African migrants that appear a month later will have largely moulted into summer plumage.

Flight identification Completely different from Black-tailed: plain brown wings, a barred tail and white rump, which extends up the back in a V, with the overall pattern reminiscent of Curlew. The underwing is duller than Black-tailed, the primary and secondary tips being greyish and less contrasting. The legs do not project so far beyond the tail, producing a more truncated rear end. If the bill is not visible, distant Bar-tailed can be confused with Whimbrel *N. phaeopus* (see p. 174).

Voice Quite vocal in flight: a rather soft, mellow, rhythmic, *er–er, er–ik–ik, ik–ik–ik* or a quicker, more urgent alarm: *ik–ik–ik–ik–ik….*

Black-tailed Godwit *Limosa limosa*

Structure and bill colour Black-tailed is tall and elegant, with a long, relatively straight bill, a long neck and long legs. Note in particular that the tibia is longer than Bar-tailed's. The base of the bill is pink in winter, duller than Bar-tailed but rather more extensive; it becomes bright orangey-yellow when breeding (all-dark on Bar-tailed).

Icelandic race *islandica*

Plumage *Winter* The entire plumage is *plain* and uniformly dull brownish-grey (with very narrow and faint pale fringes to the upperpart feathers) with a dull white belly; there is a whitish supercilium up to the eye but it is fainter behind the eye than Bar-tailed. *Juvenile* Orange on the throat and breast, often very richly coloured; in fact bright juveniles are frequently mistaken for summer adults. However, juvenile plumage is very smooth, immaculate and uniform below, lacking the adult's sharp contrast between the breast and belly, and also without barring. The upperparts of juveniles differ from adult summer and from all plumages of Bar-tailed, being brownish-black, neatly and evenly patterned with thick orangey feather fringes, which produce an attractive tortoiseshell pattern. However, the orange fades as autumn progresses and as the grey first-winter feathering starts to appear. The supercilium is prominent only before the eye. ***Adult summer*** Bright orangey-chestnut down to the lower breast, with a white belly variably barred black (summer male Bar-tailed is plain reddish right down to the undertail-coverts). The upperpart pattern differs from Bar-tailed, the wing-coverts being rather uniformly grey, but the mantle and scapulars are chequered black and chestnut, mixed with variable amounts of plain grey winter feathering. The pale supercilium peters out behind the eye. ***First-summer*** Small numbers of first-summer *islandica* remain in Britain throughout the summer. Such birds do not acquire full summer plumage. Instead, they tend to be strongly washed with orange on the neck and breast (extent and intensity varies), this plumage contrasting with the plain white belly and undertail-coverts; they largely retain grey winter feathering on the upperparts, but this is variable, with many acquiring limited amounts of summer plumage. Do not assume that such birds are summer adult Continental *limosa* (see below).

Flight identification Very easily identified by the huge white wing-stripe, black tail and square white rump, combining to produce a striking pattern. The underwings are pure white, contrasting with black tips to the primaries and secondaries. The legs project much further beyond the tail than Bar-tailed which, together with the longer neck and bill, produce a strikingly attenuated look.

Voice The normal call is a dry, throaty, slightly muffled *ick* or extended *ick ick ick ick ick ick* (number of notes varies). Feeding parties may give a peculiar throaty, coughing wheeze. When breeding, it has a rhythmic *wick-a-wick-a-wick-a* call or a persistent *wik-wik-wik-wik*. It also gives an alarm call on the breeding grounds reminiscent of Lapwing *Vanellus vanellus*.

Separating Continental *limosa* from Icelandic *islandica*

Although small numbers of the Continental *limosa* breed mainly in e. England (*c.* 60 pairs in 2010; Holling *et al.* 2012) it must be stressed that this race is otherwise rare here and does not normally occur in winter. There are structural differences between the two, *limosa* being distinctly longer-billed, longer-necked, longer-legged and generally more 'gangly' than *islandica* (but this is complicated by the fact that females of both forms are larger and longer-billed than males). Differences in their summer plumage are complicated and confusing. The basic principle is that, being a more northerly and later breeder, *islandica* has more time to complete its moult into summer plumage than the mid-latitude earlier breeding *limosa*. Consequently, summer-plumaged *islandica* have *more* summer plumage than *limosa* and that plumage is *brighter and more intense in colour*. Thus, on *islandica*, at least 70% of its body feathers are in summer plumage and the orange is deeper, more 'saturated' and more rufous in tone; also, the rufous extends further down onto the flanks and it is more extensively barred with black on the belly, flanks and undertail-coverts. *Limosa* has a complicated spring moult that results in breeding birds having a mixture of barred dull orange feathers, newer *unbarred* bright orange feathers, and quite extensive grey winter feathering on the upperparts. Confusingly, those birds with a greater proportion of the newer, unbarred, bright orange feathers look more like *islandica*. A further complication is that males of both forms are brighter than females. In juvenile plumage, *limosa* is a paler, washed-out orange, with narrower and paler feather fringes to the upperparts. This, combined with the structural differences outlined above, may enable juvenile *limosa* to be picked out from flocks of brightly coloured juvenile *islandica*.

Curlew and Whimbrel

Where and when Curlew *Numenius arquata* is a widespread breeding bird, mainly on damp upland moors in the north and west, but has recently shown large declines in Britain and, more especially, in Ireland. Outside the breeding season, it remains common around the coast and inland in some areas. Whimbrel *N. phaeopus* breeds only in Shetland, Orkney, the Hebrides and n. Scotland but is a fairly common spring (April/May) and autumn (July to October) passage migrant. Quite large numbers pass through western areas en route to and from Iceland; very small numbers winter in s. Britain and Ireland. Habitat similar to Curlew and both species often feed in fields.

Size and structure Whimbrel appears smaller and stockier than Curlew, with a short neck and short legs; the bill is less evenly curved, appearing comparatively straight but markedly downcurved towards the tip (almost appearing down-kinked). Curlew has a longer, more evenly curved bill (females have longer bills than males); juveniles are slightly straighter-billed than adults.

Plumage Whimbrel's striped head pattern is an obvious difference: it has a pale crown-stripe, dark lateral crown-stripes, a pale supercilium and a dark eye-stripe. Whimbrel's overall brown plumage tone is distinctly darker than Curlew's. Juveniles of both species are similar to adults but have fresher, buffer plumage.

Pitfalls Be wary of small Curlews (no doubt males) that have a rather short, relatively straight bill and a hint of dark lateral crown-stripes; such birds can be easily mistaken for Whimbrel.

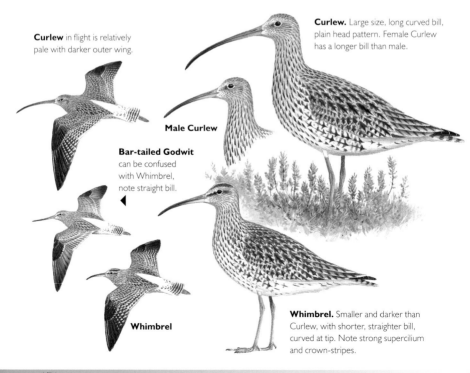

Curlew in flight is relatively pale with darker outer wing.

Curlew. Large size, long curved bill, plain head pattern. Female Curlew has a longer bill than male.

Male Curlew

Bar-tailed Godwit can be confused with Whimbrel, note straight bill.

Whimbrel

Whimbrel. Smaller and darker than Curlew, with shorter, straighter bill, curved at tip. Note strong supercilium and crown-stripes.

In addition, there have been occasional sightings of aberrant Curlews showing an abnormally short, under-developed bill.

Flight identification Curlew has relatively pale, mottled plumage with contrastingly dark outer primaries and primary coverts. Whimbrel has darker and more uniform wings. Whimbrel's wings are shorter and rather more triangular in shape, producing quicker wingbeats than Curlew, whose longer wings produce a slower, more languid, gull-like flight. Distant Whimbrels can be difficult to separate from Bar-tailed Godwits *Limosa lapponica*, the problem being particularly acute in spring when large numbers of 'Bar-wits' move up the English Channel, often well out to sea. In such circumstances, godwits often form large, purposeful flocks, whereas Whimbrels tend to form smaller, straggly lines. Whereas Whimbrel tends to give a 'head-up, crop-out' impression in flight, Bar-tailed Godwit has a long, almost drooping neck. When seen together, Bar-tailed is slightly but noticeably smaller, slimmer-bodied, slimmer-winged and paler.

Calls The easiest way to separate Curlew and Whimbrel is by call. Curlew has a familiar loud, mournful *cur-lew*, *cur-loo* or variations thereof, often drawn into a very distinctive soft, bubbling call. It also gives a rapid *kvi-kvi-kvi-kvi* alarm call. Whimbrel has a very distinctive, easily imitated seven-note whistle. Just to confuse matters, however, migrant Whimbrel rarely give a rather thin, high-pitched Curlew-like *cur-eeeee*, which may be tagged on to the conventional call. Curlew has a well-known, haunting, bubbling song. Whimbrel's song is a similar liquid rippling, somewhat like the call but longer, more liquid and interspersed with a rather shrill *oo-eeep*.

Hudsonian Whimbrel *Numenius hudsonicus*

There are seven British records (1955–2009) of this closely related North American species. The most obvious difference is its lack of a white rump, the lower back and rump being uniformly brown. However, it also has paler, buffier and more contrasting plumage with a more contrasting head pattern (the dark stripes darker and the pale stripes paler) producing a pattern reminiscent of Aquatic Warbler *Acrocephalus paludicola*. Also obvious in flight are uniformly dark underwings, with cinnamon-brown underwing-coverts.

Common and Spotted Sandpipers

Where and when Common Sandpiper *Actitis hypoleucos* is a common summer visitor, mainly to upland rivers and streams in Scotland, n. England and Wales (also in Ireland). It is more widespread on migration, frequenting both freshwater and coastal habitats. Small numbers winter, mainly in southern coastal areas. Spotted Sandpiper *A. macularius*, its North American counterpart, is a rare vagrant, mainly in autumn, but northward-bound vagrants also occur in spring and it sometimes overwinters; there are currently about six records a year (a pair nested in Scotland in 1975).

General Both species bob their rear ends and fly low over the water on bowed wings with a flicking flight action. Both show a prominent white wing-bar and Common makes very distinctive high-pitched *pee pee pee pee pee* calls. In summer, both species have their upperparts barred and mottled with black (juveniles lack these markings); summer Spotted has heavy thrush-like underpart spotting.

Structure Structural differences may attract initial attention. Spotted is rather pot-bellied, flat-backed and long-legged compared to Common (with a bit of imagination, its shape can recall a Dipper *Cinclus cinclus*). Most important is tail length: Spotted has a shorter tail which usually protrudes only just beyond the closed wings (maybe 0.5cm). Common has a longer tail which projects well beyond the wing-tips. Spotted's shorter tail produces a more truncated

Juvenile Spotted Sandpiper. Note short tail, well-patterned wing-coverts, stronger supercilium, two-toned bill and, usually, yellower legs.

Juvenile Common Sandpiper. Note long tail, less contrasting wing-coverts, duller bill and dull legs.

Winter Spotted Sandpiper. Wing-coverts plainer than juvenile. Bill, supercilium and leg differences as for juvenile.

Winter Common Sandpiper. Note long tail.

Primaries often completely cloak tail. Looks pot-bellied.

Note difference in tertial pattern. Common Sandpiper (top) has notched edges. Spotted Sandpiper's are plain.

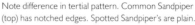

Spotted Sandpiper (left) and **Common Sandpiper** (right). Spotted Sandpiper has more black on inner secondaries and a strongly barred tail.

rear end, but beware of occasional shorter-tailed Commons (perhaps mainly autumn juveniles that have not fully grown their tail feathers).

Plumage *Juvenile* Most autumn Spotteds are juveniles, which are unlikely to occur before late August. Very detailed field notes and/or photographs are essential to ensure acceptance. Close inspection should reveal the following differences from juvenile Common (in approximate order of significance): **1** TERTIALS Differences diagnostic. On Common, the edges of the tertials have buff notching; on Spotted, they are plain, with barring confined to the tip. **2** WING-COVERTS On Spotted, prominently barred pale buff, black and brown, forming a large barred patch at a distance; juvenile Common is also barred, but the bars are much closer in colour and, therefore, less striking (note that this area can be obscured by the breast and scapular feathers). **3** BILL Spotted's bill is usually flesh-coloured with a dark tip; Common's is greyer and plainer. Some Spotteds, however, have a plain bill with an inconspicuous horn-coloured base to the lower mandible. **4** SUPERCILIUM AND EYE-RING The eye-ring is usually distinctly whiter and more obvious on Spotted and the supercilium is often better defined, producing an obviously 'sharper' facial expression. **5** LEGS Usually much yellower on Spotted, often strikingly so, but a minority have greener or greyer legs with little yellow apparent and, therefore, more similar to Common (which usually has dull green or greenish-yellow legs). **6** PLUMAGE TONE Spotted is generally slightly greyer than Common, especially on the head, neck and breast-sides. **7** BREAST PATCHES Smaller, neater, plainer and less extensive on Spotted (but can be larger on some); therefore, Spotted looks whiter on the breast when viewed front-on, lacking Common's more extensive breastband. Note that summer-plumaged adult Common has paler brown upperparts than juvenile, delicately patterned with black (some much more patterned than others) and is readily distinguishable from juvenile Spotted. *Adult summer* Adult Spotteds may start to moult during autumn migration, but the moult is not completed until arrival in the winter quarters and many reach there still in full summer plumage. This means that autumn adults in this country should retain all or most of their spotted summer plumage and are easily identifiable. *Adult winter* Spotted's winter plumage is usually acquired in October or early November; in spring, summer plumage is acquired later than Common, by mid April/early May. Adult winter appears very plain above but shows dark brown and buff barring on the wing-coverts and also somewhat randomly on the scapulars; however the wing-coverts are not as contrastingly barred as on juveniles, so this area does not really stand out as a paler, barred patch. Winter Spotteds are best identified by structure, bill colour, facial pattern (note especially the strong whitish eye-ring) and leg colour (all as juvenile). Note that many show small black flecks on the belly, rear flanks and undertail-coverts throughout the winter.

Calls Strangely, Spotted is far less vocal than Common and vagrants are often almost completely silent. Calls include a repeated loud, slightly disyllabic *tui tui* ('fuller' and mellower than Common; several calls are sometimes strung together) or a high-pitched *teep* note, sometimes quiet, occasionally loud and penetrating.

Song Common's song has a similar quality to its call: a fast, rhythmic and pulsating *see–SEE–see–see, see–SEE–see–see, see–SEE–see–see....* Short bursts are often given on spring migration.

Green and Wood Sandpipers

Where and when Green Sandpiper *Tringa ochropus* is a fairly common and widespread autumn passage migrant from mid June, with a peak in August; small numbers winter and there is a small spring passage, with a few present into May. Wood Sandpiper *T. glareola* is a spring and autumn passage migrant, mainly in May and from late June to September; largest numbers tend to occur in eastern counties. Both species are found almost exclusively in freshwater habitats. Very small numbers of Wood Sandpipers regularly breed in Scotland, with tiny numbers of Green Sandpipers also recorded in recent years.

Structure and behaviour Although superficially similar in plumage pattern, Green and Wood Sandpipers are easily separated. At rest, Green is a rather hunched, short-legged, dumpy wader whose structure tends to recall a giant Common Sandpiper *Actitis hypoleucos*; it is sleeker when active. Wood is slim, long-necked, small-headed and rather long-legged. It is in fact an exquisitely delicate and elegant wader that is likely to be seen striding along an open shoreline like a Redshank *T. totanus*, rather than feeding secretively on a small pool or muddy stream, like Green.

Plumage Green is very dark with a *prominent* breast-band and a pale eye-ring but, although it has white before the eye, *it lacks a prominent supercilium*. The upperparts spotting is quite small, but summer adults are more coarsely spotted, so beware of misidentifying such birds as Wood Sandpiper. The latter is much paler and browner, with an obvious *white supercilium* (contributing to a more capped appearance). It has white underparts with a delicately streaked breast (unlike Green, appearing pale at any distance) and *heavily spotted or chequered upper-parts*. Its greenish-yellow legs are paler than Green Sandpiper's.

Ageing Autumn juveniles of both species are immaculate and delicately patterned, while late summer adults are scruffier and more coarsely patterned. Some adult Greens suspend wing moult during autumn migration, so obvious gaps in their inner primaries permit late summer adults to be easily aged in flight.

Flight identification Although both have plain wings and a square white rump, Green is a very contrasting black-and-white bird, whose overall appearance suggests a giant House Martin *Delichon urbicum*; as well as the contrasting white rump, note in particular its *blackish underwings*. Even in flight, Green looks rather stocky and the legs do not project noticeably beyond the tail. It tends to tower when flushed, calling noisily (see below). Wood Sandpiper looks much browner in flight, and its slightly smaller white rump contrasts less with the browner plumage; the key feature is *the pale whitish underwings*. Wood looks slimmer in flight, with the feet projecting well beyond the tail; it thus has a slim-winged, elongated appearance.

Calls Totally different. Green is a noisy, excitable bird and, when flushed, typically gives a loud, ringing, *too-leet* or *too-leet . . . too-leet too-leet* (or variations thereof). Wood has a quick, shrill, high pitched *chiff-if* or *chiff-if-if*.

Song On the breeding grounds, Wood Sandpiper gives a fast, repetitive, musical *wiloo wiloo wiloo wiloo....* Green's song is very different: a fast and rather squeaky *ik-u-wee ik-u-wee ik-u-wee* (the last syllable higher-pitched and inflected upwards).

Juvenile Wood Sandpiper. Note browner plumage with strongly chequered upperparts and prominent supercilium.

Juvenile Green Sandpiper. Plumper and shorted-legged, darker above and lightly spotted with a strong breast-band. Note short supercilium.

Green Sandpipers. Very black and white, legs short, not projecting far beyond tail. Note blackish underwings.

Wood Sandpipers. Slimmer and slighter with legs projecting beyond tail. Note pale underwings, less contrasting rump and brown upperparts.

Adult Wood Sandpipers. Note slimmer shape recalling a small 'shank'. Upperparts well chequered, streaked breast-band and strong supercilium. ▼

Wood Sandpiper. Note longer tibia than Green Sandpiper, legs typically paler and yellower.

Adult Green Sandpipers. Spotting stronger than on juveniles. Note stockier shape than Wood Sandpiper, recalling large Common Sandpiper. ▼

Other confusion species

Common Sandpiper

At rest, distant Green can be confused with Common Sandpiper (see p. 175) and reports of inland wintering Commons often prove to be Greens. Common is smaller and browner, and the tail protrudes well beyond the wing-tips; in addition, brown on the breast tends to be confined to patches, while a small strip of white protrudes up and around the bend of the wing. Differences in flight are obvious: Common has a white wing-bar and a dark rump, and flies low with bowed wings and a flicking flight action.

Common Redshank

In late summer, Wood Sandpiper can be confused with juvenile Common Redshank, which looks surprisingly different from the adults (pp. 181 and 182) being much paler and browner. The upperparts are heavily spangled with buff, the head and breast are streaked brown and the bill is dull. However, it is larger and heavier than Wood Sandpiper, it lacks a supercilium and has a darker breast and orange legs; in flight, it shows a V-shaped white rump and *prominent white secondaries.*

Lesser Yellowlegs

Wood has similar plumage to the vagrant Lesser Yellowlegs *T. flavipes*, particularly when juvenile. If in doubt, the latter's larger size, long primary projection, relatively plain face, long yellow legs and *tiu … tiu* call should enable easy separation (see pp. 184 and 185).

Solitary Sandpiper

T. solitaria A very rare North American vagrant, mainly to western areas in September/ October. Similar to Green Sandpiper but slimmer with longer legs. Usually quite tame. The plumage is browner, the breast-band less obvious and it has a prominent white eye-ring. Most obviously, it has (1) a long primary projection at rest (primaries almost equal to overlying tertial length) and (2) in flight it has an all-dark rump with strong black barring on the white tail-sides.

Common and Spotted Redshanks, Greenshank and Marsh Sandpiper

Where and when Common Redshank is common on coastal marshes and estuaries, but also occurs inland, especially on migration; it breeds on both coastal and inland marshes. Greenshank breeds in n. Scotland, but both it and Spotted Redshank occur more widely on passage (inland and on the coast); relatively small numbers of both species winter, mainly on English estuaries. Marsh Sandpiper breeds in e. Europe and Asia. It currently averages about two records a year, in April to June and July to October, mainly in freshwater marshes in e. England.

Common Redshank *Tringa totanus*

General features A lively and excitable bird, typically seen flying off noisily when flushed from a saltmarsh creek; loose flocks also form on open mudflats, packing together when roosting. At rest, a rather dumpy, featureless, medium-sized wader, uniformly grey-brown from a distance, with a fairly obvious white eye-ring. At close range, bright red legs and bill base are obvious on adults (orange on juveniles – see below). It feeds with a brisk walk, stopping every few steps to pick or probe.

Plumage *Adult summer* Highly variable: grey-brown to cinnamon-brown, with variable amounts of black streaking, mottling and barring. **Winter** Plain grey-brown with a white eye-ring; lacks obvious markings at any distance. ***Juvenile*** Potentially more confusing. Paler and browner than adult; the upperparts are heavily chequered with pale buff, but the head and breast are streaked brown. The legs are orange and the bill base is grey at first, then orange. Juvenile body plumage is moulted in autumn, so first-winter plumage is much more adult-like (although the orangey bill base and legs may persist into winter). Juvenile can be confused with other species, notably Wood Sandpiper *T. glareola* (see p. 178); if in doubt, wait until it flies.

Flight identification Easily identified by its conspicuous and striking white secondaries. The lower back and rump are also white.

Calls Extremely noisy. When flushed, gives a loud, panicking *tee-dee tee-dee tee-dee tee-dee*; the normal flight call is a more subdued, melancholy and slightly ringing *tew* or *pit-ew*, mellower and less incisive than the corresponding call of Greenshank. Other calls include a mellow *tew du du, tew du* or a more abrupt alarm: *tyip tyip tyip tyip*. The song is an almost pulsating, musical *tulu-tulu-tulu-tulu....*

Spotted Redshank *Tringa erythropus*

General features Superficially similar to Common Redshank, but slightly larger and much more elegant, with a noticeably long *fine* bill and longer legs. More lively in pursuit of prey, typically wading up to the belly, immersing the head and sometimes virtually upending; frequently swims. It may occur in small parties, which usually feed in shallow water and are particularly fond of estuarine creeks and channels.

Plumage *Adult summer* Male easily identified by its solidly black head and underparts, and black upperparts, spotted with white. Females are duller and less evenly coloured. Moulting adults are patchy and blotched white. **Winter** Very pale and readily separated from the more uniform and darker grey-brown Common Redshank. The upperparts are *pale* grey, with small white notching on the feather fringes; the underparts are strikingly white (with a pale grey wash mainly on the breast-sides). The head pattern is distinctive, with a dark stripe between the bill and the eye and a conspicuous white supercilium above that, curving up and over the eye, but usually petering out behind. The legs are bright red or bright orange but, unlike Common Redshank, the bill base is red *only on the basal third of the lower mandible.* ***Juvenile*** Quite different from adult: dusky, smoky-brown, evenly coloured on the underparts but with diffuse barring on the flanks and sides of the vent, most obvious at close range. Neat pale spotting on the upperparts. A white supercilium before and over the eye contrasts markedly with the dusky plumage. The legs and base of the lower mandible are bright orange. Juvenile plumage is lost by late autumn (following body moult).

Adult summer Redshank. Variable plumage tone patterned with black.

Juvenile Redshank. Well spotted above. Orange legs and bill base.

Adult winter Redshank. Plainer grey. Red legs and bill base.

Juvenile Spotted Redshank. Dusky plumage, white supercilium, long orange legs.

Redshank. Prominent white secondaries.

Spotted Redshank. Narrow white back patch, plain wings, long legs.

Moulting adult Spotted Redshank, (July onwards) very patchy.

Winter Spotted Redshank. Pale grey above, white below. Strong supercilium.

Adult summer Greenshank. Greyish head and neck. Upperparts patterned with black.

Adult winter Greenshank. Frosty white head, pale grey above. Note green legs and uptilted bill.

Juvenile Greenshank

Adult spring Greater Yellowlegs. Barred flanks.

Juvenile Greater Yellowlegs. Boldly speckled upperparts, hint of flank barring. Yellow legs.

Juvenile Greater Yellowlegs. Size of Greenshank, note square white rump.

Flight identification Easily identified as it lacks Common Redshank's prominent white secondaries. There is a narrow white patch on the back (unlike Greenshank's large white V) but the rump and tail appear greyish. More attenuated than Common Redshank, with a long bill and long legs that project noticeably beyond the tail.

Call Diagnostic and distinctive: a loud, disyllabic *tchoo-ik*. It occasionally gives other less distinctive calls, such as a softer *cru-cru*, reminiscent of Greenshank.

Greenshank *Tringa nebularia*

General features Easily separated from both redshanks. From a distance, generally looks very pale, with a frosty white head, pale grey upperparts and very white underparts. The legs and bill base are grey-green (former occasionally yellowish); the bill is slightly but distinctly upcurved (straighter on juveniles). Impetuous, energetic and noisy; often seen walking briskly along the shoreline of a lake or reservoir, or chasing small fish through estuarine shallows with a lolloping gait and erratic changes in direction.

Plumage *Adult summer* Grey-brown with variable and irregular black blotching on the upperparts, particularly the scapulars. The head and breast are greyer than in winter, coarsely streaked and spotted with black. *Winter* Very pale with a very white-looking head, coarsely streaked with grey. The upperparts are a washed-out pale grey-brown, all the feathers neatly and narrowly fringed with white; very white underparts. *Juvenile* Pale grey-brown above, with all the feathers neatly fringed and notched with white. The head and breast are not as white as in winter and are more coarsely streaked grey-brown; there is a white eye-ring. Looks smooth, immaculate and evenly patterned compared to late summer adults, which normally retain some black upperparts feathering, appearing rather messy and coarsely streaked in comparison. Juvenile plumage is lost during the late autumn/early winter body moult.

Flight identification Distinctive: dark grey wings, whitish tail and conspicuous sharp white V extending up the back.

Call Usually wary, flying off with a penetrating loud, ringing *tew tew tew* which, when really spooked, becomes hoarse and angry. Also gives an excited *chip*.

Song A pulsating *tew-u tew-u tew-u tew-u...* quite similar to Common Redshank but sharper, more urgent and somewhat more penetrating.

Marsh Sandpiper *Tringa stagnatilis*

General features In many ways recalls a miniature Greenshank. A small, delicate, dainty wader, about half the size of a Greenshank (body size smaller than Green Sandpiper *T. ochropus*) with a long fine bill and very long legs. Feeds on soft mud or in shallow water, with a dainty walk and a rather deliberate downward dabbing movement of the bill. The legs are greenish (sometimes yellowish or orangey on breeding adults). The bill is very fine and all dark.

Plumage *Adult summer* Rather nondescript. The head, breast and upperparts are brownish, with profuse but delicate black streaking and mottling. *Winter* Starts to moult into winter plumage shortly after breeding, so some late summer adults are essentially pale grey above and very white below. The head is very pale, with a white forehead extending back to form a noticeable white wedge-shaped supercilium; there is a slightly darker patch behind the eye. May recall a winter Wilson's Phalarope *Phalaropus tricolor*, but the latter has shorter yellow legs, an uptilted 'gravy boat' shape and a *square* white rump (hardly contrasting with the pale grey tail). *Juvenile* Plumage very similar to juvenile Greenshank. The upperparts are dark grey with narrow white feather fringes, and neat grey streaking on the breast-sides. The head is not as pale as winter adult, but it has the same clear-cut supercilium and darker ear-coverts.

Flight identification Very much recalls a miniature long-legged Greenshank. The upperparts are greyish or brownish, with a very pale tail and a sharply pointed white V extending up the back. The legs project well beyond the tail, producing a very attenuated appearance. The tips of the secondaries may appear slightly paler.

Call A thin, abrupt, slightly subdued *teur* or *tiur* which can recall a rather 'full' Little Ringed Plover *Charadrius dubius*. At other times it becomes a rather more 'chipping' *tyip tyip*.

Lesser and Greater Yellowlegs

Where and when Lesser Yellowlegs is a rare North American vagrant, with widely scattered occurrences in spring, autumn and winter (averages about ten a year). Also from North America, Greater Yellowlegs is an extreme rarity (31 records to 2011).

Lesser Yellowlegs *Tringa flavipes*

General features About two-thirds to three-quarters the size of a Common Redshank but slimmer, more elegant and more delicate, with a small head, a longish neck and an attenuated rear end with long primaries often extending scissor-like beyond the tail. The bill is fine and relatively short, but it has distinctive long bright ochre-yellow legs with long tibia (note that Greenshank occasionally shows yellowish legs). All plumages show a plain head with a narrow white eye-ring and a white supercilium that peters out behind the eye.

Plumage *Adult summer* Grey-brown with blackish streaking on the breast. The upperparts show irregular black markings and random white spotting, particularly on the scapulars (plumage varies, and wears considerably by late summer). It usually shows dark vertical barring on the flanks. *Winter* Greyish breast and rather grey above, with light speckling on the wing-coverts. *Juvenile* Most autumn vagrants are in this plumage. Resembles an oversized juvenile Wood Sandpiper, but note the plainer head, lacking Wood's prominent white supercilium behind the eye (see p. 178). Like juvenile Wood Sandpiper, it has neat and regular white spotting on the upperparts, and neat streaking on the breast, but easily separated by its larger size, long yellow legs, long primary projection and call. Juvenile plumage is gradually lost during the late autumn/early winter body moult. Note that distant Wood Sandpiper *can* be confused with Lesser Yellowlegs, particularly if the plumage is sleeked down in hot weather.

Flight identification Easily identified by its *squared-off* white rump. Overall appearance thus recalls a giant Wood Sandpiper, but note that the long yellow legs protrude conspicuously beyond the tail.

Call A soft *tiu* or a subdued *tiur tiur*. When alarmed, may give a more anxious *ti ti tur tur tur* or even a rather quick, shrill *kyip kyip* (Wood Sandpiper gives a quick, shrill, high-pitched *chiff-if-if*).

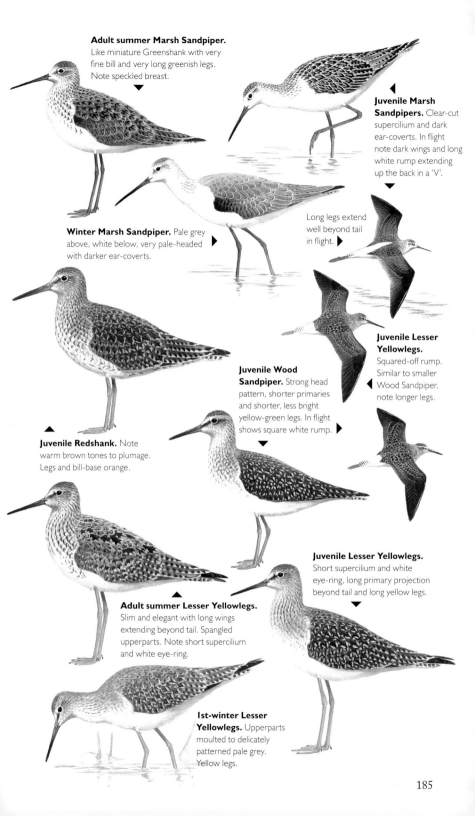

Adult summer Marsh Sandpiper. Like miniature Greenshank with very fine bill and very long greenish legs. Note speckled breast. ▼

Juvenile Marsh Sandpipers. Clear-cut supercilium and dark ear-coverts. In flight note dark wings and long white rump extending up the back in a 'V'. ▼

Winter Marsh Sandpiper. Pale grey above, white below, very pale-headed with darker ear-coverts. ▶

Long legs extend well beyond tail in flight. ▶

Juvenile Lesser Yellowlegs. Squared-off rump. Similar to smaller Wood Sandpiper, note longer legs. ◀

Juvenile Wood Sandpiper. Strong head pattern, shorter primaries and shorter, less bright yellow-green legs. In flight shows square white rump. ▶

▲ **Juvenile Redshank.** Note warm brown tones to plumage. Legs and bill-base orange.

Juvenile Lesser Yellowlegs. Short supercilium and white eye-ring, long primary projection beyond tail and long yellow legs. ▼

Adult summer Lesser Yellowlegs. Slim and elegant with long wings extending beyond tail. Spangled upperparts. Note short supercilium and white eye-ring.

1st-winter Lesser Yellowlegs. Upperparts moulted to delicately patterned pale grey. Yellow legs.

Greater Yellowlegs *Tringa melanoleuca*

Plumages resemble Lesser Yellowlegs but similar in size to Greenshank, with a similarly long, gently upcurved bill (often with extensive dull green at the base) and a more prominent 'Adam's apple' on the foreneck. Juvenile more coarsely streaked on the head and breast than Lesser, with a more distinct eye-ring, producing a more Greenshank-like appearance. The rear end is somewhat less attenuated than Lesser, with a shorter primary projection beyond the tail. The call is a loud *tu tu tu tu* (number of notes varies), much more Greenshank-like than Lesser. Feeding behaviour may also recall Greenshank, sometimes chasing fish in the shallows (sometimes with a side-to-side sweeping action like an Avocet *Recurvirostra avosetta*).

Phalaropes

Where and when Grey Phalarope breeds in the Arctic and winters off W. Africa (as well as off the Pacific coast of South America). Autumn migrants passing through the Atlantic are sometimes hit by westerly gales, and large numbers are occasionally seen off SW England (particularly Cornwall). Smaller numbers occur elsewhere, often inland. The peak month is September, but records continue well into winter. Very rare in spring and summer. It currently averages 285 a year, with a peak of 1,125 in 2001 (but several thousand in 1960; *BWP*). Small numbers of Red-necked Phalaropes breed in Shetland and the Hebrides, but virtually all European breeders migrate south-east in autumn to the Caspian and Black Seas, then on to the Arabian Sea, so they do not usually occur in western areas after gales. It is more likely to be seen in eastern counties in spring (May/June) and early autumn (peak in August, much earlier than Grey). Thus, occurrences are not usually associated with adverse weather. It currently averages 31 records a year, with a peak of 71 in 1999. Wilson's Phalarope breeds in North America and is a vagrant, currently averaging about three records a year, mostly young birds in September. There are also records of adults in spring and late summer, often on the English east coast.

Grey *Phalaropus fulicarius* and Red-necked Phalaropes *P. lobatus*

General features The two commoner species are easily separated from other waders by their persistent swimming, constantly picking at the surface in a quick, hyper-active manner, often spinning and snapping at passing insects. Both show a prominent black patch through the eye. In Britain and Ireland, Grey Phalarope is markedly pelagic, but storm-driven migrants occur in all types of saline and freshwater environments. However, most Red-necked occur in coastal marshes or freshwater habitats.

Structure Grey is a slightly larger, bulkier bird, but concentrate in particular on the bill: Red-necked's is very fine and needle-like, but Grey's is thicker, blunter and rather 'tubular' in shape. As both are invariably tame, this is not difficult to evaluate, but note that the differences are less apparent at a distance. Red-necked sits lower in the water and looks flatter-backed.

Plumage *Summer adults* Totally different: Red-necked has red confined to the neck; Grey has the entire underparts red, except for a contrasting white face; it also has a conspicuous

Juvenile Red-necked Phalarope.
Smaller than Grey Phalarope, fine bill. Dark above with buffish lines on mantle and scapulars. White patch at bend of wing.

Juvenile Red-necked Phalarope.
Dark with white wing-stripe.

Winter Red-necked Phalarope.
Grey above with darker feather centres. Note fine bill.

Juvenile Grey Phalarope looks dark-winged in flight with strong wing-bar. As moult to winter plumage progresses becomes pale-bodied and resembles Sanderling.

Juvenile Grey Phalarope moulting to 1st-winter.
Unlike juvenile Red-necked, soon acquires extensive grey on upperparts.

Juvenile Grey Phalarope moulting to 1st-winter. Dark above like Red-necked Phalarope but lacks strong buff 'V's and usually shows grey patches on back and scapulars. Note larger size and thicker bill.

Adult winter Grey Phalarope. Plain pale grey above. Note thick bill.

Phalaropes are tiny; seen here with a 1st-winter Black-headed Gull.

Juvenile Wilson's Phalarope.
Soon acquires grey feathers above.

Juvenile Wilson's Phalarope, moulting to 1st-winter. Larger than other phalaropes, more terrestrial. Note yellow legs, white rump and plain wings in flight.

yellow base to the bill; pure white underwing-coverts contrast with the dark body in flight. As their sexual roles are reversed, in both species the female is brighter than the male. Adults start to moult in late summer so, by August, they should show a scruffy mixture of summer and winter body plumage. ***Juveniles and first-winters*** Most autumn phalaropes seen in this country are young birds, but there is an important difference in their moult timings. August and September Red-necked are usually in full juvenile plumage with little or no sign of moult. However, even by late August, juvenile Greys will have already acquired at least some grey first-winter feathering on the back and scapulars and, on most, this is extensive; thus, they show a patchy two-toned appearance to the upperparts (consequently, they are best aged as 'juvenile/first-winter'). Initially, the scapulars stand out as a pale grey patch but, eventually, most of the blackish juvenile mantle feathering is also lost. By late autumn, the upperparts are usually plain pale grey, with only the wing-coverts and tertials remaining black (with whitish feather fringes). Juvenile Grey initially shows a strong pinkish-buff wash to the neck, which usually fades as autumn advances. Since August/September Red-neckeds retain juvenile plumage, they are blacker above than Grey and show two well-defined buffy-white mantle and scapular lines, which are usually prominent. They also show a small white patch at the bend of the wing. There is often a variable buffy-grey wash to the neck and breast. Whereas Grey Phalarope has a square ear-patch that extends back horizontally from the eye, on Red-necked the ear-patch *tends* to curve down behind the eye and end in more of a point. By late autumn (October onwards) even juvenile Red-neckeds start to show grey patches on the back and scapulars. ***Winter*** Note that winter Red-neckeds are extremely rare in this country, whereas small numbers of adult and first-winter Greys occur fairly regularly. As both species are essentially grey and white in winter, they are best separated by structural differences. However, even in winter plumage Red-necked is slightly darker and shows subdued pale mantle and scapular lines. Differences in the shape of the eye-patch may also be useful (see above). Note that Grey Phalaropes can be aged well into the winter as young birds retain blackish tertials (fringed white) whereas adults acquire pale grey tertials, concolorous with the rest of the upperparts. Both adult and late autumn first-winter Greys may also show yellow at the base of the bill, which Red-necked never has.

Flight identification Difficult to separate, but Red-necked is small, compact, shorter-winged and shorter-tailed than Grey. As the vast majority of phalaropes seen on seawatches are Grey, it would be extremely inadvisable to claim a fly-by Red-necked unless seen exceptionally well. In flight, Grey is similar to Sanderling *Calidris alba*, appearing pale grey and rather white-headed, with a prominent white wing-bar, but it is long-winged, rather full-breasted and has a rather long, full, dark tail. Also distinctive is a side-to-side jinking flight action. When feeding, Grey frequently flies short distances but then abruptly stalls and drops vertically to crash-land on the water.

Calls Both give a thin, monosyllabic *chit* or *tit*, Red-necked slightly lower and huskier than Grey.

Wilson's Phalarope *Phalaropus tricolor*

General features Completely different from Grey and Red-necked. A freshwater species that is much more at home on land, where confusion with Wood Sandpiper *Tringa glareola*, Lesser Yellowlegs *T. flavipes* and Marsh Sandpiper *T. stagnatilis* is more likely (see pp. 178,

180 and 184). Although Wilson's often swims, vagrants do not usually do so persistently. Conversely, it should also be remembered that most common waders, such as Spotted Redshank *T. erythropus*, will also swim occasionally. Wilson's has a fine black bill and a distinctive shape, with a small head and long neck. The whole bird often appears to be tipped forward when feeding, rather like an uptilted 'gravy boat'. Feeding action either quick and erratic, or slow and methodical, sometimes stalking flies with the head and bill stretched out parallel to the ground.

Plumage *Adult summer* Females are stunning, easily identified by their pale grey crown and nape, and a thick black stripe on the neck-sides, merging into bright chestnut and pale grey upperparts. The foreneck is strongly tinged rich apricot. Males are markedly duller, with brown and orange-brown replacing the bright head and neck colours of the female, while the upperparts are more uniformly grey-brown, lacking bright chestnut. *Juvenile, first-winter and winter adult* Juvenile and winter-plumaged Wilson's usually look very pale. The underparts are very white and there is a dark line through the eye that extends obscurely down the neck-sides. Juvenile is brown above, with pale feather fringes, but most vagrants arriving in Europe have already started to acquire extensive pale grey first-winter plumage on the back and scapulars. Winter adults are entirely pale grey above and lack juvenile/first-winter's dark brown wing-coverts and tertials. Unlike other phalaropes, the legs are yellow in autumn. In flight, it has plain wings, a square white rump and pale grey tail that hardly contrasts with the rump.

Call Generally silent, but occasionally gives an unremarkable, soft, nasal *chu* or *yup*.

Arctic, Pomarine and Long-tailed Skuas

Where and when Arctic Skua breeds in n. and w. Scotland but is a widespread coastal migrant, mainly in April/May and August to October (with a few lingering into winter). Pomarine Skua is an uncommon migrant, mainly in late April/May and August to November but, unlike the other two species, small numbers regularly occur throughout the winter, particularly in the North Sea. Most are seen on spring passage, when variable numbers move up the English Channel and also north through the Irish Sea. The vast majority, however, move up the Atlantic and, during strong north-westerly winds, large numbers may be seen off the Outer Hebrides and nw. Ireland. Long-tailed is by far the rarest skua (about 370 records a year in 1981–90) but breeding has been attempted. Like Pomarine, the vast majority move up the Atlantic, and north-westerly winds may also produce a large spring passage off the Outer Hebrides. Otherwise, small numbers are most likely to be encountered in August/September, mainly off North Sea coasts (it is very rare in the English Channel). A huge autumn movement in 1991 produced a British record for that year of 5,350, mainly in the North Sea (Fraser & Ryan 1994). All three species can appear inland; in fact some routinely migrate overland, appearing on fresh water and even in fields (the latter mostly Long-tailed) particularly during gloomy anticyclonic weather. Such birds may be incredibly tame.

General approach Skuas are exciting birds that always enliven a sea watch but, with the exception of adults in summer plumage, the three smaller species are notoriously difficult to identify. Any discussion of their identification is complicated by their variability. Firstly, adult Arctics and Pomarines have pale and dark plumage morphs (as well as intermediates). Secondly, juvenile plumages of all three species are similar, and individual variation at this age is considerable. Thirdly, skuas do not reach maturity until about three to five years old. As they usually remain in their winter quarters during their first summer, their immature plumages are unfamiliar to Northern Hemisphere birders. The identification of immatures can also be complicated by bleaching and wear. Fourthly, judging their size is complicated by the fact that females are significantly larger than males (averaging *c.* 12–15% heavier). Although experienced seawatchers may confidently identify skuas at some distance (largely by 'jizz') less-experienced birders should exercise caution and identify only those that are seen well; be prepared to log some as 'skua sp.'. It is an odd paradox that distant skuas at sea are routinely identified with confidence, even by inexperienced birders, yet close-range birds inland often prove controversial. In view of the complexities of identifying winter adults and immatures, only summer adults and juveniles are dealt with in depth; other plumages are outlined on p. 196. Finally, note that adults retain their summer tail projections throughout the autumn until they moult in their winter quarters, but the projections may be susceptible to loss or damage.

Arctic Skua *Stercorarius parasiticus*

Structure Generally the commonest small skua, this species should act as the yardstick when identifying the other two species. Size intermediate between Pomarine and Long-tailed, being similar to Common Gull *Larus canus*, but much sturdier. Structural differences from Pomarine and Long-tailed are dealt with under those species but, compared to Pomarine, note Arctic's medium build, being smaller and slimmer with a longer-looking head, narrower wings and tapered rear end; adults have an obviously pointed tail projection, which may be as long as 10.5cm (4.5 inches) suggesting Long-tailed. Conversely, skuas of all species and ages can occasionally lack central tail feathers, either through moult or damage. At close range, the bill is rather slim and slender, lacking Pomarine's more obvious gonydeal angle.

Flight Migrating flight is steady, but less heavy and ponderous than Pomarine. All three species glide on distinctly arched wings. In strong winds, Arctic may adopt a shearwater-like flight, rising and falling above the waves in a series of long arcs, but the wings are held more arched than most shearwaters and the arcs tend to be flatter.

Plumage *Adult summer* Dark-phase adults are commonest and they appear completely dark fulvous-brown; a yellowish shade to the cheeks and ear-coverts is not obvious at any distance. Pale-phase adults have a blackish cap and are mainly white below, usually with a pale yellow face; they often have a brown breast-band of variable width and extent, but generally weaker than Pomarine. At close range, three or four white shafts on the upperwing, at the base of the primaries, are noticeable (*cf.* two on Long-tailed) and these show as a pale crescent on the underwing. The upperparts are brown; pale and intermediate birds are warmer-toned than Pomarine or Long-tailed and the upperwing-coverts show little or no contrast with the black secondaries (*cf.* Long-tailed, which shows marked contrast). Intermediate adults vary between dark and light phases. *Juvenile* Except for some very dark birds, juvenile skuas can usually be aged by their barred underwing-coverts and axillaries, blue-grey to pinkish-grey bill base and blue-grey to whitish legs. Plumage tone varies, Arctic and Long-tailed being more variable than Pomarine. The underparts of Arctic vary from uniformly blackish through brown to greyish-white, narrowly barred brownish. Differences from Pomarine and Long-tailed are outlined under those species, but the following are the main characteristics of Arctic. **1** DARK PHASE Note that very dark juvenile Arctics are solidly blackish-brown throughout, lacking any obvious barring (but a few dark Long-taileds can look similar, especially at a distance). **2** FEATHER FRINGES The pale feather fringes on pale and intermediate phases are warm in tone (rufous or buff) and do not contrast with the richer brown background plumage; this renders their entire plumage warmer-toned than Pomarine or Long-tailed (thus appearing 'foxy-toned': yellow-brown or rufous-brown). **3** HINDNECK Unlike Pomarine, most have a contrastingly pale hindneck, which is often warm-toned. **4** UPPER- AND UNDER-TAIL-COVERTS Although often thickly barred brown and buff, the barring is not especially contrasting so, unlike Pomarine and Long-tailed, they fail to show obviously paler upper- or undertail-coverts. **5** UPPER-PRIMARY PATCH The shafts on the bases of the first three or four primaries are white (only two are white on Long-tailed). **6** UNDER-PRIMARY PATCH There is a *single* white flash on the under-primaries so, unlike Pomarine, it does not generally show a narrow pale crescent in front of the large pale patch. **7** PRIMARIES At rest it shows noticeable pale tips to the primaries (not shown by Pomarine or Long-tailed). **8** TAIL PROJECTIONS Short and pointed (always blunt on Pomarine; blunt and usually longer on Long-tailed). **9** BILL Usually rather uniformly dark, lacking an obviously paler base.

Pomarine Skua *Stercorarius pomarinus*

Structure Summer-plumaged adults are most easily identified by their blunt, twisted central tail feathers, which look like spoons or legs/feet trailing out behind. These are retained throughout the autumn until their winter moult. However, the 'spoons' may be abraded or even broken off by late autumn and even in spring a minority show only slight protuberances. Juveniles lack 'spoons' but they have a short, blunt projection (pointed on Arctic); this is quite noticeable in closer views (down to *c.* 400m) but difficult to see at any distance (and may even be absent). In winter plumage, adults show a short projection that is hardly twisted. With Pomarines lacking 'spoons', it is essential to concentrate on overall size and structure. Although size evaluation may be difficult with lone birds, Pomarine is a large skua, approaching Lesser Black-backed Gull *L. fuscus* (about four-fifths the size in direct comparison) or *c.* 10% larger than Common Gull (Arctic is more similar in size to the latter). It is a sturdy, thickset, powerful and 'meaty' bird with a large head, a chunky body and a shorter-looking rear end; note also the heavy, hooked bill. Most importantly, the wings are broad-based and it has a sturdy, robust appearance in flight (in direct comparison, Arctic Skua is smaller, *noticeably slimmer* and flatter-chested); on summer adults, the deep-chested appearance may be emphasised by a thick dark breast-band. They may also show a more ragged appearance to the vent. Juveniles or dark-phase adults can be confused with Bonxie *S. skua*, a mistake unlikely with Arctic. On the ground, they again look big and bulky with a relatively short primary projection.

Flight Even in direct comparison there may little difference in their flight action. However, Pomarine generally has a slightly slower, steadier, more ponderous and more lumbering flight than Arctic, with continuous gull-like flapping low over the waves – often 'hugging' the sea – interspersed with short glides on bowed wings. Thus, the flight is not as 'fast and dashing' as

Adult summer Pomarine Skua. Pale phase.

Adult summer Arctic Skua. Pale phase.

Adult summer Long-tailed Skua. Little white in wing, dark belly, long tail.

Adult summer Long-tailed Skua. Note upperwing contra not shown in Arctic Sku

Adult summe Arctic Skua. Pale phase.

Compare keel shapes, wing proportions and tail lengths and shapes.

Adult summer Arctic Skua. Dark phase.

Adult summer Pomarine Skua. Dark phase.

Juvenile Long-tailed Skua. Note cold plumage tones, dark pectoral band, pale nape and elongated, rounded central tail feathers.

Juvenile Arctic Skuas. Note warm plumage tones, pale nape, streaked upper-breast, pale tipped primaries an pointed central tail feathers. Dark birds (behind) more uniform.

Adult winter Pomarine Skua. Large and bulky. Barred uppertail-coverts. Short 'spoons' (may be lost when moulting).

Juvenile Pomarine Skua. Note large two-toned bill, greyish tone to head, which is more uniform and barred rather than streaked. Evenly barred breast and underparts. Usually lacks pale primary tips.

Juvenile Long-tailed Skua. Cold, greyish tones. Neat, even barring to underwing, tail-coverts and rump. Dark breast, pale belly.

Juvenile Long-tailed Skua. Note dark secondary bar, lacks obvious white in primaries.

Juvenile Long-tailed Skua, dark phase. Note blunt tail projection. Size of Kittiwake.

Juvenile Long-tailed Skua, pale phase.

Juvenile Arctic Skua. Note pointed tail projection, white primary bases, warm plumage tones, less distinct barring on rump and undertail-coverts.

Juvenile Arctic Skua, pale phase. Note gingery plumage tones, indistinct barring beneath.

Juvenile Arctic Skua, dark phase. Bill looks uniform, undertail-coverts may appear plain.

Juvenile Pomarine Skua. Note double white wing patch on underwing. Heavy chest, broad-winged.

Juvenile Pomarine Skua. Note size and bulk. Two-toned bill, heavily barred rump and undertail-coverts. Tail projection, if present, blunt.

Adult Bonxie. Short tail, large white wing flashes. Large size, heavy body, black bill.

Arctic. Like that species, it shears in long arcs during strong winds. When pursuing prey, the wingbeats are deep and bowed. Unlike the smaller skuas, it habitually kills and eats quite large birds, such as smaller gulls. It also chases larger birds for food, such as Herring *L. argentatus* and Lesser Black-backed Gulls, and may even force them into the sea. Like Bonxie, it often feeds on scraps.

Plumage *Adult summer* Approximately 90% of adults are pale phase. Apart from the tail 'spoons' no single character separates Pomarine from Arctic, but 'Pom' is *generally* darker, blacker-brown above, and usually has a prominent dark breast-band (but some, mainly males, just show patches on the breast-sides). It also has larger whitish wing patches as well as a greater incidence of flank barring (but beware of the superficial similarity between adult Pomarine and subadult Arctic). The extensive white face (extending onto the nape) is usually obvious at all ranges, and this is usually (but not always) washed pale yellow, visible at some distance. Dark-phase birds are completely dark brown and very striking, but their whitish wing flashes may appear less obvious than on pale birds. Many 'Poms' retain traces of winter plumage well into summer and such intermediate individuals appear much duskier below, especially at a distance. *Juvenile* The main differences from Arctic are as follows. **1** PLUMAGE TONE Pomarine is fairly consistent in plumage tone, being generally rather dark brown, with no warmth to its plumage. The body plumage is variably barred buff (strongest on pale-phase individuals) and the barring is better defined than on Arctic. **2** UNDERWING CRESCENT One of the best plumage features is a whitish crescent in front of the large and very obvious silvery-white patch at the base of the under-primaries; the crescent is formed by pale bases to the greater under-primary coverts and, in good light, is visible at long range. Arctic may show a faint, diffuse crescent in front of the under-primary patch, but this is hardly visible in the field. The white flash on the upperwing is less obvious. **3** TAIL-COVERT BARRING Strong brown-and-whitish barring on the uppertail-coverts forms a noticeable pale 'rump' in flight; the undertail-coverts, vent, lower belly and, sometimes, the flanks, are similarly barred. The equivalent barring is less contrasting and less obvious on Arctic, and note that dark Arctics usually lack barring altogether. **4** BILL Longer and heavier than Arctic, with a more prominent gonydeal angle. The basal two-thirds are pale bluish, olive or sandy, contrasting with the dark tip (recalling juvenile Glaucous Gull *Larus hyperboreus*); the pale base may 'flash' paler at a distance. Arctic's bill base is less prominent because (a) it is usually slightly darker, (b) the bill is smaller and (c) the adjacent head feathers are paler and do not contrast with the base. **5** HEAD Drab grey-brown, rarely showing Arctic's contrastingly pale hindneck (but may show a paler grey wash). Thus, wholly brown-headed birds with barred underparts are almost certainly Pomarine. The few Arctics that lack a contrasting light hindneck are usually solidly blackish-brown, with unbarred underparts. Dark streaking on the head is typical of Arctic and is never present on Pomarine, which instead is lightly *barred*. **6** TAIL PROJECTIONS Short and rounded (always short and pointed on Arctic); some Pomarines lack projections altogether. **7** LEG COLOUR There is overlap in leg colour: bluish-grey, and all three species may show whitish legs. *Second-winter* See 'Appendix' p. 196.

Long-tailed Skua *Stercorarius longicaudus*

Structure Although there is size overlap, Long-tailed is as different from Arctic as Arctic is from Pomarine. A small skua, similar in size to Black-headed Gull *Chroicocephalus ridibundus*.

Adults are readily identified by their incredibly long central tail feathers which waver in flight. They can have streamers up to 18cm (7 inches) in length but note the overlap with Arctic, which can have projections up to 14cm (5.5 inches). However, Arctic's tail tends to appear thick and tapered, whereas Long-tailed's projections are obviously long and thin (the two separate feathers are often clearly visible). Juvenile Long-tailed has a short to medium *blunt* tail projection (see below). Unlike the other skuas, it has no bulk to its body, being slimmer with a shallower breast; on the water, it appears slim and elongated. It has a smaller bill than Arctic, with less of a gonydeal angle, and a smaller, more rounded head. This combines to produce a much gentler character and appearance, perhaps recalling Common Gull.

Flight Narrower wings than Arctic, especially at the base, and often appears light, slim and agile, the whole effect being more tern-like. It tends to have a more continuously flapping flight, with little gliding, and may even feed with small gulls, dropping down to the water's surface to pick up food. Inland birds may pick insects off the water or even hawk them in the air; others have fed on earthworms in ploughed fields.

Plumage *Adult summer* Do not rely solely on tail length, but concentrate on structure and plumage. Long-tailed is more consistent in its appearance than Arctic and even Pomarine: dark-phase birds are very rare and intermediates virtually unknown. Typical adults differ from Arctic in the following respects. **1** CAP Neat, clear-cut and black, contrasting sharply and smartly with the white face (often washed pale primrose-yellow). **2** UNDERPARTS White, lacking a breast-band, *but lower belly and vent obviously dark* (ashy-grey, like upperparts) merging but contrasting with the obvious white upper breast and face (sometimes the dark belly extends up to the lower breast). Therefore, the front end looks white, the rear end dark. However, some are much whiter-bellied (mainly from Greenland, North America and E. Siberian populations, race *pallescens*). **3** UPPERPARTS *Cold ashy-grey*, not as dark as Arctic. The primaries and secondaries are black, the latter contrasting with the wing-coverts to form a noticeable dark trailing edge to the wing (Arctic appears plain brown above). The tail is black, contrasting with the paler rump and uppertail-coverts. **4** UPPER-PRIMARY PATCHES Significantly, there is little or no white in the upperwing: usually just two white primary shafts (Arctic has white bases to three or four, forming a definite patch). **5** UNDERWINGS Dark silvery-grey with a contrasting black border to the front and rear of the wing; as on the upperwing, just two pale shaft-streaks on the outer primaries. *Juvenile* Exhibits a variety of plumage tones, from pale through intermediate to dark. Paler individuals can be separated from Arctic by the following differences. **1** PLUMAGE TONE Generally colder and greyer-looking than Arctic. **2** UPPERPART BARRING Upperparts show clearly defined cream or whitish barring, contrasting with the grey-brown background colour, producing a neat, scaly effect at a distance (Arctic has darker, buffier barring that contrasts less with the browner plumage, producing a warmer tone to the upperparts). **3** UPPERTAIL-COVERTS In flight, heavy brown and whitish barring produces a noticeable whitish 'rump' (duller and less obvious on Arctic, some having plain upper- and undertail-coverts, never found on Long-tailed). **4** TAIL PROJECTIONS *Blunt-tipped*; length varies from short to medium. Always short and pointed on Arctic, so those Long-tailed that show longer projections are quite distinctive, although the projection can be difficult to make out at a distance. Beware of Arctic Skuas that lack central tail feathers, the two adjacent ones then appear to be blunt central feathers. **5** PRIMARY PATCHES *Only one or two white primary shafts* (although white on the bases of the third and fourth may be perceptible at point-blank range). On Arctic, three or four show obvious white. Consequently, Long-tailed shows little white on the upperwing, but has a larger broad pale whitish-grey crescent-shaped patch on underwing. **6 UNDERPARTS** Typically greyish, with finely barred flanks; many show a darker head and neck, with a large whitish area immediately below. Undertail-coverts strongly barred. **7 HEAD** A pale greyish area on the sides of the head and nape shows to a greater or lesser degree. Some pale individuals are strikingly white-headed. **8 UNDERWING-COVERTS** Heavily barred, especially on the axillaries (some darker Arctics have uniform underwing-coverts, never found on Long-tailed). **9 BILL** Generally more black at tip (40–50% of bill is black, compared with 25–30% on Arctic) and the black usually extends back past the gonydeal angle and frequently tapers along the cutting edge, about halfway into grey base. **10 PRIMARIES** Unlike Arctic, it lacks buff fringes to the tips of the closed primaries, which appear plain black at rest. *Dark juveniles* Much more similar to dark Arctic but they show more contrastingly pale buff feather fringes to the upperparts (although some are very finely patterned and look wholly dark at a distance). However, on closer birds, quite striking dark brown-and-white barring on the undertail-coverts should be obvious, both at rest and in flight.

Other plumages of skuas

Owing to the complexities of identifying winter adults and immatures, it must be stressed that the following details are generalised. *Adult winter* On failed breeders, winter plumage starts to appear from July onwards and may be complete by August, but for most the moult starts from late August and is completed in the winter quarters. Unlike most juveniles, winter adults lack underwing-covert and axillary barring, and have a black bill and legs. On pale-phase birds, the cap becomes less distinct and the throat and neck duskier. The upperparts feathers show pale fringes and the tail-coverts are barred, as on juveniles. Dark-phase individuals are more similar to summer adults, but may acquire indistinct barring on the tail-coverts. All adults have shorter tail projections in winter, while moulting individuals may temporarily lose them altogether. *Immatures* Owing to our incomplete knowledge, the following gives only an outline and it should be stressed that immature skuas are notoriously variable, a problem exacerbated by wear and bleaching. The following details relate to Arctic Skua (from *BWP*), but the *sequence* appears similar for all species, although Pomarine appears to take a year longer to reach maturity. Juvenile plumage is moulted in midwinter and 'first-winter' plumage is characterised by a mixture of adult winter (including slightly longer tail projection) and juvenile characters (such as pale legs and, on pale-phase birds, barred underwing-coverts and axillaries). Dark-phase 'first-winters' are more similar to juveniles, but on average less heavily barred. 'First-winter' plumage is retained until late summer, when it is replaced directly by second-winter, so there is no first-summer plumage. In their second summer, some skuas arrive at the breeding colonies with variable amounts of adult-like summer plumage mixed with second-winter plumage. From then on, the moults gradually produce a more adult-like plumage until maturity, and traces of winter plumage are no longer retained in summer. However, some fully mature adults (mainly Pomarines) retain traces of winter plumage on arrival on the breeding grounds. Some mature earlier, so ageing is very difficult once juvenile characters are lost (such as the pale legs and partially barred underwing-coverts and axillaries).

Appendix *Second-winter Pomarine Skua* Given Pomarine's propensity to winter around our coasts, the following details may be useful, based on a well-studied November individual. Uniformly dark brown upperparts, head and breast, the latter producing a hooded effect in flight (with a slight dark cap). Belly silvery-white but flanks and undertail-coverts heavily barred pale buff. Pale tips to the uppertail-coverts formed a noticeable pale 'rump' in flight. Noticeable blunt tail projection. Underwing-coverts *plain brown* (barred on juvenile) and legs and feet largely *blackish* (obviously pale bluish-grey on juvenile, sometimes whitish). Plumage less immaculate than juvenile, with clear signs of inner primary moult. The narrow whitish crescent in front of the large whitish under-primary flash was faint. Bill appeared 'blob-ended', the black tip contrasting with a slightly paler base.

References Broome (1987), Davenport (1987), Fraser & Ryan (1994), Jonsson (1984), Mather (1981), Olsen & Christensen (1984), Stoddart (2012), Ullman (1984).

Immature Kittiwake and Little and Sabine's Gulls

Where and when Kittiwake is a common coastal species throughout the year, but rarer in winter; it also migrates overland and small numbers (occasionally large flocks) may appear on inland lakes and reservoirs, mainly in March/April and September/October. Little Gull is a scarce passage migrant, mainly from March to May and August to October, but smaller numbers are encountered throughout the year; it is much more likely to be seen inland than Kittiwake. Large numbers occur at selected coastal sites, notably in the North Sea, with over 6,000 sometimes gathering off the Yorkshire coast. Sabine's Gull is a rare autumn passage migrant from Canada, mainly in September/October, with occasional spring occurrences from March to June; it is extremely rare in winter. It currently averages 165 records a year, with a peak of 710 in 1987. Most records are from the west coast, particularly Cornwall (and w. Ireland) but it can occur in any coastal county and even inland; numbers depend on the prevalence of westerly gales and large movements or 'wrecks' sometimes occur.

Kittiwake *Rissa tridactyla*

Flight identification Juvenile and first-winter are superficially similar to the equivalent plumages of Little Gull, both species showing a large black 'W' across the wings in flight. Kittiwake is much larger than Little Gull, being slightly larger than Black-headed Gull *Chroicocephalus ridibundus*. The grey mantle, scapulars and leading wing-coverts *are quite dark, highlighting the extreme whiteness of the inner primaries and secondaries*. Also, the black 'W' is better defined, particularly on the primaries: this combination produces a smarter, more contrasting 'grey-black-white' pattern than first-winter Little Gull. Juvenile and first-winter Kittiwakes are similar, but the juvenile has a prominent black collar across the nape, which is usually lost in first-winter plumage (when it may show a grey shawl instead). A thick black bill contrasts strongly with the white head, which has a black spot or smudge behind the eye. The underwing is snowy white, with contrasting black tips to the under-primaries; the very white primaries and secondaries appear translucent from below. By spring, it lacks the black collar and the plumage wears and fades considerably, so that the outer primaries are not as black, the mantle and wing-coverts are paler grey and the general appearance is whiter and less contrasting; later in summer, some may become severely bleached and abraded.

Identification at rest In fresh plumage, looks smart and contrasting. Identified by its thick black bill, predominantly white head, black collar (when present), rather dark grey mantle and scapulars, and large black bar across the base of the wing-coverts, extending onto the tertials. Note also the black legs, which are very short for a gull.

Little Gull *Hydrocoloeus minutus*

Flight identification A tiny gull, about two-thirds the size of Kittiwake; its small size is usually obvious, even without other species for comparison. Its feeding behaviour is remarkably tern-like, flying back and forth and dipping down to the water like a Black Tern *Chlidonias niger* (Kittiwake is more typically 'gull-like'). Juvenile is easily separated from Kittiwake as the *back and scapulars are completely blackish* (the feathers edged pale) and this coloration extends onto the nape and neck-sides; the crown and ear-coverts are also black, quite unlike Kittiwake. As autumn progresses, however, it moults into first-winter plumage, the black feathering on the mantle and scapulars being replaced by grey, while the nape becomes white; at certain stages of moult, it can show the effect of a dark collar, suggesting Kittiwake, but this is rarely as clear-cut. First-winter *retains the distinctive dark crown and ear-coverts* that Kittiwake lacks; in flight, it looks less contrasting than Kittiwake, the greys contrasting less with the whites, and the black on the primaries is less clear-cut (the underwing too is a less pure white). First-summer may acquire a partial black hood.

Identification at rest Juvenile is easily identified by its predominantly black upperparts, crown and ear-coverts; this attractive black-and-white plumage suggests a large juvenile phalarope *Phalaropus*. First-winter is more like Kittiwake but, again, is easily separated by its black crown and ear-coverts, less contrasting upperparts, the small, delicate bill, and short pinkish-red legs.

Call A quick, throaty, tern-like *ar – akar akar akar akar*, recalling a squeaky toy (first-year Kittiwakes are relatively silent).

Black-winged Little Gulls There are records of atypical first-year Little Gulls with the whole upperwing black (with paler feather fringes) leaving a prominent white trailing edge; in first-winter, these show a grey mantle and scapulars that stand out as a pale 'saddle'.

Sabine's Gull *Xema sabini*

A very distinctive gull, easily identified if seen well. Despite this, many claimed Sabine's, particularly distant individuals on sea watches, are misidentified Kittiwakes. Caution is therefore essential.

Size and structure In direct comparison, Sabine's is noticeably smaller than Black-headed Gull and has long, thin, pointed wings; it has a tail fork but this is difficult to see at any distance. On the ground, short legs produce a pigeon-like gait. ***Adult summer*** In early autumn (August/early September) most Sabine's passing offshore are adults and most retain a full black hood (by late autumn, this is often reduced to a large black smudge across the rear of the head). The *combination* of black hood, black primaries, dark grey mantle/wing-coverts and the huge white triangle on the inner primaries/secondaries permits instant recognition. The underwings are pure white, with dark tips to the primaries (often with a thick greyish bar

Juvenile Kittiwake. Note clean, contrasting plumage, dark collar.

1st-winter Little Gull. Note small size and dark cap.

Juvenile Little Gull. Dark 'saddle'.

1st-winter Little Gull. Rare plumage variant with all dark wings.

1st-winter Sabine's Gull. Similar to juvenile but grey mantle and black rear crown.

Adult Sabine's Gull, late summer plumage. Note combination of black head and striking wing pattern.

Juvenile Sabine's Gull. Note brown head, mantle and wing-coverts. White triangle on rear of wing.

Juvenile Little Gull. Very dark above.

1st-winter Little Gull. Very small, note dark cap.

Juvenile Sabine's Gull. Note brown head, nape and mantle.

Juvenile Kittiwake. Note black line on wing and black collar.

1st-winter Sabine's Gull. Grey mantle and scapulars. Black area on rear of crown. This plumage is acquired on wintering grounds, thus very rare in northern hemisphere.

across the greater underwing-coverts). ***Juvenile*** Dark grey-brown mantle and wing-coverts (with noticeable scaling at close range, caused by narrow whitish feather fringes) and extensive washed-out grey-brown on the nape and breast-sides. The latter two areas produce an obvious dark 'front-end' to the bird in flight. The brown areas, along with the black primaries, contrast strongly with the white triangle on the rear of the wing. The underwing is white but has noticeable grey shading on the under-primaries and a broad dark greyish bar across the greater underwing-coverts. Juvenile Sabine's therefore lacks the black 'W', grey mantle, black collar and predominantly white head of juvenile/first-winter Kittiwake. Note that juvenile Sabine's do not moult until they arrive in their winter quarters. ***Winter adult and first-summer*** Sabine's Gulls do not normally occur in the Northern Hemisphere in winter, so observers should not claim a winter Sabine's unless it is seen exceptionally well. Winter adults resemble summer birds but show a greyish or blackish 'half-hood' from the eye over the rear of the crown. First-years usually remain in the winter quarters during their first summer, but individuals occasionally move north in spring. At this time, such birds have a grey back and scapulars (forming a grey 'saddle') contrasting with the worn brown wing-coverts; they also show brown smudging on the rear crown and nape (plus a brown spot behind the eye) and pinkish legs. The overall appearance at rest may suggest a huge winter-plumaged Grey Phalarope *Phalaropus fulicarius*. By spring, such birds have already started wing and tail moult. By late summer, first-years are adult-like but have variable messy grey shading over the crown and/or nape, advanced birds showing a 'moth-eaten' hood.

Behaviour Sabine's often flies on noticeably bowed wings. When feeding over the water, it has a flapping flight before stalling and dropping to the surface with the wings raised almost vertically.

Mediterranean Gull

Where and when Formerly an irregular vagrant, Mediterranean Gull *Larus melanocephalus* has spread across n. Europe and is increasing. In Britain, over 1,000 pairs bred at c. 34 sites in 2010, the largest colonies in Hampshire, Kent and Sussex (Holling *et al.* 2012). It has also bred in Ireland since 1995. Small numbers are widespread outside the breeding season, both on the coast and inland, but it is rarer in n. England and Scotland. Adults disperse from their colonies in late June and return in March or April. Juveniles appear later, generally in July/August, and first-years linger into the summer. Large gatherings occur in some areas, with over 800 recently recorded on the Fleet in Dorset and over 600 in Southampton Water.

Size, structure and behaviour 'Med Gull', as it is universally known, is slightly larger, distinctly heavier, chunkier and squarer-headed than Black-headed Gull *Chroicocephalus ridibundus* (although small individuals, no doubt females, may be similar in size). On the water it looks rather neckless, flat-backed and somewhat less attenuated, while a thick, rather blunt bill is apparent at surprisingly long range. In flight it looks bull-necked and deep-chested with stiffer, less pointed wings, the latter effect perhaps emphasised on adults by their lack of a white primary wedge. It has rather a smooth, high-stepping, plover-like gait and is often markedly aggressive to other small gulls. In spring, it has a distinctive low, soft, deep but far-carrying call: *eeuurr* or *a-ahar*, rising and falling slightly (the rhythm vaguely suggesting the call of male Eurasian Wigeon *Anas penelope*).

Plumage *Adult* AT REST Easily separated from Black-headed Gull by its prominently white primaries and (in winter plumage) by a large black, wedge-shaped ear-covert patch that often extends as a narrow grey shawl over the back of the head. The head pattern, however, is variable at all ages, some showing less extensive markings, while a small minority lack obvious markings altogether, looking peculiarly white-headed. In summer plumage (usually attained in March) the hood is black (brown on Black-headed) and extends further down the nape than on Black-headed (but this varies with posture). There is also a prominent broken white eye-ring. The thick, blunt bill is usually bright red and close views should reveal a black subterminal band and small yellow tip, but it may fade to orangey-red or even dull orange in autumn. IN FLIGHT A beautiful, ghostly white bird. Unlike Black-headed Gull, the underwings are pure white, while the upperwings shade from pearly grey on the mantle and wing-coverts to pure white on the primaries, lacking both the white primary wedge and black primary tips of Black-headed. The only real pitfall is the very occasional aberrant white Black-headed Gull, Common Gull *L. canus* or Kittiwake *Rissa tridactyla* (see below). *Second-year* As adult, but with variable amounts of black on the primaries. Most have relatively small subterminal markings, often showing as black arrowheads on the closed wing, but others have larger black primary wedges and are less easy to pick out at rest from Black-headed Gulls; however, unlike that species, they usually show prominent white within the black. *First-year* At rest, does not always stand out from Black-headed Gulls, but look for the combination of the black ear-covert wedge, heavy blunt bill and thickset appearance. The bill is black at first, gradually acquiring a pinkish, orangey or reddish base as winter progresses. Compared to first-winter Black-headed, the closed wing shows browner coverts, solidly dark tertials (only narrowly fringed white) and solidly dark primaries; these plumage differences, combined with the structural ones, produce a unique 'jizz' that is very distinctive once learnt. In first-summer plumage, they may gain at least a partial black hood (even full) and the upperparts become pale grey as the dark-centred wing-coverts and dark tertials are replaced. In flight, it has a similar pattern to first-year Common Gull and, when seen with that species, it can be surprisingly difficult to pick out but, again, look for the black ear-covert patch. The wings are, however, cleaner-looking and more contrasting than Common: the primary wedge and secondary bar are blacker, and the mid-wing panel (greater coverts and inner primaries) is clean grey; also, it lacks Common Gull's obvious dark grey 'saddle' (back and scapulars), the mantle being a pale pearly grey. Most distinctive are the underwings: unlike Common, the underwings lack brown markings and are cleanly white, the only real dark being at the under-primary tips. Structural differences, particularly the bull-neck and shorter, stiffer wings, are also useful. *Juvenile* Even for observers familiar with first-winter Mediterranean Gulls, their first juvenile may come as a surprise. Quite unique, being closest to juvenile Common Gull at rest, as well as in flight. The whole of the upperparts, including the hindneck, are dark chocolate-brown, each feather neatly and cleanly fringed with white, producing an attractively scalloped appearance. The greater coverts, however, are plain grey, standing out as a broad, pale unmarked strip along the base of the closed wing. The breast has brown mottling, concentrated mainly at the sides. Also of note is that, unlike first-winters, the

2nd-winter Mediterranean Gull. Similar to adult (right), but variable amount of black on primaries. Bill variable.

Adult winter Mediterranean Gull. White primaries, heavy bill and, usually, dark wedge behind eye.

Adult winter Black-headed Gull

1st-summer Mediterranean Gull moulting into 2nd-winter plumage.

▲ **2nd-winter Mediterranean Gull**

▼ **1st-winter Black-headed Gull**

1st-winter Mediterranean Gull. ▲
Note heavy bill, black face mask, pale mantle.

Juvenile Common Gull. Weak bill, pale legs. Browner and less contrasting, lacking strong greater covert panel of Mediterranean Gull.

Juvenile Mediterranean Gull. Very dark, strongly scalloped upperparts, pale greater covert panel, whitish head, dark bill and legs.

Juvenile Black-headed Gull

Adult winter Black-headed Gull.
White primary wedges, black tips.

Adult winter Mediterranean Gull shows striking white underwing.

Adult winter Mediterranean Gull shows white primaries.

◀ **Black-headed Gull** has dark under primaries.

◀ **2nd-winter Mediterranean Gulls** show variable amounts of black in wing-tip. ▼

Adult Kittiwake, in moult may show very little black in wing-tips. ▼

1st-winter Mediterranean Gull. Note black eye patch and contrasting wings. ▼

1st-winter Mediterranean Gull

1st-winter ▶ **Common Gull** shows less contrasting pattern. Note dark 'saddle'.

1st-winter Common Gull. Underwing is less 'clean'.

white head is relatively unmarked, with no real wedge and only a faint grey suffusion behind the eye and over the rear crown. The thick black bill is prominent against the featureless head, while the legs are also noticeably dark, reddish-black. The flight pattern is contrasting and similar to first-winter, except that the mantle is blackish, not grey. In comparison, juvenile Common is browner and less contrasting: it lacks the pale greater-covert panel, has extensive brown underpart mottling, a weak bill (with at least some pale at the base), round head, gentle expression and, most importantly, pale greyish or flesh-coloured legs. At rest, Common also looks long-winged and short-legged. Juvenile plumage of both species is quickly lost from early August onwards, during the post-juvenile body moult.

Pitfalls *Hybrids* When identifying second-years, always bear in mind the remote possibility of hybrid Black-headed × Mediterranean Gull, which shows characters intermediate between the two and may be confusable in a cursory glance: note particularly the hybrid's slimmer bill and slighter build, as well as traces of Black-headed Gull plumage (such as a hint of a white primary wedge and black tips to the trailing edge of the primaries). *Leucistic gulls* All-white Black-headed and Common Gulls, and Kittiwakes are not unusual but their sheer whiteness, particularly across the mantle and wing-coverts, instantly separates them from Mediterranean; structural differences and bare-parts coloration are also helpful. *Moulting Kittiwakes* Another potential pitfall is late summer adult Kittiwakes: when still growing their outer primaries, such birds have shorter, more rounded wings than usual, with the amount of black at the tip severely reduced (perhaps suggesting second-winter Mediterranean).

Ring-billed Gull

Where and when A North American gull, first recorded in Britain in 1973, but currently averaging *c.* 60 records a year, with a peak of 108 in 1992. Most are in seen in western areas, but records have occurred throughout the country. Some have returned to the same wintering sites for many years. Although records have occurred in every month, first-years appear mainly from November to February, often remaining to summer (note that first-years are extremely unlikely *before* November). Adults and second-years occur mainly from November to April. All age groups also show a marked spring passage, with a peak in March and April; these are thought to be northward-bound birds that have wintered further south.

General approach Ring-billed Gull *Larus delawarensis* should be identified with caution, especially in first-year plumages; most are found by experienced observers who habitually scrutinise their local gull flocks. A thorough understanding of all plumages of Common Gull *L. canus* and Herring Gull *L. argentatus*, including their abnormalities and idiosyncrasies, is essential. The following details outline its separation from the similar Common Gull; differences from Herring are summarised at the end.

Size, structure and behaviour Ring-billed is always conspicuously smaller than Herring Gull, and basically resembles a large Common Gull. However, the size of all three species varies. Some male Ring-billed are noticeably larger than most Common Gulls, while small females are about the same size (comparisons should always be made with *several* individuals of the commoner species). Ring-billed is slightly different structurally, looking stockier, bulkier and deeper-chested, while on the water it looks flat-backed, sleek and attenuated compared to Common. The most obvious structural difference is the bill, which looks longer, noticeably thicker and more 'parallel': this effect is apparent even at a distance, when a thick black band (or tip) makes the bill appear rather blunt. The head is slightly more angular, less rounded than Common, but this has been over-emphasised and the head shape depends largely on attitude: when relaxed, Ring-billed can look quite round-headed. The size difference may be more obvious in flight, when Ring-billed looks distinctly longer- and broader-winged than Common. The wing-tips appear more pointed than those of Common but on adults and second-years this is emphasised by differences in the wing-tip pattern (see below). The legs are often noticeably longer than Common, resulting in a strutting walk. Ring-billed Gulls are often attracted to man and may become very tame.

Plumage *Adults at rest* Always remember that winter adult and second-year Common Gulls, and second- and third-year Herring Gulls, often show a prominent, clear-cut ring on the bill. The best way to pick out an adult or a second-year Ring-billed at rest is by a combination of mantle colour and tertial and wing-tip patterns. **1** MANTLE Noticeably paler than that of Common, being closer in shade to that of Black-headed Gull *Chroicocephalus ridibundus*. **2** TERTIALS Rather square and lack Common's conspicuous broad white crescent. At close range, the tertial tips on Ring-billed *are* whiter, but they are narrow and do not contrast with the paler mantle. **3** PRIMARIES The closed primaries look uniformly black, with three inconspicuous white primary tips that decrease in size towards the wing-tips (unlike Common, the large white mirrors on the outer primaries are not readily apparent at rest). The pale mantle and black primaries, unrelieved by an obvious white tertial crescent, produce a pattern quite distinct from adult Common, and quite similar to adult Black-headed Gull. **4** PITFALLS Two pitfalls need to be considered: (1) unusually pale Common Gulls do exist, and (2) second-year Commons often show a narrow tertial crescent and little white in the primaries. Make sure that your 'adult Ring-billed' is not a second-year Common. To avoid such pitfalls, it is absolutely essential for the identification to be confirmed by other features. Structural differences, outlined above, are especially important, and pay particular attention to the bill. **5** BILL The black band should stand out clearly and cleanly, and contrast with the pale yellow base, even in winter. **6** EYE COLOUR A key difference: Ring-billed has pale irides (as well as a narrow orange orbital-ring) but the pale eye is difficult to detect at any distance. However, it usually produces a squint-eyed expression, in contrast to the dark-eyed, open-faced look of Common. **7** HEAD STREAKING Ring-billed tends to have paler, mottled head streaking, but some show quite dark streaking (this is so variable on Common as to render it of limited value in the field). **8** LEGS Often yellower.

Adults in flight If a suspected Ring-billed flies or wing-flaps, concentrate on the wing-tip pattern: Common has two large, conspicuous white mirrors right across the wing-tip, but on Ring-billed the mirrors are small, relatively inconspicuous and often confined to just one mirror on the inner web of the outer primary (the relative lack of white emphasises the more pointed wing shape). The pale mantle and wings contrast strongly with the black primary wedges so that, in flight, Ring-billed's pattern looks surprisingly similar to that of Herring Gull; very white underwings reinforce this impression.

Second-years Similar to adult, but easily aged by the presence of dark feathering on the

Adult winter Common Gull. Note thin bill, dark eye, dark grey upperparts and thick white tertial crescent.

Adult winter Ring-billed Gull has thick bill and pale eye. Best picked out by pale upperparts and thin white tertial crescent.

1st-winter Common Gull. Thin bill, lightly spotted head and nape, dark back and scapulars, wide pale edges to tertials.

Common Gulls (upper two), **Ring-billed Gull** (lower one). All 1st-winter and show great variation in upperpart moult and wear.

Note median-covert pattern in fresh plumage, The dark centres are spade-shaped on Common, arrow-shaped on Ring-billed Gull.

1st-summer Ring-billed Gull. Concolorous upperparts.

1st-winter Ring-billed Gull. Heavy bill, usually heavy spotting on head, neck and breast. Pale back and scapulars, pale greater coverts and narrow pale edges to tertials.

2nd-winter Herring Gull. Note barred greater coverts and tertials, heavier build. Often shows a pale eye. Note short primary projection and often a pronounced 'tertial step'.

1st-summer Common Gull. Note dark 'saddle'.

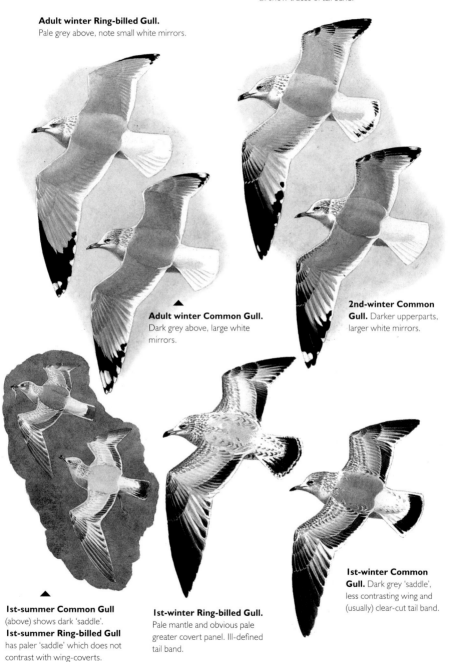

2nd-winter Ring-billed Gull. Contrasting wing pattern, one small mirror. Most but not all show traces of tail band.

Adult winter Ring-billed Gull. Pale grey above, note small white mirrors.

Adult winter Common Gull. Dark grey above, large white mirrors.

2nd-winter Common Gull. Darker upperparts, larger white mirrors.

1st-summer Common Gull (above) shows dark 'saddle'. **1st-summer Ring-billed Gull** has paler 'saddle' which does not contrast with wing-coverts.

1st-winter Ring-billed Gull. Pale mantle and obvious pale greater covert panel. Ill-defined tail band.

1st-winter Common Gull. Dark grey 'saddle', less contrasting wing and (usually) clear-cut tail band.

primary coverts. Most also show vestigial black markings on the tail and, sometimes, the secondaries (although many lack them). Conversely, some second-year Commons also show them (but usually only on the tertials). Ring-billed shows only a small white mirror on the inner web of the outer primary (often difficult to see), whereas second-year Common shows one or two obvious white mirrors. The age at which Ring-billed develops adult bare-part colouring varies: most acquire a complete black bill band and a yellow base by their first summer, although some still retain a black tip and/or a greenish base a year later; and some remain dark-eyed into their second summer.

First-years The most difficult age to identify, as many of the subtle differences are incon-sistent. First-year Ring-billed has a distinctive 'jizz' *once learnt*, but it should always be identi-fied by a combination of minor differences. Close views and detailed notes or photographs are essential, and observers should always bear in mind the possible occurrence of odd Common Gulls (for example, unusually pale individuals). All the following features (in rough order of significance) should be checked. **1** BILL The best character, being heavy, thick, 'parallel' and blunt-ended (Common's bill looks slender, pointed and weedy). Usually pale orangey-pink with a prominent black tip, strangely reminiscent of the bill of first-winter Glaucous Gull *L. hyperboreus* (some Commons have a similar bill colour, but most have a duller, grey or greenish base). **2** TERTIALS Solidly dark brown, *narrowly* fringed white (on Common, paler brown, with broad white fringes, but beware the effects of abrasion). **3** MANTLE AND SCAPULARS *Pale* grey, lacking the dark 'saddle' effect of Common. Whitish tips to many of the scapulars and the retention of some dark juvenile feathering may create a more variegated pattern than Common, but the dark feathers are moulted and pale tips wear off as winter progresses. **4** GREATER COVERTS Usually appear pale grey, sometimes barred on the inners (unlike Common), producing a pale strip along the base of the closed wing and forming a noticeable pale mid-wing panel in flight. **5** HEAD AND UNDERPARTS Usually well mottled and spotted about the head (first-winters with head streaking may show a white eye-ringed effect) and more heavily mottled or scalloped below. There is often heavy dark barring or spotting on both the upper- and undertail-coverts, which Common usually lacks. Both species, however, are variable. **6** TAIL The dark of the tail-band usually extends up the outer web of each tail feather to intrude into the tail base, which usually shows delicate greyish mottling or shading (occasionally almost an all-dark tail); the tail therefore looks messy compared to Common, which shows a *clear-cut* band and a white base. On some Commons, however, dark also intrudes into the white, while a minority also show grey mottling at the base, so the difference is not absolute. Ring-billed has dark mottling or barring on the outer web of the outer tail feather, which Common seems to lack. **7** MEDIAN COVERTS In fresh plumage, the brown centres to the median coverts are pointed on Ring-billed, rounded on Common, but this distinction breaks down with wear and fading, and is of little use in worn plumage. **8** LEGS Sometimes quite pink on first-year Ring-billed (but may be pale greyish).

First-summer Both species fade and bleach, and eventually replace their wing-coverts and tertials with grey second-winter plumage. First-summer Commons look washed-out and pale, their outer primaries and secondaries fading to brown and the rest of the wing becoming creamy and worn, contrasting conspicuously with the dark grey 'saddle', particularly in flight. Ring-billed also fades but, because it lacks the dark 'saddle', the mantle and wing-coverts look uniformly pale grey and concolorous. Unlike Common, first-year Ring-billed soon gains a pale bill tip and, by their first summer, the bill pattern is usually similar to that of the adult. *Juveniles* Full juvenile has never been recorded in Britain and Ireland. Similar to first-winter, but mantle and scapulars brown, fringed white, and the head and underparts are also heavily marked.

The Herring Gull problem

First-year and second-year Ring-billed may be confused with second- and third-year Herring Gulls respectively, both of which can show a prominent bill band. The easiest and most obvious difference is their size: most Herring should appear large, bulky, angular-headed, heavy-billed and meaner-looking. If in doubt, check the wing-coverts: second-year Herring shows noticeable brown *barring* across the wing-coverts (including the greater coverts), which first-year Ring-billed lacks; in addition, second-year Herring shows rather mottled tertials and, usually, a pale eye. In flight, second-year Herring shows fairly uniform grey inner primaries, producing a pale grey 'window' extending to the tips of the feathers; first-year Ring-billed has dark subterminal marks on these feathers. Third-year Herring is also easily separated as it retains traces of dark mottling on the wing-coverts, obvious vestiges of immaturity that second-year Ring-billed would never exhibit (although *small* amounts of brown may be retained on the leading lesser coverts); third-year Herring also has pinkish legs, whereas second-year Ring-billed usually has greenish or yellowish legs.

Herring, Lesser Black-backed and Great Black-backed Gulls

General approach Many birders find the identification of large gulls daunting. The best approach is to familiarise yourself with the Herring Gull, which should act as the yardstick. Start with adults, gradually moving on to third- and second-years, before looking at juveniles and first-years. The first real identification challenge is the separation of juvenile and first-year Herring Gulls from similarly aged Lesser Black-backed Gulls.

Ageing immature large gulls – general principles Aside from hybridisation, the reason why immature large gulls are so difficult to identify is related to three main factors: (1) moult, (2) individual variation and (3) the effects of wear and bleaching.

Moults *Juvenile* A large gull's first plumage is 'juvenile'. This is brown, heavily patterned on the upperparts and always neat and immaculate. The mantle and scapulars have brown feathers with white fringes, forming a scalloped appearance. *First-year* In late summer and autumn, most juveniles begin moulting into *first-winter* plumage. This involves just the back and scapulars, and variable numbers of head and body feathers. The back and scapulars then show anchor-shaped marks and/or cross-barring. In their *first-summer*, many species acquire extensive but variable amounts of greyer second-winter feathering on their back and scapulars (the shade depending on the species) while simultaneously their old juvenile or first-winter feathers become very worn and often bleached (the new greyer feathers help considerably in the identification process). Strictly speaking, however, 'first-summer' is not a discrete plumage, but simply the early stages of the bird's moult into second-winter. **Second-year** The sequence in their second-year is similar to the first, gradually acquiring more adult-like grey feathering on their back and scapulars by their second-summer, and later the wing-coverts. Second-years otherwise retain extensive black in their primaries, primary coverts, secondaries and tail. **Third-year** Third-winter plumage is similar to adult, but variable amounts of brown immature feathering remain (particularly on the wing-coverts, primary coverts and tail), but this becomes increasingly limited; the bare parts may also retain signs of immaturity. Owing to their relatively consistent appearance, third-years are not usually covered in the individual species accounts below. **Fourth-year** Although essentially adult, many retain subtle traces of immaturity (as indeed can a few even older individuals).

Useful tips 1. *Shape of juvenile primary tips* A useful point to remember is that both juvenile and first-year large gulls have slightly more pointed primary tips than older birds and, when worn, these are often obviously pointed, contrasting with any newly growing, rounded, second-year feathers. This may help to age puzzling individuals.

2. Effect of moult timing on subsequent plumage type The earlier an immature gull moults, the more 'immature-like' the replacement feathers will be; the later it moults, the more 'adult-like' they will be. Thus, a first-year gull moulting in May will replace the old feathers with more immature-like feathers than a gull replacing those same feathers in August, which will grow more adult-like feathers. This explains much of the individual variation (and is presumably related to hormone levels).

3. *Effect of latitude on moult timing* Southerly species breed earlier than northern ones and they subsequently moult earlier; consequently, they are usually more advanced in their moults and in their acquisition of the next age of feathering. Thus, Yellow-legged and Caspian Gulls acquire significant amounts of plain grey adult-like feathers in their first-summer, whereas 'first-summer' Herrings, Lesser Black-backs and Great Black-backs do not. This is significant for identification.

4. *Variation* As indicated above, the plumage of large gulls is extremely variable and some defy accurate ageing. For example, colour ringing has revealed that a few second-winter Herring Gulls look just like third-winters.

Aberrant birds and hybrids If a gull doesn't look right, check for anomalous features – both plumage and structural – that may suggest aberration (albinism, leucism, etc.) or hybridisation. There are no easy answers to this, except logical thought and an open mind. Olsen & Larsson (2003) and Howell & Dunn (2007) are recommended for further reading.

Herring Gull *Larus argentatus*

Where and when Herring Gull – the familiar 'seagull' of seaside towns – is a common breeding bird around our coasts and also inland on city buildings. It is more widespread in winter, when Scandinavian immigrants of the nominate race *argentatus* are numerous in some northern and eastern areas.

Size and structure Generally slightly larger and stockier than Lesser Black-back, with a shorter, less attenuated rear end at rest. Being a more sedentary species, the wings are proportionately slightly broader, shorter and less pointed in flight, and the body appears slightly bulkier. **Plumage** *Adult* Easily identified by its pale grey mantle, black wing-tips and pale pink legs (note that the black in the wing-tips may become severely reduced during autumn primary moult, when the wings are much more rounded). Beware of confusion with the much smaller Common Gull (see p. 207). *Juvenile* Similar to juvenile Lesser Black-back (see p. 212) but plumage is slightly but distinctly paler, being more grey-brown, less smoky looking. The feather fringes are also generally paler. At rest, the most important differences are: **1 TERTIALS** Herrings show obvious whitish notching around the fringes and tip (although there is individual variation in the exact pattern). **2 GREATER COVERTS** Obviously chequered, normally lacking the thick dark bar across the inner greater coverts of most Lessers (at the base of the closed wing, nearest the wing bend). **3 HEAD** Subtly paler, plainer and more bland-looking than Lesser Black-back. **4 IN FLIGHT** Usually appears paler with the following differences from Lesser: (1) more obvious whitish inner primary 'windows', both above and below (the latter obviously translucent against the light); (2) paler brown underwing-coverts (not dark chocolate-brown), making the underwing appear paler and more uniform and (3) a slightly narrower tail-band with more white at the base; also, the tail-band does not contrast as strongly with the uppertail-coverts and rump, which are not as white looking as Lesser, being more heavily mottled. The structural differences outlined above are also useful. *First-winter* By late August/September, a partial post-juvenile body moult gradually introduces new mantle and scapulars feathers, which have brown anchor-shaped markings or more distinct cross-barring; when fresh, these feathers have a buff background, which soon whitens. The head and underparts also become whiter as winter progresses. *First-summer* Unlike Lesser Black-back and Yellow-legged, most show little if any adult-like pale grey feathering on the back and scapulars (if present, it is usually confined to a limited area on the upper back). Instead, any new feathers resemble first-winter and, as these are mixed with old, worn, first-winter feathering, individuals in active moult often look very scruffy. Note that, on the scapulars, the new feathers may be heavily barred dark brown on a whitish background. New second-winter tertials may show extensive white at the tips. The head and underparts may also retain scruffy grey streaking and mottling. By late summer, however, a minority show extensive pale grey on the back and scapulars, but these feathers usually have brown shaft-streaks and limited cross-barring. *Second-year* Very variable. At first surprisingly similar to first-year, with extensively scalloped or barred upperparts (but new brown tertials and greater coverts are often lightly peppered with white). Pale grey feathering gradually appears on the back and scapulars, and usually dominates by midwinter, providing an obvious difference from similarly aged Lessers. The underwings remain paler brown than equivalent Lessers, some being quite white. They gradually acquire a pale eye and an extensive pale base to the bill. If in doubt about ageing, the shape of the primary tips may be useful.

Scandinavian Herring Gulls *L. a. argentatus*

Herring Gulls breeding in Britain and Ireland are of the race *argenteus*. In Scandinavia, nominate *argentatus* takes over, becoming larger and darker towards the north, with less black in the primaries. In many parts of n. and e. Britain, most winter Herring Gulls are of Scandinavian origin. Whilst many are not safely separable in the field from *argenteus*, the more extreme examples are markedly different (see below). The more distinctive adults (apparently from N. Norway and the White Sea) have a darker mantle, which also tends to be colder grey. Such birds generally show reduced black in the wing-tip, large white tips to the outer two primaries, larger white primary spots at rest and a broader white tertial crescent (see pp. 230, 231 and 233). In flight, some of these dark birds are predominantly white on the underwing, often showing little black on the under-primaries. In addition, they tend to show a very heavy and angular head, heavy neck streaking in winter, a large, pale, washed-out bill and a particularly 'mean' expression. They also tend to be bulkier and have a pronounced tertial step. A few Baltic *argentatus* (formerly known as '*omissus*') have yellow legs, perhaps birds with Yellow-legged or Caspian Gulls in their ancestry. ***Immatures*** Some, presumably more northern juvenile/first-winters, may appear rather large, with paler plumage than *argenteus* (almost suggesting juvenile Glaucous *L. hyperboreus* or Iceland Gulls *L. glaucoides*). More distinctive individuals have obviously pale brown primaries with noticeable creamy-white fringes to the individual feathers.

Glaucous × Herring Gull hybrids

It is possible that some such birds have Glaucous Gull in their ancestry and apparent first-generation Herring × Glaucous Gull hybrids have been seen in Britain (often called 'Nelson's Gull' in North America). Smaller individuals may also suggest Thayer's Gull (see p. 233).

Lesser Black-backed Gull *Larus fuscus*

Where and when Breeds around much of the British and Irish coast (rarer on English east coast) and nests abundantly on buildings in many inland cities. Also widespread in winter (rarer in n. Scotland) with most inland, often on farmland. Lesser Black-back is strongly migratory and many head south to winter on coasts of Iberia and NW Africa (some reaching W. Africa). The darker Scandinavian race *intermedius* also occurs in small numbers, mainly in winter. The Baltic race *fuscus* – or 'Baltic Gull' – is on the British List by virtue of two recoveries ringed as chicks in Finland. Given that it migrates south or south-east to winter on the coasts of E. Africa, it is a relatively unlikely vagrant to Britain. However, a number of recent well-documented claims suggest that small numbers do occur, but the Rarities Committee will currently accept only records of birds ringed within the breeding range (Kehoe 2006).

Size and structure Similar size to Herring Gull, but marginally smaller and slighter, with a slightly smaller bill. Being a longer-range migrant, it has long wings, so at rest Lesser looks 'long and low' with long pointed primaries and a tapered rear end. These structural differences are particularly useful in flight, especially when identifying first-years or birds high overhead: compared to Herring, Lesser Black-back looks slim, with proportionately longer, narrower and more pointed wings.

Calls Adult's calls are distinctly deeper than those of Herring Gull.

Plumage *Adult British race graellsii* For differences from Great Black-backed Gull, see p. 215. **1 UPPERPARTS** Very dark blackish-grey above, obviously paler than Great Black-back. In winter, the head and neck show heavy grey streaking, but a pre-breeding body moult from January onwards soon produces a white head. **2 UNDERWINGS** Viewed from below, has dark grey across the inner primaries and secondaries (Herring is completely and obviously

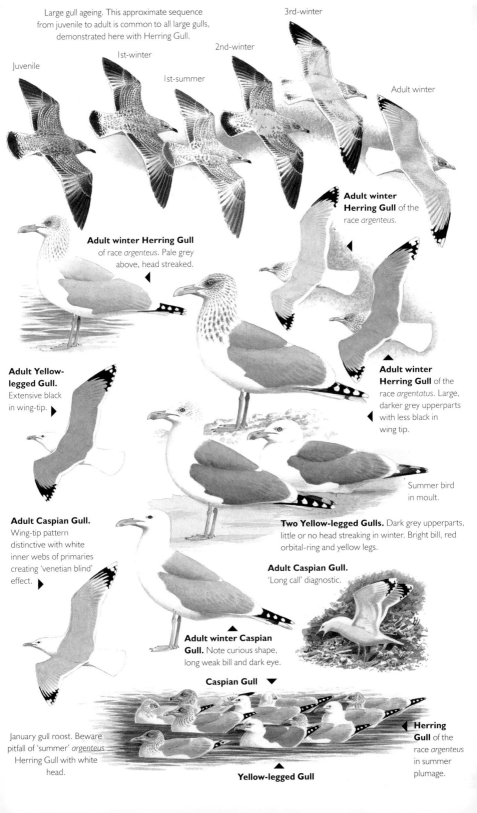

Large gull ageing. This approximate sequence from juvenile to adult is common to all large gulls, demonstrated here with Herring Gull.

3rd-winter

2nd-winter

1st-summer

1st-winter

Juvenile

Adult winter

Adult winter Herring Gull of the race *argenteus*.

Adult winter Herring Gull of race *argenteus*. Pale grey above, head streaked.

Adult Yellow-legged Gull. Extensive black in wing-tip. ▶

Adult winter Herring Gull of the race *argentatus*. Large, darker grey upperparts with less black in wing tip.

Summer bird in moult.

Adult Caspian Gull. Wing-tip pattern distinctive with white inner webs of primaries creating 'venetian blind' effect. ▶

Two Yellow-legged Gulls. Dark grey upperparts, little or no head streaking in winter. Bright bill, red orbital-ring and yellow legs.

Adult Caspian Gull. 'Long call' diagnostic.

Adult winter Caspian Gull. Note curious shape, long weak bill and dark eye.

Caspian Gull ▼

January gull roost. Beware pitfall of 'summer' *argenteus* Herring Gull with white head.

Yellow-legged Gull ▲

◀ **Herring Gull** of the race *argenteus* in summer plumage.

white in this area). **3 BARE PARTS** Legs bright yellow in summer but dull creamy-yellow in winter (pale pink on Herring). In summer, the bill is often brighter than Herring's and the orbital-ring is red (orange or orange-yellow on Herring). *Juvenile* **1 UPPERPARTS** Compared to juvenile Herring Gulls, juvenile Lessers are distinctly dark and rather smoky looking, both above and below. The back, scapulars and wing-coverts appear rather chocolate-brown, the individual feathers being less contrastingly fringed grey-buff or dark buff. The dark tertials are relatively plain with a narrow whitish fringe, lacking the *obvious* notching of Herring Gull; however, many Lessers show *slight* notching, particularly towards the tip, while a minority show stronger whitish indentations at the tip, more similar to many Yellow-legged Gulls. Unlike Herring, at rest most juvenile Lessers show *a variable thick dark bar across the bases of the inner greater coverts* (i.e. those closest to the bend of the wing), this being similar to that shown by juvenile Yellow-legged Gull. **2 IN FLIGHT** If in doubt, juvenile Lessers are more easily identified in flight. The dark, smoky appearance is obvious, the wing-coverts contrasting much less strongly with the black primaries and secondaries. Most importantly, unlike Herring and Yellow-legged, on juvenile Lesser's upperwing the black of the secondaries extends right across the inner primaries, forming a solidly dark rear wing (with a narrow white trailing edge), lacking the obvious pale inner primary windows shown by Herring and, to a lesser extent, Yellow-legged Gull (Lesser shows only subdued greyish-white windows). The underwing too is dark, with chocolate-brown underwing-coverts that show relatively little contrast with the dark grey or black flight feathers. Lesser may show more black on the tail, covering all but the base; in flight this contrasts more strongly with the white rump and uppertail-coverts, which are less mottled than Herring. **3 AUTUMN JUVENILES** Note that juvenile Lessers often fade considerably by early autumn, the browns becoming paler and the feather fringes and background colour to the head and underparts becoming white. Such birds may then suggest Herring Gull or, more particularly, Yellow-legged and even Caspian Gulls; nevertheless they can be separated by most of the features outlined above. *First-winter* By late August and September, juveniles usually begin a variable and partial body moult into first-winter plumage. They gradually acquire new feathers on the back and scapulars, showing a distinct anchor-shaped pattern; the head and underparts continue to whiten. Dark chocolate underwing-coverts are retained (paler brown on Herring). *First-summer* Acquired from early spring onwards. Incredibly variable, but most look quite dark as a result of the gradual acquisition of second-winter feathering, which has a dark grey background colour (albeit with variable dark chocolate-brown shaft-streaks, anchors, cross-bars and often pale fringes). Other feathers gained at this time are more like first-winter, being dark chocolate, variably patterned with darker brown. In active moult, the mix of new second-winter feathers and old worn first-winter feathers creates an extremely scruffy appearance (many old feathers become severely bleached, especially across the wing-coverts and tertials). Significantly, first-summer Herring Gulls acquire little, if any, adult-like pale grey back and scapular feathers and so remain much as first-winter, appearing paler and browner than Lessers, but often strongly barred brown across the whitish scapulars (once the new second-winter feathering is acquired). *Second-year* Very variable, the result of a complicated mixture of adult-like dark grey and immature-like pale chocolate feathering (both feather types variably patterned with dark brown and sometimes fringed paler). While some have extensive plain, adult-like dark grey feathering on the back and scapulars, a minority remain more like first-winters, being heavily patterned with dark anchors, diamonds and cross-bars on quite a pale background. Nevertheless, all are distinctly darker above than second-winter Herring Gulls. The tertials show more extensive white at the tip, often with a large anchor shape within the white. The eye gradually turns pale, as does the bill base. In flight, the underwing-coverts remain dark chocolate-brown. By their second-summer, the mantle and scapulars become predominantly dark grey, forming a distinct 'saddle', and many also show extensive dark grey on the upper-wing-coverts, increasing considerably by late summer. The head and underparts become white and the bare parts often brighten considerably.

Scandinavian races
Adult *intermedius* Lesser Black-backs become darker towards the north and east. As its name suggests, Scandinavian *intermedius* is intermediate in shade between British *graellsii* and nominate Baltic *fuscus*, appearing distinctly darker and blacker than *graellsii*. Both its structure and the timing of its moults are similar to *graellsii*.

Adult *fuscus* Classic individuals are distinctly or even strikingly smaller and slighter than *graellsii*, with a smaller, slimmer bill, more domed head, shorter legs and long, scissor-like primaries (it has long slender wings in flight). Their jizz may be reminiscent of a smaller gull, such as Common Gull *L. canus*. The upperparts are virtually black, similar in shade to Great Black-back, showing little contrast with the black primaries, which have only a small white mirror on the inner web of the outer primary. *Fuscus* also has a whiter head in winter, with light grey streaking confined to the crown and hindneck.

Moult Differences are particularly significant: *graellsii* and *intermedius* have a complete post-breeding moult from mid May to December, with their primary moult commencing anytime between May and August. Apart from perhaps the innermost one or two primaries, *fuscus* does not commence its primary moult until arrival in its winter quarters, from October to April. Consequently, any late summer/autumn *fuscus* seen in Britain is likely to have old and worn primaries until at least October, by which time local *graellsii* will have new, fresh primaries. Winter *fuscus* should be in active primary moult, whilst in spring the primaries should be new and fresh at a time when *graellsii*'s are starting to wear. However, great caution must be taken when identifying a potential *fuscus* in Britain. In particular, the effects of the light must be carefully considered, particularly at evening roosts (in certain lights even *graellsii* can look very dark). There is also intergradation between *fuscus* and *intermedius* while, apparently, some small female *intermedius* closely resemble *fuscus*. It has to be accepted that any apparent *fuscus* seen in Britain is unlikely to make it beyond the level of 'possible' or 'probable', unless of course it carries an identifiable ring.

Great Black-backed Gull *Larus marinus*

Where and when Breeds fairly commonly, mainly on rocky coastlines, right around Britain and Ireland, except along the British east coast south of the Firth of Forth. More widespread in winter, when Norwegian immigrants are common in the east, penetrating well inland. Small numbers of local breeders may also feed throughout the year on inland lakes and reservoirs.
Size and structure Fundamental to its identification. A huge, bulky brute of a gull, with a large head, deep chest, often a pronounced tertial step and a somewhat truncated rear end. Conspicuously larger than all other gulls and always dominates them when feeding. Males are larger than females, but most are *c.* 20% larger than Lesser Black-back and, more importantly, about

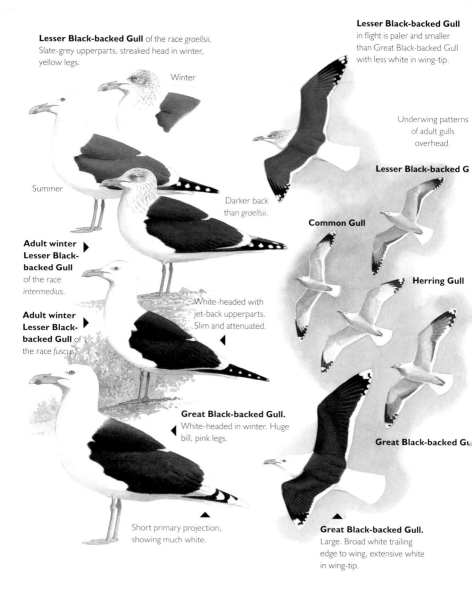

Lesser Black-backed Gull of the race *graellsii.* Slate-grey upperparts, streaked head in winter, yellow legs.

Lesser Black-backed Gull in flight is paler and smaller than Great Black-backed Gull with less white in wing-tip.

Winter

Underwing patterns of adult gulls overhead.

Summer

Lesser Black-backed G

Darker back than *graellsii.*

Common Gull

Adult winter Lesser Black-backed Gull of the race *intermedius.*

Herring Gull

Adult winter Lesser Black-backed Gull of the race *fuscus.*

White-headed with jet-back upperparts. Slim and attenuated.

Great Black-backed Gull. White-headed in winter. Huge bill, pink legs.

Great Black-backed Gu

Short primary projection, showing much white.

Great Black-backed Gull. Large. Broad white trailing edge to wing, extensive white in wing-tip.

twice as heavy. In comparison, Lesser is rather slender, with long primaries that at rest produce an obviously tapered look to the rear end (but wings appear shorter during autumn primary moult). Great Black-back's deep and powerful bill is always conspicuously larger than that of Lesser Black-backed and Herring Gulls, with a rather bulbous tip and prominent gonydeal angle. Note also the small beady eye (looks dark at any distance), set well back on the head. In flight, it appears slow, lumbering and pregnant-looking, with relatively rounded wings and slow wingbeats (its 1.5m wingspan is larger than that of Common Buzzard *Buteo buteo*). In comparison, Lesser Black-backed's wings are long, narrow and pointed although, in the absence of direct comparison, Great Black-backed's can look *proportionately* somewhat shorter.

Calls Adults' somewhat muffled calls are loud, deep, powerful and often bellowing.

Plumage *Adult* If seen well, separation from Lesser Black-backed is not difficult but, if accurate size assessment is not possible, confusion is surprisingly easy, especially if poor light affects the mantle colour. **1 UPPERPARTS** Great Black-back is virtually black above, readily separating it from the British race *graellsii* of Lesser Black-back (Scandinavian Lesser Black-backs – *intermedius* and the rare *fuscus* – are darker above). At rest, Great Black-back has a thicker and more prominent white tertial crescent. **2 LEG COLOUR** When in doubt, concentrate on size, structure and leg colour, Great Black-back having flesh-coloured legs (Lesser Black-back yellow, bright in summer but duller in winter when the yellow can be difficult to discern in weak light). **3 HEAD COLOUR IN WINTER** Great Black-back retains a largely white head in winter (any markings are faint) whereas *graellsii* Lesser has a heavily streaked head until at least December. **4 IN FLIGHT** Great Black-back is blackish right across the wings, whereas the paler wings of *graellsii* Lesser contrast with the blacker primaries. Great Black-back has *large white tips to the outer two primaries* forming a diagnostic and conspicuous *white spot* at the wing-tip (Lesser usually has much less obvious *subterminal* mirrors). Greater also has a broader white trailing edge to the wing that contrasts much more strongly with the blackish upperwings. The under-primaries and under-secondaries are noticeably dark grey. *Juvenile* Most likely to be confused with similarly aged Herring Gulls (and even more similar to juvenile Yellow-legged Gull, which see), so correct size and structural evaluation is essential. **1 BILL** Pay particular attention to the large black bill, which usually contrasts strongly with the predominantly white head (lightly streaked around and behind the eye). **2 UPPERPARTS** Plumage neat and immaculate and appears much 'cleaner' than first-year Herring, with a buffy-white background colour (gradually wearing whiter). The wing-coverts are distinctly *chequered* with chocolate-brown, whereas the mantle and scapulars are plain brown, each feather neatly fringed with white (Herring's underparts and upperparts are browner, the latter with a less contrasting, duller ground colour, but some pale individuals can look more similar to Great Black-back). The tertials show broad white tips and fringes. **3 IN FLIGHT** The rather pale upperwing-coverts contrast quite strongly with the black primaries and secondaries. Most useful in flight is a rather narrow tail-band which, at close range, often has a series of narrow black bars in front of it and thus appears ill-defined (tail-band solidly brown or black on Herring and Yellow-legged Gulls). Unlike Herring, the tail-band contrasts strongly with the white base, uppertail-coverts and rump, which are only lightly mottled. *First-year* **1 UPPERPARTS** Following partial post-juvenile body moult in autumn, the head and underparts become even whiter and the white-fringed juvenile mantle and scapular feathers are gradually replaced by first-winter feathers that show brown anchor-shaped marks on a buff background, with noticeable white fringes (producing a more barred impression). **2 BILL** Black throughout the first winter, but often acquires a pale tip and base in the first summer (Herring usually acquires a paler base during its first winter). *Second-year* **1 PLUMAGE** Becomes even whiter on head and underparts. Initially very heavily patterned above, often with obviously all-dark greater coverts. It gradually acquires obvious blackish feathering on the back and scapulars, allowing easy identification. **2 BILL** Becomes paler and pinker, with black only at the tip, and the eye gradually turns pale.

References Olsen & Larsson (2003), Howell & Dunn (2007).

Yellow-legged and Caspian Gulls

Background Some of the greatest recent advances in bird identification have related to the large gulls. It has always been known that the widespread and familiar 'Herring Gull' occurs in various forms across the Northern Hemisphere, but recent developments in the analysis of their DNA have indicated that, what was formerly considered one species, is at least six. These can be divided into two clades (groups of closely related species that share a common ancestry). The North Atlantic clade contains European Herring Gull, Yellow-legged Gull and Armenian Gull *L. armenicus* as well as Great Black-backed Gull *L. marinus*, all evolving from a common ancestor in the North Atlantic. The Aralo-Caspian clade (named after the Aral and Caspian Seas in SW Asia, a past hub of gull evolution) contains Caspian Gull, all races of Lesser Black-backed Gull *L. fuscus*, the various Siberian forms of 'Herring Gull' and, remarkably, American Herring Gull *L. smithsonianus*, the ancestors of which apparently colonised North America not from the Atlantic, but from Asia. As far as we in Britain are concerned, the most significant changes relate to Yellow-legged and Caspian Gulls. It has long been known that the former is a regular late summer and winter visitor from the Mediterranean, and the Atlantic coasts of Iberia and France. Caspian Gull, which breeds principally in SE Europe and SW Asia, was officially split as recently as 2008. In recent years it has started to spread north-west and is now breeding as close as Poland. The problem is that, as with many expanding species, the pioneers are interbreeding with a closely related species, in this case Herring Gull. Although classic Caspian Gulls are fairly distinctive, it is this hybridisation that makes it such a difficult and controversial species to identify.

Yellow-legged Gull *Larus michahellis*

Where and when Most occur in late summer (July to September) with smaller numbers remaining until late winter (by spring and early summer, only a few immatures usually remain). They occur both on the coast and inland, particularly on rubbish tips and reservoirs. Since 1995, one or two pairs have bred in s. England.

Size and structure Older birds are usually located by their darker grey mantle (see below) but structural differences are also significant. Somewhat larger and more powerful than Herring Gull and males may be particularly big, clearly intermediate in size between Herring and Great Black-back. The bill is usually longer, heavier and more bulbous than Herring's, with a deep and rather blunt tip; some large males have a particularly long and heavy bill. It also tends to be deeper-breasted and rather long-legged, sometimes producing a distinctly gangly impression (often adopts a very horizontal stance with a bulbous breast and long legs). In flight, it has noticeably long, pointed wings and glides on arched wings, rather like a skua. Those that occur in Britain are of the w. Mediterranean race *michahellis*, but some resemble so-called '*lusitanius*' from w. Iberia (see p. 233).

Plumage A useful point to remember is that, as they breed further south, Yellow-legged Gulls nest earlier than both Herring and Lesser Black-backed; consequently, at each plumage stage young birds mature significantly earlier than equivalent-aged Herring Gulls.

Adult Superficially similar to Herring Gull, but remember that it shares many features with Lesser Black-back. When separating adults, the following should be looked for (in approximate order of significance): **1** MANTLE Clearly a darker 'mid grey', intermediate between *graellsii* Lesser Black-back and *argenteus* Herring (similar, in fact, to Common Gull *L. canus*, but warmer). **2** LEGS Yellow, bright in summer, but paler and washed-out in winter (when the colour can be difficult to determine). **3** HEAD SHAPE At rest looks bulbous with a thick neck but, when alert, may show a rather small, flat head and distinctly long neck. **4** HEAD STREAKING Lacks Herring Gull's heavy head streaking in winter, instead showing only limited light grey streaking in autumn and early winter on the crown, perhaps on the nape and most obviously a stronger backward-pointed area around and behind the eye. This difference is valid only until late December/January onwards when both Yellow-legged and *argenteus* Herrings start to acquire their white summer head colour. However, early moulting Yellow-leggeds often stand out as sleek, smart and contrasting amongst winter-plumaged Herrings. **5** BILL Brightly coloured with a large red spot; it may acquire a narrow black band in winter. **6** ORBITAL-RING Red, like Lesser Black-back, producing a beady-eyed effect at a distance (yellow or orangey-yellow on Herring). **7** UPPERWING In flight, tips more extensively black than Herring; also, the black wing-tips are cut off more squarely and show little white at the tip (two mirrors on P9 and P10). **8** UNDERWING Dusky-grey across the under-secondaries and inner primaries (not as dark as Lesser Black-back); Herring is all white in this region. **9** MOULT Earlier than Herring, with extensive primary moult from July to early September. Primaries are practically fully grown by September, when local *argenteus* Herring's are usually still short and partially grown. **10** CALLS Distinctly louder and deeper than those of Herring Gull (and Lesser Black-back), often recalling Great Black-back.

Juvenile Superficially similar to juvenile Herring Gull, both at rest and in flight, but it has a distinctly white background colour, creating a smarter, cleaner impression. In consequence, Yellow-legged is oddly reminiscent of juvenile Great Black-back, being rather white-headed with rather pale, 'clean'-looking mid brown upperparts with contrasting whitish feather fringes (forming a strong contrast with the very dark tertials, primaries and secondaries). Note that juvenile Lesser Black-backs, although they start off dark, often fade considerably by early autumn, with paler brown plumage, a pure white background colour to the head and underparts, and white feather fringes to the upperparts; such birds can strongly suggest Yellow-legged. Concentrate on the following. **1** HEAD The head and neck pattern can be divided into three areas. (1) There is fine grey streaking on the crown and a large greyish smudge or 'shadow' that extends behind and below the eye to form an ill-defined but fairly noticeable mask. (2) Immediately behind this area is a large crescent-shaped swathe of white, running from the upper nape down the neck-sides, below the eye shadow. (3) Below this again, the hind-neck is strongly streaked brown, forming a 'shawl' that runs onto the lower neck-sides and to the upper breast, the streaking then extending onto the breast and belly. The ground colour to this whole area is very white, contributing to the very clean impression. **2** UPPERPARTS At rest, juveniles are mid brown, with nearly all the feathers showing contrasting buffish-white fringes that produce a clean-cut, well-patterned, scalloped appearance (little chequering, most obvious on the inner greater coverts). A useful *aide-mémoire* is that the combination of white-looking head, dark eye shadow and whitish-fringed upperpart feathering can, with some imagination, suggest a giant juvenile Mediterranean Gull *L. melanocephalus*. **3** GREATER COVERTS Also distinctive is that, like Lesser Black-back (but unlike Herring) most Yellow-leggeds show dark bases to the outer greater coverts, forming a thick dark band across the lower edge of the closed wing (closest to the wing bend),

Juvenile Herring Gull.
Note barred greater coverts and tertials patterned to base.

1st-winter Herring Gull. Head paler, upperparts moulted from juvenile. Tertials notched and greater coverts chequered. In flight note pale inner primary 'window'. ▼

1st-winter Herring Gull
of the race *argentatus*. Large, boldly patterned. ▶

1st-winter Herring Gull.
Some show less of a primary 'window'; and darker outer greater coverts. ▶

1st-summer Herring Gull. ▶
May become very worn and faded.

1st-winter Herring Gull. Mid-bro underwing, inner primary 'window'.

Juvenile Yellow-legged Gull.
White head with dark 'eye mask'. Dark bases to outer greater coverts; tertials, relatively plain with dark base.

1st-winter Yellow-legged Gu
Underwing dark.

1st-winter Yellow-legged Gull. White rump contrasts with tail-tip, poor inner primary 'window'.

1st-summer Yellow-legged Gull. Early acquisition of adult-type upperparts.

1st-winter Yellow-legged Gull.
Mantle and scapulars mostly moulted, dark tertials evident, white head.

Late 1st-winter Yellow-legged Gull.
White head and breast. Distinctions in the tertials and greater coverts gradually lost to wear.

A variety of 1st-winter gulls at roost, including a Caspian (front middle), a Great Black-back (top) and a Lesser Black-back (right) amongst Herring Gulls.

Juvenile Caspian Gull. Long bill and legs, flat back and high breast distinctive. Note dark tertials and greater covert bases.

1st-winter Capsian Gull. Very pale underwing. Note pale mantle and contrasting white head and rump.

Two 1st-winter Caspian Gulls. White-headed, coarse 'shawl' around neck sides, grey based lower scapulars.

1st-winter Lesser Black-backed and **Herring Gull** (right) from below.

Juvenile Lesser Black-backed Gull. Dark body. Tertials and greater covert bases solidly dark.

1st-winter Lesser Black-backed Gull lacks primary obvious 'window'. White rump.

Juvenile Great Black-backed Gull. Rather pale and coarsely chequered.

1st-winter Lesser Black-backed Gull. Dark-bodied, inner greater coverts dark-based, little patterning in tertials.

1st-winter Great Black-backed Gull. Pale head, boldly chequered above, huge bill.

1st-winter Great Black-backed Gull. Pale-bodied, inner primary 'window' and thin black tail band.

but note that this area may be hidden by the fluffed-up flank feathers. **4** TERTIALS Unlike Herring (which shows strong notching) the tertials are plain brown with white feather fringes (similar to Lesser Black-back). Many also show white indentations at the tips (the so-called 'oak leaf' effect), while others show a thick dark anchor at the tip with deep white indentations behind (then a solidly dark basal area). **5** IN FLIGHT Juvenile Yellow-legged shows a relatively narrow but clear-cut jet back tail-band (usually slightly browner on Herring) that contrasts strongly with the pure white base, rump and uppertail-coverts, all of which show only limited black mottling. The very dark secondary bar contrasts quite strongly with the pale wing-coverts and most show a second dark bar across the outer greater coverts, in front of the secondary bar (but this can be surprisingly difficult to confirm on the moving wing). Although it shows a pale 'window' on the inner primaries, this is much less obvious than on Herring Gull. Although somewhat variable, the underwing-coverts often stand out as rather dark chocolate-brown (like Lesser Black-back) contrasting with the predominantly greyish flight feathers (underwing paler and more uniformly brown on Herring). When flying head-on, it shows two prominent buff 'landing lights' at the bend of each wing, much more obvious than the equivalent marks on Herring and Lesser Black-back. To eliminate Great Black-back, remember that this species has an obviously narrow tail-band that is often broken into rows of thin black bars, creating an ill-defined and rather messy pattern at a distance. Notwithstanding large male Yellow-leggeds, Great Black-back should always look considerably larger and more powerful, with an extremely large and heavy-ended bill.

First-winter By late July, some start to acquire first-winter feathers on the back and scapulars, these being buff with blackish anchor-shaped markings and basal cross-bars, these predominating by early autumn.

First-summer Unlike Herring and Lesser Black-backs, most first-summer Yellow-leggeds acquire extensive plain grey second-winter feathering on the back and scapulars, often appearing very clean-cut and contrasting. This may appear from early spring and predominates by late spring and summer, allowing such birds to be readily separated from Herring (which largely retains worn first-winter plumage) and Lesser Black-backed (which shows either new strongly patterned chocolate feathers on the back and scapulars or plain *dark* grey ones). Like the adults, Yellow-legged's new feathers are a dark 'mid grey' so, as they mature, they become increasingly easy to pick out and identify. First-summers are also very white on the head and underparts but may still retain a darker shadow around the eye and an obvious shawl of heavy streaking around the hind neck. The tertials at this time often appear solidly brown, the whitish fringes having worn off. In flight, the underwing-coverts remain dark. The legs may turn yellowish.

Second-year Still shows dark primaries, secondary bar, tertials and tail-band, and also a dark band across the outer greater coverts. However, the extensive but variable 'mid-grey' that appears in their first-summer often spreads onto the wing-coverts and such birds appear much more advanced than similarly aged Herring Gulls. The shawl of dark streaking around the base of the hind-neck is often very obvious and the greyish eye shadow is often retained. The eye and bill base gradually turn pale (but variable). In flight, the underwing-coverts continue to show dark brown feathering. Following a spring body moult, the head, neck and underparts become pure white, the back and scapulars form a strong mid-grey 'saddle' and the wing-coverts are often strongly chequered (more advanced birds already have extensive grey in the wing-coverts). The bill often becomes bright yellow by late spring, with black and bright red at the tip. The legs also show a distinct yellow tone by midsummer.

Behaviour Less versatile in its feeding behaviour than Herring Gull. On the coast and at inland reservoirs, Yellow-leggeds often patrol quite high above the water (maybe 10–20m) and sometimes plunge for fish from low elevations (almost like a Gannet *Morus bassanus*). More usually, they scavenge, often hanging around behind a feeding Great Black-back, clearly taking second place in the pecking order ahead of Herrings and Lesser Black-backs. They also feed with other gulls on rubbish tips and in fields.

Pitfalls When identifying adult Yellow-legged Gulls, two particular pitfalls need to be considered. (1) Local *argenteus* Herrings can look dark in certain lights, a problem especially acute at evening gull roosts, and this can be compounded by the fact that *argenteus* regularly acquire a white head by January. (2) Confusion is also likely with Scandinavian Herring Gulls *argentatus*, which may be distinctly darker grey than local *argenteus* (see p. 211, 230, 231 and 233). However, such birds tend to have less black and more white in the primaries and also show heavy winter head streaking.

Herring × Lesser Black-back hybrids

Such hybrids are not infrequent and, inevitably, show a darker mantle than Herring Gull. However, the mantle tends to be a darker, 'flatter' shade of grey than Yellow-legged Gull. Also, unlike Yellow-legged, they show extensive head streaking in winter. Leg colour may be yellow but some show flesh-coloured legs, as well as a yellow or orangey orbital-ring (rather than Yellow-legged's red). Such birds often appear very much halfway between the two species and fail to show Yellow-legged's structural differences. If a potential Yellow-legged Gull does not look quite right, then consider a hybrid.

'*Lusitanius*' Yellow-legged Gulls

Some Yellow-legged Gulls seen in Britain resemble so-called '*lusitanius*' from w. Iberia. Such birds seem to occur mainly in south-western areas, usually from August onwards, often associating with Lesser Black-backs. They appear much more similar in shape to Lesser Black-back, being distinctly smaller than *michahellis*, with shorter legs, *a distinctly short, thick-looking, rather stubby bill*, a rounded head and a distinctly short thick neck.

Azorean Gull *L. m. atlantis*

There have been a number of British and Irish claims of gulls considered to be Azorean Gull. Winter adults resemble *michahellis* but are distinctive in that classic examples show a hood of dense dark grey streaking, whilst immatures appear very dark chocolate-brown. Any such bird should be carefully studied and photographed. For further information, see Olsen & Larson (2003) or check images on the internet.

Caspian Gull *Larus cachinnans*

Where and when Most occur in late autumn and winter in SE England, East Anglia and the East Midlands, with a few in the W. Midlands and NE England; rare elsewhere. They tend to frequent coastal areas, rubbish tips, pig fields and reservoir and gravel pit roosts.

Two **2nd-winter Herring Gulls.** Similar to 1st-winters but bill and eye paler, plumage more variegated. More advanced birds begin to show grey in mantle.

Two **2nd-winter Herring Gulls.** Inner primary 'window' evident, mantle variable.

2nd-winter Yellow-legged (top) and **Caspian Gulls**, late winter.

2nd-winter Yellow-legged Gull. Crisper than Herring Gull, whiter bodied and darker grey upperparts.

2nd-winter Yellow-legged Gull. Neat tail-band and clean grey mantle.

2nd-winter Yellow-legged Gull. Typically dark underwing.

2nd-winter Caspian Gull. White head and breast, coarse 'shawl' to sides of neck. Weak-billed and long-legged.

2nd-winter Caspian Gull. Pale underwing.

2nd-winter Lesser Black-backed Gull. Dark, typically with much adult-type upperparts.

2nd-winter Great Black-backed G[ull]. White-headed, outer wing-coverts dark innermost boldly patterned.

2nd-winter Great Black-backed Gull. Some show slaty-black mantle feathers. Bill and eye become paler.

Very narrow dark tail band.

2nd-winter Lesser Black-backed Gull. Rather uniformly dark part from white rump.

Large gulls in 3rd-winter type plumages closely resemble adults (but are very variable). Typical signs of immaturity are dark marks within the adult bill pattern, brown centred feathers within the smaller wing-coverts, inner greater coverts, tertials, and reduced, fine markings in the tail.

3rd-winter type Herring Gull. Pale grey, inner primary 'window' still evident.

3rd-winter type Herring Gull. Extensive head streaking, pale grey upperparts and pink legs.

3rd-winter type Yellow-legged Gull. Dark grey above, extensive black in wing-tip. ▶

3rd-winter type Yellow-legged Gull. White-headed, dark grey upperparts, bare parts often brightly coloured. Yellowish legs.

3rd-winter type Caspian Gull distinctive shape, mid-grey upperparts, white head. Eye dark, bill and legs often insipid. ▶

3rd-winter type Caspian Gull. Suggestion of adult pattern in wing-tip.

3rd-winter type Lesser Black-backed Gull. Densely streaked head, brownish hue to upperparts, yellow legs.

3rd-winter type Great Black-backed Gull

3rd-winter type Lesser Black-backed Gull. Tail virtually unmarked, little white in wing-tip.

3rd-winter type Great Black-backed Gull. Sparse head streaking, some browner wing-coverts, pink legs.

225

General Caspian Gull has proved to be a difficult and controversial bird to identify, especially outside its normal areas of occurrence. The following attempts to strike a balance between clarity and the remorseless complexity that bedevils the subject. As a useful *aide-mémoire*, it may be helpful to think of Caspian as resembling a giant, long-billed, long-necked, beady-eyed Common Gull.

Size and structure Although similar to both Herring and Yellow-legged Gulls, its structure is likely to attract attention and is fundamental to its identification. **1** BILL When active, it usually shows a distinctly longer, slimmer bill than Herring Gull, with a weak gonydeal angle. It may also appear more pointed, a consequence of a gently curved tip to the upper mandible. However, females have shorter bills than males (overlapping with Herring Gull) and the bill of both sexes may be less striking when the bird is at rest, the fluffed-up head feathering reducing the impact of bill length. Beware also of Herring or other gulls showing an abnormally long bill (such aberrations are not infrequent). **2** STRUCTURE WHEN ACTIVE Similar in size to Herring Gull and, in direct comparison, males can appear quite big and bulky, females obviously smaller. It is a gentler looking, more elegant bird than Herring, with a flatter back, little or no tertial step and relatively long wings. The tail tip falls about one-third of the way along the closed primaries, compared with about halfway on Herring (but this feature is not valid with birds in outer primary moult). The head is small and rather pear-shaped, often showing a low, sloping forehead which, combined with the long bill, may give a drawn-out, 'snouty' impression. The neck is long and thin, producing a distinctive 'wine bottle' shape; the lower neck often shows an additional 'bulge' at the front, suggesting a bulging crop or an 'Adam's apple'. It also has a rather long and slightly decurved gape line. Distinctly longer spindly legs often give it a gangly impression. Its overall shape may be reminiscent of a giant Slender-billed Gull *Chroicocephalus genei*. **3** STRUCTURE AT REST When relaxed with its head sunk into its shoulders, its structure is far less distinctive. With the feathers fluffed-up, the head then looks more rounded, dome-shaped, or even square, but usually with a sloping forehead, an angle before the eye and an obviously rounded or slightly angular rear crown. The fluffed-up body plumage may also make the legs look shorter. It still has a distinctly deep-breasted impression and the underparts tend to show a distinct bulge in the ventral region, immediately behind the legs. At rest with its feathers sleeked (e.g. in hot weather) a combination of the long bill, pear-shaped head, high 'bosomed' breast, the 'bulge' in the ventral area and long wings is distinctive. This may be emphasised by its habit of standing with the body tipped at an angle, with the closed primaries pointing towards the ground. In flight, appears long-winged with a more protruding bill, head and neck.

Plumage *Adult* Likely to be located by a combination of structure, white head (even in winter), dark eye and 'intermediate' mantle colour. **1** HEAD In autumn and winter, looks noticeably white-headed, showing little if any streaking, which is fine and confined to a light dusting of pale grey on the crown or a more obvious 'shadow' around the eye, the latter emphasising a subtle white eye-ring. More distinctive is a shawl of fine pale grey streaking around the lower hindneck, but this usually wears off after autumn. **2** EYE Dark, standing out against the white head to give a softer, gentler demeanour than Herring Gull, although at a distance it can look 'beady-eyed' or 'piggy-eyed'. At close range, the eye can often be seen to be ivory coloured. Beware of fourth-winter (or even adult) Herring and Yellow-legged Gulls with dark eyes. **3** BILL In winter, the bill is a rather dull, insipid, washed-out yellow, often tinged green, sometimes with small black spots or a narrow black ring crossing both mandibles, immediately behind the dull red gonydeal spot. In summer, the bill is bright yellow with a red gonydeal spot, lacking any dark. **4 LEGS** Intermediate between Herring and Yellow-legged: an insipid greyish-pink in winter, often with a yellow tint (yellower in summer). **5 MANTLE COLOUR** Paler than Yellow-legged and closer to *argenteus* Herring, being a quite a flat mid grey (also described as a neutral 'silky' grey) with less of a bluish hue than both races of Herring Gull (Gibbins *et al.* 2010), but bear in mind that the exact shade varies according to the light and angle of viewing. **6 PRIMARIES** Having found a candidate, the most important feature on which to concentrate is the primaries. Firstly, these usually have two large white mirrors at the wing-tip: the larger (on P10) reaching the tip of the feather, the smaller on P9. More importantly, there is extensive pale in the primaries, which takes the form of long pale 'tongues' or 'lobes' on the inner webs (grey on the upperside, white on the underside). These are separated from the white at the tips by a relatively narrow band of black cutting across the entire width of the feather. This pattern of pale 'tongues' on the inner webs creates what is often referred to as a 'Venetian Blind' effect. This distinctive pattern is best evaluated on the underwing at rest, when the bird is preening, or if it droops its primaries. It is also apparent in flight, significantly reducing the size of the black in the primaries, particularly on the underwing (where the black hooks back along the trailing edge).

Juvenile **1 AT REST** Very similar to juvenile Yellow-legged Gull (although rather more 'washed-out'). Consequently, structure is fundamental to its identification at this age. Like Yellow-legged, it shows a subtle dark eye-patch, a curved area of white on the rear head/ upper nape (extending onto the neck-sides) and a band of dark streaking on the hindneck (extending onto the upper breast). The mantle, scapulars and wing-coverts are 'mouse-brown' showing whitish fringes to all of the feathers, creating a scalloped impression. Like Yellow-legged and Lesser Black-backed (but unlike Herring) it has a thick dark bar across the outer greater coverts (at the base of the closed wing, closest to the wing bend). Chequering on the greater coverts is slight and, like Yellow-legged, is limited mainly to the inner feathers. It differs from Yellow-legged in showing broader, more diffuse white tips to the plain brown tertials, extending back around the outer edge of the feathers towards the base. The legs are a pale, washed-out flesh colour. Note that juvenile Lesser Black-backed Gulls, although starting off dark, often wear considerably by early autumn, with paler brown plumage and a pure white background colour to the head and underparts; they also show white feather fringes to the upperparts; such birds can suggest Caspian Gull – even their tertial pattern can be identical. **2 IN FLIGHT** The most distinctive feature is that, unlike Herring and Yellow-legged, the underwing-coverts have a white background, but with grey-brown on the axillaries and narrow bands of grey-brown across the tips of the lesser and median coverts. It should be noted, however, that a minority of Caspian Gulls show more solidly dark lesser underwing-coverts. The under-primaries and secondaries are also paler (silvery to off-white). The pale inner primary window is usually somewhere between Herring and Yellow-legged in prominence, and is not a particularly helpful feature. As Yellow-legged Gull (and Lesser Black-back), in flight it shows a dark 'bar' across the greater coverts, immediately in front of the black secondary bar. Also like Yellow-legged Gull, the black tail-band contrasts strongly with the white tail base, uppertail-coverts and rump (thus showing more contrast than Herring Gull).

First-winter A post-juvenile body moult from mid August gives the bird a distinctive snow-white head, with only limited grey shading immediately around the eye (which largely fades away by winter). The white head forms a smart contrast with the largely all-black bill and also with a shawl of grey streaking around the lower hindneck. It also emphasises the dark eye. The underparts too become extremely white. The back and scapulars are rather pale grey with dark brown anchor-shaped marks; the important point about these markings is that they are very narrow, creating a much more delicate pattern than that shown by Herring and Yellow-legged (lacking the heavy anchors and strong cross-barring of Yellow-legged). The exact pattern of the scapulars is, however, variable, some showing delicate streaks, wedges, or diamond shapes (and occasionally somewhat heavier markings). The thick dark band across the outer greater coverts is retained (visible both at rest and in flight). The tertials are solidly dark with thick white tips. Most importantly, the underwings appear increasingly white (brown on Herring and dark brown on Yellow-legged).

First-summer Similar to first-winter but, as a result of wear and bleaching, generally paler and less contrasting. From late winter onwards, variable but often significant amounts of plain pale grey second-winter feathering appears on the back and scapulars. The bill base turns increasingly pale pink or yellowish, retaining a dark tip and cutting edges.

Second-year The complete moult that begins in spring or summer is completed by early autumn. Second-winter is similar to first-winter (including on some the shawl of grey streaking on the lower hindneck) but gradually acquires more extensive plain grey feathering on the back and scapulars (forming a grey saddle) and, later, the wing-coverts. Such birds are obviously 'cleaner' and greyer than similarly aged Herring Gulls, which retain extensive brown immature plumage. Some Caspians, however, appear much more similar to first-winter, with dark streaks, blobs or delicate anchors on the back and scapulars (but lacking *strong* anchors). The underwings become even whiter in second-year (but may retain limited brown markings). Unlike the more messily patterned Herring Gull, the black tail-band continues to contrast strongly with the white rump and uppertail-coverts. Most second-winter Caspians show a small white or off-white mirror in the outer primary (rare in Herring and Yellow-legged). The bill base continues to become paler and the legs grey-pink, but the eye remains dark (unlike most Herring and Yellow-legged Gulls).

Third-year In some ways third-years are more difficult to identify, as this age has few concrete plumage differences from Herring and Yellow-legged Gulls. Structure and general 'jizz' become the best 'feature', but note also the dark eye and the shawl of dark neck streaking. The bill is usually greenish-yellow with a black subterminal band. Differences in the primary pattern may be useful. Third-winter Caspians usually show large white mirrors on P9 and P10 (larger then equivalent Yellow-leggeds) but third-winter Herring can also show these. The following esoteric features are probably best analysed from photographs: (1) Caspian shows black bands across the tip of P5 (and often P4), which are rare on Herring, and (2) some also have the effect of grey 'tongues' on the inner webs of the middle primaries. Remember that some third-year Herring Gulls have dark eyes and white heads, and can look disconcertingly similar to Caspian (concentrate on structure).

Behaviour When feeding on water, they often duck their head in and even up-end, rather like Slender-billed Gull. Also when feeding, they often raise their wings, especially in aggressive interactions.

Calls and long-call posture The calls are surprisingly distinctive. The 'long call', begins with a deep, throaty *eaah eaah* or a deep, buzzing *zeep zeep* followed by a deep, rather fast, rhythmic laughing: *ah-ah-ah-ah-ah-ah-ah-ah*, oddly reminiscent of Common Guillemot *Uria aalge*. This is given with the head progressively raised and the wings held open and pushed back in a so-called 'albatross posture' (birds squabbling over food may indulge in this behaviour). The normal call is also distinctive: a peculiar deep, rather nasal, buzzing *zeeeup*.

Further reading For further information on the complicated identification of this species (and hybrids) see Gibbins *et al.* (2010, 2011).

References Gibbins *et al.* (2010, 2011).

Glaucous and Iceland Gulls

Where and when Both are scarce winter visitors, mainly from October to May, but numbers vary considerably from year to year, with periodic influxes. They are liable to occur anywhere that attracts large gulls. Most of our Glaucous Gulls *Larus hyperboreus* originate from e. Greenland and Arctic Europe, so they tend to occur in the north-west, the north, and on the east coast. Despite their name, Iceland Gulls *L. glaucoides* originate from Greenland, so have more of a north-westerly distribution. The NE Canadian race of Iceland Gull, *kumlieni* or 'Kumlien's Gull', breeds principally on Baffin Island and winters in Iceland and NE Canada, south to the Great Lakes. It has proved to be a regular visitor to Britain and Ireland, in small numbers. The breeding range of Kumlien's is sandwiched between Iceland and Thayer's Gull *L. g. thayeri*, which breeds across the Canadian Arctic, mainly to the west of Kumlien's. Thayer's is currently treated as a full species in North America, but in Britain it is regarded as a race of Iceland Gull. There are currently no accepted records of Thayer's in Britain, although two very strong claims are under consideration (and several accepted records in Ireland).

Separating Glaucous and Iceland Gulls

Size and structure Both can be separated from Herring Gull *L. argentatus*[†] by their white primaries but, to differentiate between them, concentrate on size, structure, head and bill shape and, on juveniles, bill colour. Size varies sexually, males averaging larger than females, but Glaucous is usually (but not always) much larger than Herring, approaching Great Black-backed Gull *L. marinus* in size; Iceland is similar to or slightly smaller than Herring. Glaucous is a large, powerful, mean-looking brute, with a relatively short, blunt primary projection and often a prominent tertial step. Iceland is a rather stocky but gentler looking, less powerful gull, often lacking a tertial step and has long, tapered primaries that produce an attenuated rear end. A useful way of evaluating the difference in the primary projection is that, on Glaucous, the projection beyond the tail is the same as, or shorter than, the bill length, whereas, on Iceland, it is longer than the bill. Iceland has quite short legs that are often deep pink. A useful *aide-mémoire* is that, structurally, Iceland suggests a rather delicate Lesser Black-backed Gull *L. fuscus*, whereas Glaucous suggests a large Herring.

Head and bill Although overall size and shape may be the first clue to their identification, pay

Juvenile Iceland Gull. Small dark bill, rounded head, long primary projection.

Juvenile Glaucous Gull. Large bill, pink with black tip, angular head shape, short primary projection, obvious 'tertial step'.

2nd-winters closely resemble juveniles. Eyes become pale, Iceland Gull (above) loses dark base to bill.

Both species gradually attain grey upperpart feathering. This is a **3rd-winter Glaucous Gull.**

Adult winter Glaucous Gull. Note differences in structure.

Adult winter Iceland Gull

Herring × Glaucous Gull hybrids are not uncommon, and show intermediate characteristics.

Iceland Gull of race *kumlieni* resembles nominate race, but shows pale grey in primaries.

Herring Gulls of the race *argentatus* can be very large and can show very little black in wing-tip. 1st-winters can appear very pale.

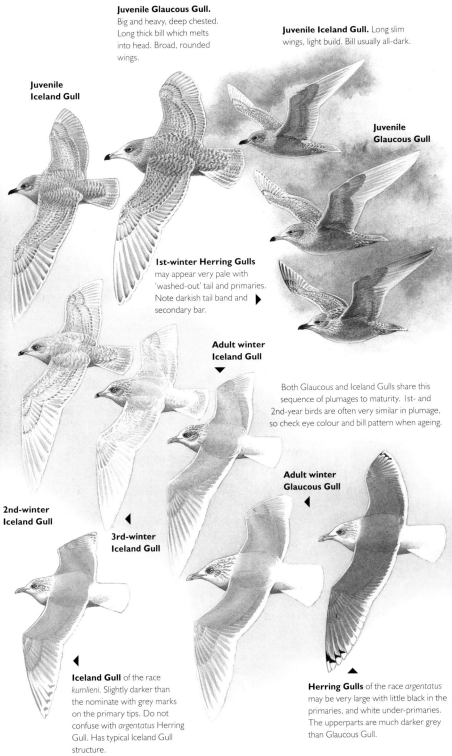

Juvenile Iceland Gull

Juvenile Glaucous Gull. Big and heavy, deep chested. Long thick bill which melts into head. Broad, rounded wings.

Juvenile Iceland Gull. Long slim wings, light build. Bill usually all-dark.

Juvenile Glaucous Gull

1st-winter Herring Gulls may appear very pale with 'washed-out' tail and primaries. Note darkish tail band and secondary bar. ▶

Adult winter Iceland Gull ▼

Both Glaucous and Iceland Gulls share this sequence of plumages to maturity. 1st- and 2nd-year birds are often very similar in plumage, so check eye colour and bill pattern when ageing.

2nd-winter Iceland Gull

Adult winter Glaucous Gull ◀

3rd-winter Iceland Gull ◀

Iceland Gull of the race *kumlieni*. Slightly darker than the nominate with grey marks on the primary tips. Do not confuse with *argentatus* Herring Gull. Has typical Iceland Gull structure. ◀

Herring Gulls of the race *argentatus* may be very large with little black in the primaries, and white under-primaries. The upperparts are much darker grey than Glaucous Gull. ▲

particular attention to head and bill shape. Glaucous has a long, thick, heavy, rather 'parallel' bill, longer and thicker than Herring's, but (unlike Great Black-back) lacking a prominent gonydeal angle. Its prominence is complemented by a low forehead, a flat crown and a peaked rear crown, all of which reinforce a general impression of severity. Iceland's bill is shorter, slimmer, stubbier and more pointed, being similar to, or slightly shorter than, that of Lesser Black-back; its steeper forehead and rounded head create a gentler expression, reminiscent of Common Gull *L. canus*. (However, head shape differences require cautious evaluation as both species can vary according to attitude.)

Flight Identification Glaucous is a large, heavy, 'lumbering' gull, with broad wings and a lazy flight that recalls Great Black-backed. Iceland is smaller, but has a rather stocky body and proportionately longer, narrower wings; its tail appears slightly, but distinctly shorter than Herring's and slightly more wedge-shaped. Its flight tends to be more energetic than that of Glaucous.

Plumage *Juveniles and first-years* Note that, unlike our common large gulls, juveniles of both species show very little if any post-juvenile moult until late winter so, to all intents and purposes, they remain 'juvenile' until then. Both are essentially pale, biscuit-brown, intricately and delicately mottled and barred, but with whiter primaries. Plumage tone varies individually, but most become paler through wear and bleaching. Bill colour is diagnostic: Glaucous has a pink bill with a prominent, clear-cut 'dipped-in-ink' black tip. Iceland's is usually black with a small, inconspicuous pale area at the base (the pale becomes more extensive as winter progresses) but the black tip is not clear-cut and extends back along the cutting edge so that, at any distance, the tip appears to merge with the pale base. *Second-years* Usually paler and less well patterned than first-years; they may look very creamy when worn. Grey feathering on the mantle and scapulars is gradually acquired, but some second-years show very little grey, if any. Again, plumage tone varies individually, so concentrate on bill and eye colour for accurate ageing. Most significantly, both species acquire pale irides during their second year. On Iceland, the bill base turns creamy, and the black tip becomes sharply defined so that is more similar to first-winter Glaucous but, unlike Glaucous, the black tends to protrude back along the cutting edge so that the tip is less clear-cut. Second-year Glaucous usually acquires a pale tip to the bill and, sometimes, a yellowish tint to the base. *Third-years* Both gain a grey mantle, scapulars and wing-coverts, and are much more adult-like, but retain traces of immaturity on the wings and tail. They show dark terminal or subterminal markings on a yellow bill, but Iceland may acquire a greenish or greyish tint to the base; bill colour may brighten towards spring. *Adults* Both are similar. Fourth-years may retain hints of immaturity, including small dark markings on the bill.

Kumlien's Gull *L. g. kumlieni*

General approach Any apparent Iceland Gull should be checked for features indicative of Kumlien's Gull. The basic difference from nominate Iceland is that adults show variable amounts of grey in the primary tips, while juveniles and other immatures show pale brown in both the primaries and tail. There is, however, much individual variation in immatures, to the point where pale Kumlien's will be impossible to separate from slightly darker nominate Iceland. Before identifying this race, first establish beyond doubt that the bird concerned is an Iceland Gull, mainly by the structural characteristics outlined above.

Plumage *Adult*. Adult Kumlien's shows grey in the outer primaries, usually along the outer webs and subterminally across the tips. At rest, this usually appears as grey along the lower edge of the closed primaries, with two or three grey chevrons towards the tip. Although fairly obvious at rest, this may be surprisingly difficult to see in flight. The shade and precise extent of the grey varies individually, some birds looking, or appearing to look, blackish. *Juvenile* Kumlien's resembles nominate *glaucoides* Iceland except they show distinctly browner primaries and tail (sometimes also a browner secondary bar). The darkness and extent of the brown varies individually; on some paler birds the dark in the primaries and tail may be apparent only in flight. Older immatures are similarly dark in the primaries and tail.

Thayer's Gull *L. g. thayeri*

Thayer's shares Kumlien's Gull's 'Iceland-like' structure and jizz, but adult Thayer's has black primaries and, superficially, appears much more similar to Herring Gull. Note, however, that adult Thayer's has a heavily streaked hood and breast-band in winter, and short, deep pink legs. Similarly, juvenile and other immature plumages of Thayer's are darker than nominate Iceland or Kumlien's. They also have black or blackish primaries, with noticeable whitish chevron-shaped tips to the individual feathers, as well as a blackish tail-band and secondary bar. The problem is that there is a certain amount of intergradation between Kumlien's and Thayer's, to the point where intermediate and 'non-classic' individuals will not be identifiable to race. Thayer's and darker Kumlien's also have to be separated from Herring Gulls, particularly those northern *argentatus* that show reduced black in the primaries (see below). If you come across a potential Thayer's, take photographs, seek expert advice and consult detailed literature, such as Olsen & Larsson (2003) and Howell & Dunn (2007). There are also large numbers of useful gull photographs on the internet.

Pitfalls

The following pitfalls should be considered when identifying any of the above.

Glaucous × Herring hybrids
Occasionally seen in Britain and Ireland (in North America, often referred to as 'Nelson's Gull'), such birds may show characters intermediate between the two species, but others may be very similar to Glaucous, revealing their true identity by traces of black in the primary tips (adults) or traces of dark brown on the outer primaries, secondaries and tail (immatures).

Northern Herring Gulls
Adults Dark Scandinavian Herring Gulls of the nominate race *argentatus* are common winter visitors to n. and e. Britain. Some, probably from the extreme north, are large and dark and show little black in the wing-tips. In flight, the black may be very difficult to detect at any distance, while the underwings are almost entirely white, often with faint grey shading only at the primary tips. As they can approach Great Black-backed in size, distant birds can be misidentified as Glaucous; they are, however, much darker above (similar to or even darker than Common Gull) while on the water they show a pronounced tertial step, a broad white tertial crescent and black primaries with very large white mirrors and primary tips. *Immatures* It seems increasingly likely that the large, pale, 'Glaucous-like' immature Herring Gulls that occur occasionally in northern and eastern areas are these birds' offspring. It may

be difficult, however, to decide what is a Glaucous × Herring hybrid and what is an immature Northern Herring, but unless an immature gull shows obvious pro-Glaucous characters, such as a Glaucous-like bill, consider the possibility of Northern Herring Gull (see p. 211). First-years seem to be fairly consistent in their appearance, being large (sometimes approaching Great Black-backed), heavy-billed with their shape being typical of *argentatus* Herring Gull (see p. 212). However, they are very pale brown on a whitish background and the primaries are brown (much paler than first-winter *argenteus*) with white feather fringes and pale submarginal markings; the bill has a pronounced gonydeal angle and is mainly black. In flight, they resemble very pale and severely washed-out *argenteus* Herring Gulls.

Leucism Older field guides may state that second-year Glaucous and Iceland Gulls are completely white. This is not so: even very faded second-years usually retain a creamy look, and close inspection will reveal traces of delicate patterning on the wings and tail. A pure white gull will almost certainly be aberrant, and structure and bare-part colours provide the best clues to its identity. (Note: leucistic or albino Lesser Black-backs are closer to Iceland in shape than leucistic or albino Herrings.) Second- or even third-summer Glaucous and Iceland may look very white at a distance, but closer views should reveal a grey 'saddle' (mantle and scapulars) and the other features outlined above.

Moulting Herring Gulls In autumn, moulting adult Herrings may show severely reduced areas of black in the wing-tips, which appear noticeably rounded as a result of partially grown outer primaries. In flight or at distance, such birds may suggest Glaucous or Iceland. Fortunately, with local *argenteus* this phenomenon occurs prior to the main arrival of Glaucous and Iceland Gulls (but *argentatus* moults later, finishing in November–January).

[†]All references to Herring Gull are to British race *argenteus*, unless otherwise stated.

References Howell & Dunn (2007), Olsen & Larsson (2003).

Sandwich and Gull-billed Terns

Where and when Sandwich Tern *Thalasseus sandvicensis* is a common summer visitor, likely to be seen on coasts from March to October (very small numbers winter, mainly on the south coast); small parties may appear briefly inland on migration. Gull-billed Tern *Gelochelidon nilotica* is a very rare vagrant, currently averaging about three records a year. Most occur in May, with fewer in summer and autumn (a pair bred in Essex in 1950). Numbers occurring here have declined in recent decades, in line with a significant decrease in breeding populations in Denmark and Germany. Many recent records have been in western areas, suggesting over-shooting from Iberia, where significant populations persist. Historically, this species has a very high rejection rate, but this belies the fact that, if seen well, it is not difficult to identify.

Habitat and behaviour Unlike Sandwich Tern, Gull-billed is not a bird of the open sea and birders should be especially wary of claiming the species on sea watches. It typically occurs on coastal marshes, lagoons and estuaries, and also hawks insects over dry land; it may even be seen patrolling ploughed fields. It also catches larger prey, such as frogs, and one Norfolk vagrant developed a liking for Little Tern *Sternula albifrons* chicks. Unlike Sandwich Tern, it rarely plunge-dives.

Structure and flight An awareness of the structural differences between the two species is an essential first step to identification. Sandwich is a large tern, similar in size to Kittiwake *Rissa tridactyla*, and it lacks the long tail-streamers of Common *S. hirundo*, Arctic *S. paradisaea* and Roseate Terns *S. dougallii*. Compared to Gull-billed, it is a *noticeably slim, rakish-looking bird* with long, *narrow*, rather angled wings. The head and neck protrude noticeably and the bill is long and slender with a yellow tip (adults only). Although similar in size, Gull-billed is distinctive. In flight, it shows quite broad-based wings that are also long and pointed but tend to be swept back at the carpals. The overall effect may recall a small gull, while the slow, easy, rather languid flight also suggests Common Tern (indeed, the whole effect in flight may recall juvenile or winter-plumaged Common, rather than Sandwich). Rather bull-necked in flight. On the ground, it looks long-legged for a tern, with a rather horizontal carriage, but the most important feature is the *thick and stubby black bill*. It also tends to look short-necked and round-headed with a rather rounded belly and deep-based primaries.

Calls The calls of adults are totally different: Sandwich has a well-known shrill, guttural, rasping *kerrr-ick*, whereas Gull-billed has a low, deep *ger-erk* or *ger-vik* (quite unmistakable). Juveniles' calls are quite different from adults'. Sandwich has a high-pitched, squeaky and rather sibilant *sreee sree sree sree…* or *pree-peep pree-peep*, often persistent and rather irritating. Gull-billed has a similar high, but soft *pe-eep* or a quick *pe-pe-eep*.

Plumage *Adult summer* In flight, Sandwich is a very whitish looking tern, with contrasting blackish outer primaries (these darken with wear as summer progresses). Adults may show a strong pink or a weak salmon tinge to the underparts. In flight, Gull-billed looks very pale, uniform whitish-grey above, including the rump and tail (Sandwich has a white rump and tail). However, in bright light, the upperpart colours of the two species can look surprisingly similar. Unlike Sandwich Tern, the underwings show an obvious blackish trailing edge to the primaries, similar to that shown by Common Tern (Sandwich has a more diffuse and much less obvious greyish trailing edge). Other plumage differences are less significant: when the head is raised, the black extends down the nape on Gull-billed (on Sandwich, the black is confined to the cap and the nape is white). Both species undergo a complete post-breeding moult. The primaries start with the inners in late summer, finishing with the outers in early to midwinter (on reaching the winter quarters); in late winter, the inner primaries are moulted a second time, prior to spring migration. The result of this moult is that, while in Europe, both species have old outer and new inner primaries. When new, the primaries have a pale grey bloom (radii) which is steadily lost with wear, revealing progressively more of the blackish base colour (rami). This produces a contrast between the old dark outer primaries and new grey inners, so that both species often possess noticeable dark outer primary 'wedges'. In Europe, the wings look most uniform in spring, but dark outer wedges will be most pronounced in late summer and autumn, when the outer primaries are oldest. It seems, however, that, on average, Gull-billed wears less dark than Sandwich and usually shows less obvious primary wedges. Even as late as early September, some adults still look uniformly pale whitish-grey across the entire upperwing; at a distance, such birds look rather concolorous, showing little contrast between the upperwing and the underwing (according to *BWP*, about one-third of European Gull-billed do not replace their inner primaries a second time in late winter and such birds may account for at least some of these plain-winged individuals). *Adult winter* Adult Sandwich starts to lose the black on the lores, forehead and crown from mid June

Adult summer Sandwich Tern.
Note protruding head, slender pointed wings, dark primary wedges and white rump and tail.

Adult summer Gull-billed Tern.
Thick-set and gull-like. Note primary pattern, grey rump and tail, thick bill.

Juvenile Gull-billed Tern. Black 'face' patc recalls Mediterranean Gull. Relatively plain above, buff mottling on back and wings.

Juvenile Sandwich Tern.
Bill shorter than adult. All dark crown, strongly patterned back and wings. Tail marked black.

Adult Sandwich Tern moulting in late summer. White forehead, dark primary wedges.

1st-winter Sandwich Tern. Darkish carpal bar, secondaries and outer tail. Black 'shawl' over head.

1st-winter Gull-billed Tern. Plainer above, grey rump and tail. Head mainly white.

Adult Gull-billed Tern, moulting (September). Black crown becomes peppered with white. May show darker primary wedges.

Adult Summer Sandwich Tern. Greyish under primary tips.

Adult summer Gull-billed Tern. Thick-set and neckless. Dark trailing edge to under primaries.

Juvenile Roseate Tern. Superficially similar to juvenile Sandwich Tern.

Adult winter Forster's Tern. Head pattern suggests Gull-billed Tern. Bill shape, size and structure more similar to Common Tern. Note red legs.

Adult winter Gull-billed Tern. Thick bill, white head with variable black eye patch and long legs.

Juvenile Sandwich Tern. Chequered mantle and carpal bar. Black crown. Note bill is shorter than adult's.

1st-winter Sandwich Tern. Forehead white, chequered mantle replaced by grey feathers. Tertials retained.

Adult winter Sandwich Tern. Slender bill, black 'shawl' over head and spiky crest.

1st-winter Gull-billed Tern. White head with black eye patch. Stubby bill. Plain grey above, poorly marked tertials.

onwards so that, by late August/September, it has a prominent white forehead and black 'shawl' from the eye back over the nape. Individuals in full winter plumage by late July are probably non-breeders (*BWP*). Adult Gull-billed starts its moult later than Sandwich, from late July to mid August, although some still have a full black crown in early September. A significant difference between moulting adults is that Gull-billed *does not gradually acquire a white forehead*; instead, the whole cap is moulted at once so that transitional individuals have the entire cap peppered with white. When moult is completed, Gull-billed has a completely different head pattern from Sandwich: instead of a black shawl around the nape, it has a white head with a variable grey or blackish patch immediately in front of and behind the eye, strongly reminiscent of first-year Mediterranean Gull *Larus melanocephalus*; this is faint on some, so these birds may look very white-headed from a distance. Some have very fine black streaking on the crown and nape (difficult to see in the field) while others have a grey wash to the rear crown (it seems that Gull-billed never has a black shawl). ***Juvenile Sandwich Tern*** Easily separated. Juvenile Sandwich has an all-black crown (including the forehead). However, it starts body moult soon after fledging so by autumn it acquires a white forecrown, a streaked rear crown and a black shawl around the nape (like winter adults). Initially, the upperparts are strongly patterned with dark brown and white (note in particular that this pattern occurs along the leading wing-coverts, unlike Gull-billed Tern; see below). However, during the late summer post-juvenile moult, much of the upperpart patterning is lost so that, by autumn, the mantle and scapulars are plain grey with variable traces of immaturity confined to the wing-coverts, tertials and tail (latter is mainly black). Note that the bill of juvenile Sandwich is often markedly shorter than the adult's and lacks a yellow tip, thus looking stubbier; to the inexperienced, this could suggest Gull-billed so attention to plumage detail is essential. Whereas late summer adults have prominent black wedges in their outer primaries, juveniles have fresh pale grey primaries, the feathers narrowly fringed with white. Also note that juvenile terns can fly before their outer primaries are fully grown, so that recently fledged individuals have noticeably shorter, more rounded wings than adults; on Sandwich this, combined with the shorter bill, can suggest something unusual. *Juvenile Gull-billed Tern* Totally different from juvenile Sandwich. It lacks a black crown or shawl but instead has a small black ear-covert patch, like winter adult (very close inspection may reveal brown shaft-streaks on the crown and nape). Instead of being strongly patterned on the upperparts, the mantle and scapulars are relatively plain, usually appearing *uniformly ginger* in the field; by September, this colour fades to cream and is soon replaced by grey first-winter feathering. Note in particular the fairly plain wing-coverts, lacking Sandwich Tern's dark feathering along the leading coverts. The overall effect is that, from a distance, juvenile Gull-billed looks very plain above and in flight can be difficult to separate from winter adult. Other differences include: the tertials are ginger with a brown feather centre and tip (juvenile Sandwich has the tertials strongly patterned with dark brown) and the tail feathers are grey with a pale tip and dark subterminal patch (on Sandwich, black, thickly edged with white). Note that, like adult, juvenile Gull-billed has a prominent dark trailing edge to the under-primaries. It may also show orange on the base of the lower mandible. *First-year* In early winter, towards the end of the post-juvenile moult, both species resemble winter adult, but are easily aged by retained juvenile tertial and tail feathers. First-winter Sandwich also has a dark secondary bar (absent on winter adult). By late winter, even these feathers are lost and both resemble winter adult. First-summer birds do not normally return to their breeding areas. *Second-year* Subsequently, both resemble adult but second-summer Sandwich does not attain full summer plumage, retaining white flecking on the forehead, lores and crown. Second-summer Gull-billed may show narrow white fringes to the black feathers of the cap and some vagrants, perhaps second-summers, have shown uneven wing-coverts with a slight brownish wash.

Other confusion species

Roseate Tern S. *dougallii*
Note the similarity between juvenile plumages of Roseate (p. 245) and Sandwich Terns. Roseate is easily separated by size and shape, which are more similar to Common and Arctic Terns.

Forster's Tern S. *forsteri*
Winter and juvenile Gull-billed may be confused with winter or juvenile Forster's Tern, which also has a black ear-covert patch (20 British records 1980–2003, mainly in winter). Forster's is easily separated from Gull-billed by the following features. **1 BILL** Long and thin.

2 SIZE Much smaller (slightly larger than Common Tern). **3** STRUCTURE More similar to Common (in particular, it has a longer tail). **4** LEG COLOUR Pale (dull orange on first-winter, bright scarlet on adult). **5** PRIMARIES Forster's Tern usually has whiter primaries than Gull-billed.

Winter and first-year Common Tern

Beware also of first-summer and winter-plumaged Common Terns (latter can occur from July onwards). These lack long tail-streamers, have an all-dark bill and a grey tint to the rump and tail. However, they are easily separated by size and shape, dark carpal bar and (when adult) red legs. First-summer Common Terns are similar to winter adults, but often show great contrast between the old, dark outer primaries/inner secondaries and new, grey inner primaries and outer secondaries.

Orange-billed terns

Lesser Crested Tern *Thalasseus bengalensis* has occurred in Britain and Ireland, as have what are thought to be Elegant Terns *T. elegans* from W. North America. Any 'Sandwich Tern' with an orange bill should be carefully studied, photographed and expert advice sought. This is a very difficult subject, not least because hybrids may also complicate the issue.

Marsh terns: Black, White-winged Black, Whiskered and American Black Terns

Where and when Black Tern is a spring and autumn passage migrant, commonest in s. England, occurring at lakes, reservoirs, gravel pits and on the coast; large flocks sometimes appear, particularly during gloomy, anticyclonic weather. White-winged Black Tern is a very rare but annual spring (May–June) and autumn (mainly August–September) migrant, currently averaging 18 records a year, with a peak of 49 in 1992. Whiskered Tern is a vagrant, with records currently about eight a year, usually in April–June but rarely in autumn. Occasionally seen in small flocks, the largest being 11 in Derbyshire in April 2009. Its occurrences are normally independent of the other two species.

General features Small, compact terns and, compared to the larger Common *Sterna hirundo* and Arctic Terns *S. paradisaea*, they are rather stiff-winged, short-tailed and lack a prominent tail fork. They usually feed over fresh water and do not plunge for food (although Whiskered may belly-plunge on occasion). However, over fresh water Common and Arctic Terns also usually feed by surface-picking, just like marsh terns.

Black *Chlidonias niger* and White-winged Black Terns *C. leucopterus*

Structure White-winged Black is slightly shorter-billed than Black (recalling Little Gull *Hydrocoloeus minutus*) and has slightly shorter and rounder wings with perceptibly stiffer wingbeats.

Plumage *Adult summer* Easily separated. Black Tern has a black head and body, grey upper- and underwings and a grey rump and tail. In comparison, White-winged Black is a stunning bird: the head and body are black, but the forewing is white and the underwing-coverts are black (contrasting strongly with the pale grey under-primaries and secondaries). The rump and tail are strikingly white. On the upperwing, the outer primaries and inner secondaries are blackish, contrasting with the rest of wing. Its legs are red (blackish on Black Tern). *Winter adult* Autumn adult White-winged Blacks nearly always retain remnants of summer plumage, particularly traces of black on the underwing-coverts, white on the upperwing-coverts and the white rump and tail, making them readily identifiable. Individuals in full winter plumage are very rare in this country, but may be distinguished from Black Tern by the following features. **1** BREAST PATCHES Lacks Black Tern's dark breast patches (often referred to as 'breast pegs') and the absence of this feature should be very carefully checked. **2** UPPERWING Paler than Black and, in autumn and early winter, has contrasting blackish outer primaries and inner secondaries, a result of feather darkening caused by wear; it also has a less obvious dark carpal bar on the leading edge of the wing-coverts. **3** RUMP Noticeably whitish compared to the grey rump of Black Tern. **4** HEAD The dark on the head is less extensive than on Black, producing a rather white-headed appearance in comparison: the black is confined to a small patch behind the eye and streaking over the rear crown (on Black Tern, the black is more solid and more extensive). *Juvenile* Black Tern has entirely plain grey upperparts with a browner mantle and scapulars, and a broad, dark carpal bar; from below, the 'breast pegs' are obvious. It is essential to check all of the following features before claiming a juvenile White-winged Black. **1** 'SADDLE' The dark brown 'saddle' (mantle and scapulars) contrasts sharply with the pale grey wings. Note, however, that in late autumn the dark 'saddle' may be less obvious owing to a combination of fading and the acquisition of the first grey winter back feathering; conversely, juvenile Black can look quite contrasting in certain lights. **2** WINGS Distinctly paler than juvenile Black Tern. **3** RUMP The 'saddle' also contrasts strongly with the white rump, which is obvious even at a distance, particularly when flying away. **4** CARPAL BAR Narrower and less conspicuous. **5** 'BREAST PEGS' Completely lacking. **6** HEAD PATTERN Like winter adult, the black on the head is less extensive, being confined to the ear-coverts and rear crown. *First-summer* Both species remain in their winter quarters during their first summer, but occasional individuals may move north with adults. Any Black or White-winged Black largely in winter plumage during the summer months is likely to be a first-summer, which often show strongly variegated plumage owing to the contrast between new and old feathering.

Hybrids Although hybrid Black × White-winged Blacks are rare, any White-winged Black should be checked for evidence of hybridisation. A juvenile hybrid in Somerset had the white rump, pale wings and dark 'saddle' of a White-winged Black, but the 'breast pegs' and structure of Black (Vinicombe 1980).

Whiskered Tern *Chlidonias hybrida*

Structure Although obviously a marsh tern, Whiskered is larger and stockier than the other two which, along with its generally pale plumage, means that winter adults, juveniles and first-summers are far more likely to be confused with Common and Arctic Terns (see p. 240). Structure is important: it has the typical *Chlidonias* shape, being rather stocky, neck-less and short-tailed (with only a shallow fork) with shorter, broad-based, straighter wings than Common. The bill is rather stubby, but it varies sexually, males having longer bills than females (no overlap). When perched, the legs are rather long and spindly.

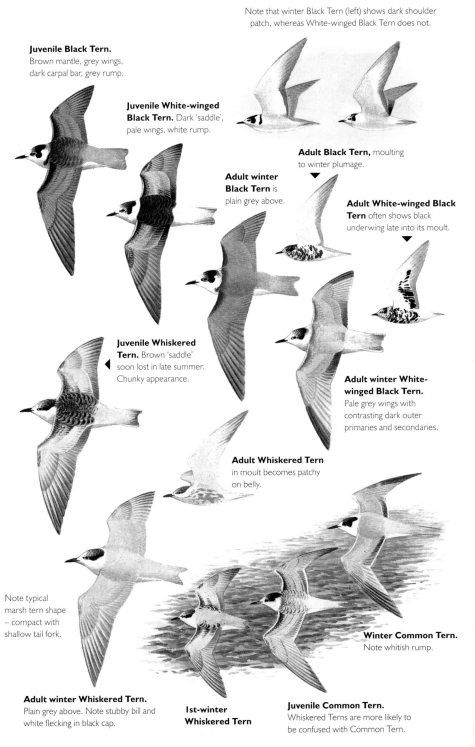

Note that winter Black Tern (left) shows dark shoulder patch, whereas White-winged Black Tern does not.

Juvenile Black Tern. Brown mantle, grey wings, dark carpal bar, grey rump.

Juvenile White-winged Black Tern. Dark 'saddle', pale wings, white rump.

Adult Black Tern, moulting to winter plumage.

Adult winter Black Tern is plain grey above.

Adult White-winged Black Tern often shows black underwing late into its moult.

Juvenile Whiskered Tern. Brown 'saddle' soon lost in late summer. Chunky appearance.

Adult winter White-winged Black Tern. Pale grey wings with contrasting dark outer primaries and secondaries.

Adult Whiskered Tern in moult becomes patchy on belly.

Note typical marsh tern shape – compact with shallow tail fork.

Winter Common Tern. Note whitish rump.

Adult winter Whiskered Tern. Plain grey above. Note stubby bill and white flecking in black cap.

1st-winter Whiskered Tern

Juvenile Common Tern. Whiskered Terns are more likely to be confused with Common Tern.

241

Plumage *Adult summer* Easily identified. Very uniformly pale grey across the wings, rump and tail; it has a black cap and an obvious white face that contrasts with the blackish or dark grey underparts, which in turn also contrast with the white undertail-coverts. The bill is blood-red. Beware of confusion with darker-bellied Common and Arctic, or even with oiled individuals. *Adult autumn* Variable. Post-breeding moult commences as early as late June so autumn migrants usually show a patchy mixture of summer and winter plumage. Many soon acquire a white forehead or crown and gradually lose the dark grey underparts. Others may retain a complete black crown while losing the dark underparts. *Adult winter* White instead of dark underparts and the black on the head is reduced to a line through the eye and across the nape. The bill is black. *Juvenile/first-winter* Birds of this age are surprisingly rare in Britain. Juvenile has a dark brown 'saddle' (mantle and scapulars) with the individual feathers fringed buff (wearing paler). A rather restricted dark 'breast peg' extends onto the neck sides. However, unlike the other two species, Whiskered starts to moult soon after fledging, so *full* juvenile plumage is rare here. By the time they migrate, most have lost or started to lose much of their dark juvenile mantle and scapular feathering, so late autumn juveniles/first-winters look rather plain grey above but often have some dark brown on the scapulars. Juveniles have a dark bill and legs, and small areas of black confined largely to the ear-coverts and rear crown. *First-summer* First-summer Whiskered seem to return north more regularly than other terns. Although some are apparently adult-like, others have the underparts mainly white, with only a few dark feathers (individually variable).

Eliminating Common and Arctic Terns Whiskered in non-breeding plumage is easily identified by its typical *Chlidonias* shape and pale grey upperparts. Common and Arctic Terns of all ages have a longer, more deeply forked tail and long wings that, in the case of Common, produce a more languid flight action. For plumage and other details, see pp. 242–249.

American Black Tern *Chlidonias niger surinamensis*

First recorded in Britain in 1999, this potential split has now occurred here five times (1999–2012) as well as four times in Ireland (to 2007). It is likely to have been overlooked previously. All British records have related to late autumn juveniles which can be distinguished from juvenile Black Tern by (1) extensive grey on the breast-sides and flanks; (2) greyish underwing-coverts (as opposed to white); (3) darker and more uniform upperparts (including the rump) and (4) a paler grey crown (with white flecks) contrasting with black ear-coverts (Andrews *et al.* 2006). Any juvenile Black Tern showing a combination of such characters should be photographed and expert advice sought.

References Andrews *et al.* (2006), Vinicombe (1980).

Common, Arctic and Roseate Terns

Where and when Common Tern is a widespread coastal and inland-breeding species, with a protracted spring and autumn migration; in summer, non-breeders may occur far from traditional nesting areas. Arctic Terns predominate in n. Britain, passing through southern areas mainly in late April and early May and again from late July to October (with a peak in September). It is more pelagic than Common, with the bulk of the passage apparently in the

Atlantic; it would appear that relatively few use the English Channel as a migration route. It is, however, frequent inland on passage, although the largest numbers tend to follow westerly gales. Otherwise, only in late September and October (when most are juveniles) do Arctics usually outnumber Commons in southern areas. Roseate Tern currently breeds mainly on Coquet Island, Northumberland (80 pairs in 2010) with a few pairs scattered elsewhere. A much larger population occurs in Ireland, on Rockabill Island, Co. Dublin (1,080 pairs in 2011). Significant post-breeding gatherings may occur near the breeding colonies, otherwise, very small numbers are recorded around the coast on migration, mainly in eastern and southern counties; it is exceedingly rare inland.

Common *Sterna hirundo* and Arctic Terns *S. paradisaea*

Approach to identification Because of identification difficulties, many observers lump Common and Arctic as 'Commic' Terns but, with the exception of distant birds, the two species are readily separable, although thorough practice is essential. It must be stressed that in most of England and Wales, Common Tern is *far more numerous* than Arctic, both as a breeding bird and on passage, and it should not be assumed that both species occur in equal numbers.

Structure At rest, Common is slightly but distinctly larger and sturdier than Arctic, with a longer bill, neck and legs; in flight, adults are distinctly longer-winged and proportionately shorter-tailed. At rest, Arctic is more delicate and in many ways its structure is reminiscent of a marsh tern *Chlidonias*. Note that resting birds appear neckless, a rounded head merging seamlessly with the body; they also have a shorter bill and *noticeably short legs*. In flight, Arctic has shorter, more slender wings, a consequence of a shorter 'arm' and narrower, more pointed primaries; it also has a strikingly long tail. It is useful to remember that Arctic's overall shape is reminiscent of a Swallow *Hirundo rustica*, with a similar 'short-headed/narrow-winged/long-tailed' shape. From below, its wings often form an inverted W shape, with the 'arm' pushed forward and the sharply pointed primaries angled back, more so than Common. Although lacking long tail-streamers, juvenile Arctic is nevertheless distinctly long-tailed compared to the more evenly proportioned juvenile Common. Note that recently fledged juveniles of both species fly before their outer primaries are fully grown, so young juveniles close to their breeding sites may have shorter, more rounded wings than adults (but their wings are fully grown by the time they migrate).

Flight Common Tern's long wings produce an easy, languid flight action, and the wings usually look quite bowed when head-on. Arctic has slightly but distinctly shallower, quicker, wingbeats. When surface-feeding inland, Commons have a steady flapping flight, continually dipping to the water's surface, whereas Arctics tend to fly higher (maybe 3–5m), stall, and then stoop steeply to the surface; however, both also patrol lower over the water, rapidly picking off insects with a downward head movement. Note that migrating birds of both species frequently fly low and fast over the sea, often in large groups, with angled-back wings and 'whippy' wingbeats; such birds are often impossible to identify, especially at a distance (see 'Distant birds' p. 246).

Plumage *Adult summer* Pay particular attention to the following features. **1** UPPER PRIMARIES Common moults its inner primaries twice a year (in late summer and late winter) but the outers are moulted only once (in early winter). Consequently, in Europe there

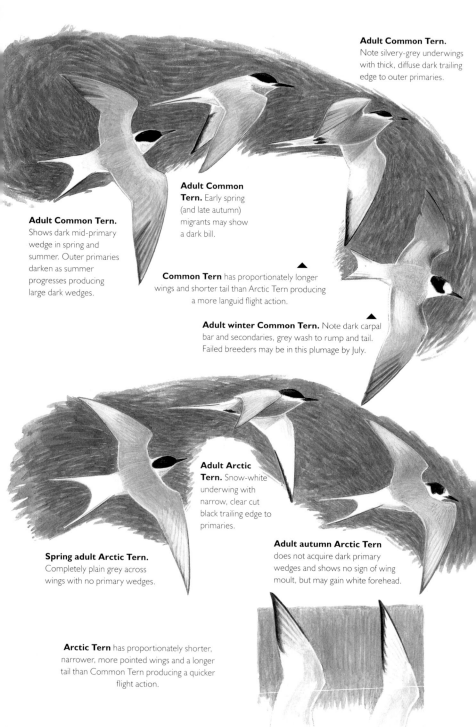

Adult Common Tern. Note silvery-grey underwings with thick, diffuse dark trailing edge to outer primaries.

Adult Common Tern. Early spring (and late autumn) migrants may show a dark bill.

Adult Common Tern. Shows dark mid-primary wedge in spring and summer. Outer primaries darken as summer progresses producing large dark wedges.

Common Tern has proportionately longer wings and shorter tail than Arctic Tern producing a more languid flight action.

Adult winter Common Tern. Note dark carpal bar and secondaries, grey wash to rump and tail. Failed breeders may be in this plumage by July.

Adult Arctic Tern. Snow-white underwing with narrow, clear cut black trailing edge to primaries.

Spring adult Arctic Tern. Completely plain grey across wings with no primary wedges.

Adult autumn Arctic Tern does not acquire dark primary wedges and shows no sign of wing moult, but may gain white forehead.

Arctic Tern has proportionately shorter, narrower, more pointed wings and a longer tail than Common Tern producing a quicker flight action.

Note differences in wing shape and pattern of black. Common Tern (left) and Arctic Tern (right).

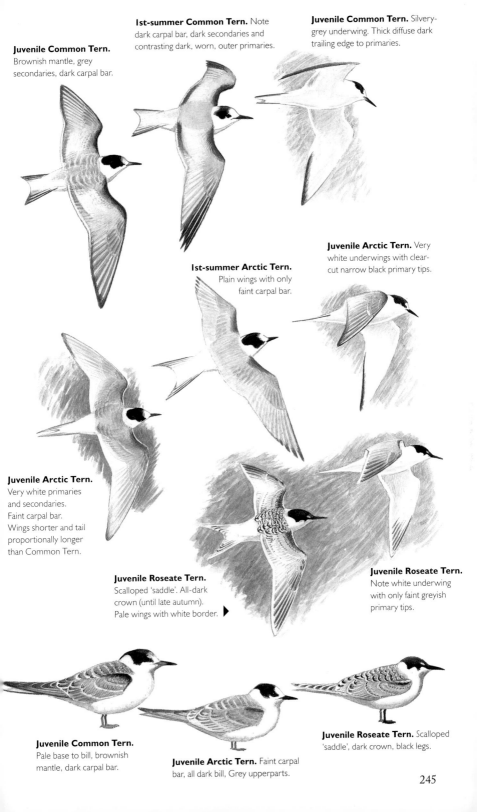

Juvenile Common Tern. Brownish mantle, grey secondaries, dark carpal bar.

1st-summer Common Tern. Note dark carpal bar, dark secondaries and contrasting dark, worn, outer primaries.

Juvenile Common Tern. Silvery-grey underwing. Thick diffuse dark trailing edge to primaries.

1st-summer Arctic Tern. Plain wings with only faint carpal bar.

Juvenile Arctic Tern. Very white underwings with clear-cut narrow black primary tips.

Juvenile Arctic Tern. Very white primaries and secondaries. Faint carpal bar. Wings shorter and tail proportionally longer than Common Tern.

Juvenile Roseate Tern. Scalloped 'saddle'. All-dark crown (until late autumn). Pale wings with white border. ▶

Juvenile Roseate Tern. Note white underwing with only faint greyish primary tips.

Juvenile Common Tern. Pale base to bill, brownish mantle, dark carpal bar.

Juvenile Arctic Tern. Faint carpal bar, all dark bill, Grey upperparts.

Juvenile Roseate Tern. Scalloped 'saddle', dark crown, black legs.

is usually a contrast between the new inner primaries and the old outers. When new, the feathers have a pale grey bloom (radii) that is steadily lost with wear, revealing progressively more of a blackish base colour (rami). Consequently, the older outer primaries are darker than the newer inners, producing a large dark wedge that is obvious at some distance. However, it is important to remember that the contrast is most obvious in late summer and autumn, and least obvious in spring, particularly at a distance. It must be stressed that, in spring and summer *many lack a dark wedge* while many others show only a small one confined to the *middle primaries* and this can be difficult to detect at any distance; as a consequence, such birds are routinely misidentified as Arctics (see 'Distant birds' below). Arctic Tern has a *complete* wing moult once a year (in late winter, prior to spring migration) so the primaries are always similarly aged. Consequently, they always appear uniformly pale grey throughout the species' stay in the Northern Hemisphere, never showing Common Tern's dark primary wedges. **2 UNDERWINGS** Common has rather *silvery-grey underwings with a broad, diffuse dark grey trailing edge to the under-primaries* (visible at some distance); also, when viewed against the light from below, only the inner primaries appear translucent. Arctic has *very white underwings* (recalling Kittiwake *Rissa tridactyla*) *with a narrow, neat, clear-cut black trailing edge to the under-primaries.* When viewed against the light, *the whole of the under-primaries appear translucent.* **3 BILL** Bright orange-red with a black tip on summer Common, the red being obvious at a considerable distance. In contrast, Arctic has a shorter, stubbier and darker blood-red bill. It must be stressed, however, that both have black bills in winter, and some spring and autumn Commons, as well as second-summers, show a black bill (or dark red with a black tip) as well as a white forehead. Similarly, some spring Arctics have dark red merging into an inconspicuous black tip. Nevertheless, any 'Commic' Tern with a long, bright orange-red bill with clear-cut black tip is safely identifiable as a Common, but it is not safe to identify darker-billed individuals without reference to other characters. **4 UNDERPARTS** Spring Arctic tends to have darker grey underparts, but some Commons are also quite dark (individual variation and vagaries in the light make this a difficult character to evaluate). **5 HEAD** When perched, the black on the head of adult Common extends further down the nape, whereas the shorter-necked Arctic has a more rounded black 'skull cap'. **6 DISTANT BIRDS** It must be stressed that distant birds, particularly in spring, need to be identified with caution as the pro-Common features, such as the dark primary wedges, may be very difficult to see or may even be absent. The best way to separate such birds in spring is by a combination of shape, underwing colour and the pattern of the under-primary tips, but this requires practice. Distant spring migrants on sea watches are particularly problematic and are usually best logged as 'Commic Terns' (see also 'Flight' p. 243). **Juveniles** Easily separated, even at a distance. In southern areas, the first migrant juvenile Arctics (mainly late August to October) really stand out from the more familiar juvenile Commons, which pass through from mid July onwards. As well as the structural differences outlined above, the following plumage features are significant. **1 WINGS** On the upperwing, Common has a thick, blackish carpal bar, a grey secondary bar (narrowly tipped white) and grey primaries. In contrast, Arctic lacks an *obvious* carpal bar but, most importantly, it has essentially white upper-secondaries and paler, whiter primaries; therefore, *the entire rear wing often looks strikingly white, even at a distance.* This impression of whiteness is further emphasised by the 'Kittiwake-white' underwings, with a *narrow, clear-cut black trailing edge to the primaries.* When viewed against the light, the underwings of juvenile Arctic often appear strikingly translucent. **2** UPPERPARTS Common shows a distinct ginger-buff tone to the upperparts, with narrow dark scalloping, both of which soon fade. Young Arctics show fainter pale buff tones to the upperparts and delicate dark scalloping, but these too also fade, so that autumn Arctics have pale grey upperparts that appear much 'cleaner' than those of Common, forming a contrast with the much whiter primaries and, particularly, secondaries. Common also has a greyer rump. **3** BILL Common usually shows more orange at the base of the bill (bill all black on Arctic, although recently fledged juveniles can show some red). ***Adult winter*** Structural and underwing differences remain the same as in summer. Certainly by August, and sometimes even in July, adult-type Commons may show signs of winter plumage; indeed, some attain full winter plumage by time they head south. As well as acquiring a dark bill and a white forehead, winter adults have a dark carpal bar, traces of a dark secondary bar, a grey tail and grey-washed rump. Moulting individuals often look rather worn and tatty. Winter-plumaged Commons may confuse beginners and suggest marsh terns, particularly on inland fresh water where, like marsh terns, they feed mostly by surface-picking, rather than plunge-diving (several erroneous claims of Whiskered Terns *C. hybrida* have proved to be winter adult, or even juvenile Commons). Before leaving our shores, adult Arctics may acquire a black bill and white forehead but otherwise show very little sign of moult until their arrival in the winter quarters; consequently, any 'Commic' Tern showing signs of wing moult in autumn should be a Common. ***Immatures*** Unlike gulls, immature Common Terns do not show clear-cut age related plumages during the breeding season, so their accurate ageing is not usually possible (White and Kehoe 2001, upon which much of the following is based). The vast majority of first-summer Common and Arctic Terns remain in their winter quarters but a few head north. First-summer Common is similar to winter adult and individuals undergoing active wing moult may have worn one-year old dark outer primaries that contrast with fresh grey inners; it is only these moulting birds that can be accurately aged. This is because many first-summers have already completed their moult and replaced all their juvenile primaries, rendering them indistinguishable from second-summers. The great majority of second-summer Commons return to the breeding grounds but most are indistinguishable from the adults (although some show white on the forehead and an all-dark bill). However, most adult-like second-summer Commons show little contrast in their primaries, with some showing evenly pale grey upper primaries with only a faintly darker grey inner primary wedge. Such birds may suggest Arctic Tern, particularly at a distance. Some immature Arctic Terns also return north in summer and most seem to have uniformly fresh, adult-like pale grey primaries and also new tail feathers (but they lack the very long streamers of adults). Otherwise, they are similar to winter adult, with a black bill, a large white forehead and a faint dark carpal bar (smaller and less obvious than on Common); maybe also with the odd grey feather in the underparts. Whether such birds are first-summer or second-summer is not clear but they lack the strongly patterned plumage of first-summer Common, appearing more uniform and more adult-like.

Calls Common's various calls include a muffled *kik*, a familiar rolling *keea-ya* and a characteristic, grumpy *keeyar*. Common's calls are fuller and throatier than those of Arctic, whose equivalent calls tend to be slightly thinner and less rasping. Juvenile Commons have a repetitive guttural begging call *sri-sri-sri-sri-sri…*, often rapidly alternated with the adult's answering *kik*. Begging juvenile Arctic gives a thin, high-pitched, musical *pi-pi-pi-pip*.

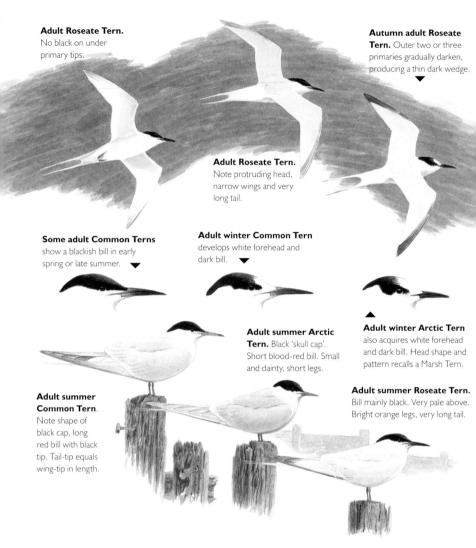

Adult Roseate Tern. No black on under primary tips.

Autumn adult Roseate Tern. Outer two or three primaries gradually darken, producing a thin dark wedge. ▼

Adult Roseate Tern. Note protruding head, narrow wings and very long tail.

Some adult Common Terns show a blackish bill in early spring or late summer. ▼

Adult winter Common Tern develops white forehead and dark bill. ▼

Adult summer Arctic Tern. Black 'skull cap'. Short blood-red bill. Small and dainty, short legs.

Adult winter Arctic Tern also acquires white forehead and dark bill. Head shape and pattern recalls a Marsh Tern. ▲

Adult summer Common Tern. Note shape of black cap, long red bill with black tip. Tail-tip equals wing-tip in length.

Adult summer Roseate Tern. Bill mainly black. Very pale above. Bright orange legs, very long tail.

Roseate Tern *Sterna dougallii*

Structure Slightly smaller and more delicate than Common, with shorter, narrower wings that appear more centrally placed on the body. Extremely long tail-streamers (even longer than Arctic's) which may quiver up and down in flight. The head and neck project slightly further than the other two species.

Flight Stiffer flight action than Common. When feeding, it may fly around in tight circles with quick, fluttery wingbeats, followed by a steep dive towards the water.

Plumage *Adult summer* Remember that Roseate shares many plumage characters with the larger Sandwich Tern *Thalasseus sandvicensis*. Like that species, the upperparts are very white, but a key feature is that, unlike Common and Arctic, Roseate *lacks* a dark border to the under-primaries, showing at best only a slight darkening (difficult to see in the field). Like Sandwich, the underparts show a delicate rose-pink flush in summer. Like Common, it acquires dark blackish primary wedges as summer advances, these contrasting conspicuously with the very pale upperwings. Any very white '*Commic*'-*like tern that shows blackish primary wedges should prove to be a Roseate* (but beware of confusion with the much larger Sandwich Tern). The bill is slender and black (usually with variable amounts of red at the base) but remember that both Common and Arctic can show black bills, even in midsummer. Other differences include: (1) the cap tends to go straight back from the bill at rest (rather than curving up behind the eye like Common) and (2) the feet and longer legs are extremely bright orange (recalling Puffin *Fratercula arctica*). They may acquire a white forehead before migrating in autumn. *Juvenile* Easily separated from Common and Arctic, and bears an uncanny resemblance to a miniature juvenile Sandwich Tern. It has a completely dark head with an all-dark forehead (but the crown is sometimes slightly paler and some start to acquire a white forehead by late September). The mantle and scapulars are strongly scalloped with dark brown, forming a distinctive 'saddle' in flight. The wings are similar to those of juvenile Common, with a variable dark carpal bar, but it lacks the obvious dark tips to the under-primaries. Also, unlike Common and Arctic, the legs are black. *First-summer* Usually remains in the winter quarters, so very rare in Britain. It resembles adult (also shows dark outer primaries) but has a white forehead, an obvious dark carpal bar, a black bill and blackish legs.

Calls Adult has a diagnostic, high-pitched, shrill, slightly rasping *tch-wit* or *tchu-weet*, in quality somewhere between Sandwich Tern and Spotted Redshank *Tringa erythropus*. Also, an equally characteristic harsh, rasping *zraak*. Begging calls of juvenile are a thin, squeaky *zi-zi-zi-zi-zi-zi...*; later calls are similar to the adult's but shriller and squeakier.

References Scott & Grant (1969), Grant & Scott (1969), Grant *et al.* (1971), Hume & Grant (1974), Hume (1993), White & Kehoe (2001).

Auks

Where and when Guillemot and Razorbill breed on rocky coastlines around Britain and Ireland, but are mainly absent on the English coast from Lincolnshire to Hampshire; Guillemot is much more numerous. Both disperse widely in winter and can be encountered almost anywhere around the coast. Puffin has a more northerly breeding distribution, although there are significant colonies in Wales and Ireland; very small colonies persist along the English Channel. In winter it is markedly pelagic, with few seen from shore except during wrecks. Little Auk is a rare winter visitor, but thousands occasionally occur in late autumn in the North Sea during persistent northerly winds, when some may be blown well inland. Brünnich's Guillemot is a very rare winter visitor, not annual, mainly to Shetland and n. Scotland; most records relate to tideline corpses, live birds being extremely rare.

Guillemot *Uria aalge* and Razorbill *Alca torda*

Bill When separating Guillemot and Razorbill, pay particular attention to the bill. Guillemot's is slim and pointed, obvious at some distance, even in flight, giving it a distinctly pointed head/bill profile. Razorbill's diagnostic oblong bill is readily apparent in flight, giving it a characteristic blunt-headed impression; note, however, that first-winter Razorbill has a much smaller, less blunt bill than the adult, but it is nevertheless obviously short and thick compared to Guillemot (full bill size is not acquired until about three years old). White lines on the bill and face (vertical on bill and horizontal across the top of the lores) are noticeable in summer, even at a distance. In winter, adults basically show only the vertical white bill line (the white loral line is very faint); first-winters lack any white.

Size and structure Guillemot is slightly smaller and slimmer and, on the water, shows a more rounded head than Razorbill. The latter is sturdier and more thickset with a larger, heavier and squarer head. It also has a noticeably long, pointed tail (can be obvious when diving); at rest on the sea, the tail is cocked to give a more 'banana-like' shape than Guillemot, which has a shorter tail and a more truncated rear end. Structural differences are not obvious in flight, although Razorbill appears heavier at the front, mainly because of its large bill. Guillemot's legs trail behind its body whereas Razorbill's are cloaked by its longer pointed tail. This difference is, however, extremely difficult to discern at any distance.

Plumage *Adults summer* Guillemot's upperparts are obviously very brown, whereas Razorbill's are jet black, this difference easily separating them (providing light conditions are reasonable). Consequently, both the white trailing edge to the secondaries and the white sides to the rump are much more contrasting on Razorbill. Guillemot has variable dark streaking on the flanks, the smarter Razorbill having unmarked pure white flanks. ***Winter*** In winter plumage, Razorbill has extensive and solidly black ear-coverts, with a lobe of white extending up behind them, but this is often sullied with grey; thus, Razorbill has a distinctly capped appearance at any distance. Guillemot is much whiter-headed, the result of the dark brown in front of and around the eye curving up strongly behind to produce an extensive area of white on the rear of the head; in addition, a noticeable dark line curves back from the eye, into the white. Both show a partial black collar on the neck-sides, sometimes extending with varying intensity right around the foreneck, most frequently on Guillemot, particularly northern birds (Mather 1991). *'Northern and Bridled Guillemots'* Guillemots gradually become slightly larger and darker towards the Arctic, such birds looking black in the field and thus more similar to Razorbill. These birds are unlikely to be seen in southern areas. The incidence of 'bridled' Guillemots also increases towards the north. In summer plumage these birds have a narrow white eye-ring and a narrow white furrow running back from the eye; they are decidedly uncommon in the south of their range. *Juvenile to first-summer* Juveniles leave the nest sites before they are fully grown and are cared for at sea by the adults. Guillemots at this age are small, short-billed and fluffier than the adults and lack dark flank streaking; they can usually be detected by their persistent high-pitched begging calls. Juvenile Razorbills are similarly small, and have poorly developed stubby bills, lacking the adults' white markings. Juveniles of both vary, but Razorbills *tend* to be darker on the chin and throat, and seem far less vocal. Following a partial post-juvenile moult, both species resemble winter adults, but the bill remains less well developed (see 'Bill' above). First-winter Guillemots have the flanks less streaked than the adults and may have a larger area of white on the head-sides. Note that adult Guillemots and Razorbills moult early into summer plumage, so winter-plumaged individuals in late winter and spring are invariably first-winters. Even in their first summer, they may retain white feathering on the chin and throat (as can older individuals); Guillemots also have faded brown wings.

Diving action Note that all species of auk dive with a distinctive open-wing action (but do not use their legs to dive).

Razorbill's aerial display At their breeding sites, pairs of Razorbills (sometimes singles or trios) often indulge in a distinctive aerial display with slow-motion wingbeats with the wings held in a pronounced 'V' (strangely reminiscent of the display of Rock Dove/'Feral Pigeon' *Columba livia*).

Winter Guillemot. Slimmer body, pointed bill.

Winter Razorbill. Heavy body with blunt head and pointed tail. Plain underwing.

Winter Puffin. Dark underwing, triangular head.

Winter Little Auk. Starling-sized, dark underwing, rapid wingbeats.

Adult summer Puffin. All-dark rump, lacks white trailing edge to wing. White 'face'.

Summer Razorbill. Black above. Note pointed tail.

Winter Little Auk

Summer Guillemot. Southern birds are browner above than Razorbill.

Juvenile Guillemots and **Razorbills** go to sea when half grown, during July and August.

Winter Guillemot. Note head pattern, short tail, streaked flanks.

Winter Little Auk. Tiny. Note head pattern and small bill. Little Auks do not arrive in British waters until late autumn.

Winter Razorbill. Note extent of white on 'cheeks', plain flanks, pointed tail.

Juvenile Razorbills tend to be darker on the throat than Guillemots.

Winter Brünnich's Guillemot. Note thick bill, extensive black on 'face', plain flanks.

1st-winter Puffin. Smaller bill than adult, dusky face and rear flanks.

Puffin *Fratercula arctica*

Distinctive and well known, but distant individuals in flight, winter adults and young birds may be less easily identified.

Size and structure Its small size is usually apparent (about three-quarters that of Guillemot and Razorbill). In summer, the huge, brightly coloured, triangular bill and white face should be obvious; in winter, the bill is duller and the face greyer. Juveniles have a smaller, less deep and stubbier bill that is predominantly greyish. On the water, it looks stocky and large-headed, with a truncated rear end. Fluorescent orange legs and feet are obvious in summer, but are duller and yellower on juveniles and winter adults.

Flight identification Easily identified in summer, appearing small and stocky with a whitish face and a massive colourful bill, which produces a front-heavy appearance. The bill and face are darker in winter and the bill is smaller on juveniles. The body is otherwise rather stocky with a stubby rear end. Short, rounded wings are slightly angled back from the carpal. The flight action is rapid with rather flappy wingbeats and frequent shifts in body angle. The upperparts are black (like Razorbill) but it lacks Razorbill and Guillemot's prominent white sides to the rump, as well as their white trailing edge to the secondaries. The underwings are very dark, appearing black, but grey at close range.

Little Auk *Alle alle*

Size and structure A tiny auk, only two-thirds the size of a Puffin and half the size of a Razorbill or Guillemot; it is in fact similar in size to a Common Starling *Sturnus vulgaris*. On the water, it appears small, stubby, horizontal and neckless, with a small but thick black bill that produces a snub-nosed effect; it also has a short, cocked tail. It floats very low on the water, often dragging its wings between dives.

Plumage Most distinctive in winter is a prominent lobe of white extending from the throat up behind the black ear-coverts; also a narrow black half-collar around the neck-sides. It has three or four short white lines across the scapulars, usually visible at some distance. In summer plumage, the chin and throat are black and it has a narrow white half-circle over the eye (the latter sometimes present in winter). First-winters are duller and browner.

Flight identification Very distinctive, appearing small and dumpy – or 'short and stubby' – with a stubby bill. It has quite short, pointed, rather swept-back wings and rapid wingbeats. It can look almost wader-like at a distance. Very black and white and clean-looking, the black including the rump and tail. Even in flight, the head pattern is distinctive, with black extending below the eye, a lobe of white up behind the eye and a black half-collar from the upper mantle onto the neck-sides. Like Puffin, it has very dark underwings but, unlike that species, it has a white trailing edge to the secondaries.

Brünnich's Guillemot *Uria lomvia*

A very rare vagrant, which should not be identified unless all of the following differences are clearly established.

Bill Pay particular attention to the length and shape of the bill. It is noticeably thicker and stubbier than Common Guillemot's with the upper mandible downcurved towards the tip, while the lower mandible may show a slight gonydeal angle halfway along. To quantify this, the distance between the eye and the foremost extension of the feathering on the bill is twice that from the tip of feathering to the tip of the bill (about equal on Common Guillemot). Brünnich's has a narrow white stripe at the base of the upper mandible ('tomium stripe') that is obvious at close range but not at a distance. This can be faint or lacking on some whilst, conversely, some Common Guillemots may also show a suggestion of this feature. Also, beware of the effects of Common Guillemots carrying fish in the bill, especially when observed at a distance.

Size and structure Slightly larger and stockier than Common Guillemot; on cliffs it has a heavy, 'rugby ball' shape. On the water, it tends to hold its shorter and stubbier bill slightly more horizontally. It has a steeper forehead, with a stronger forehead peak. The tail is short, like Common Guillemot, but can be persistently cocked. Juvenile Common Guillemots and Razorbills have shorter bills, so particular care is required in late summer and early autumn (although Brünnich's is very unlikely to occur at this time).

Plumage Darker brown than Common Guillemot, but not as black as Razorbill. It lacks Common's flank streaking, the clean white flanks appearing much more similar to Razorbill's. Consequently, in summer it looks cleaner and more 'black and white' than Common. From the front, the white of the breast protrudes into the black of the upper neck in a sharper point than on most Common Guillemots. In flight, both the white trailing edge of the wing and the white rump-sides contrast more strongly. In winter, the most important plumage feature is that the dark of the head extends down to *include the whole of the ear-coverts*, so the black is much more extensive than on either Common Guillemot or Razorbill, forming something of a capped or hooded effect. In addition, on first-winters at least, the whole of the face is dingy grey and there is also a very thick black half-collar on the neck-sides. In consequence, the whole of the head appears darker than on Razorbill or Common Guillemot. Note also that it moults into and out of winter plumage much later than Common Guillemot, so any dark-headed guillemot in late autumn/early winter or any pale-headed guillemot in spring is worth a second look.

References van Duivendijk (2011), Grant (1981), Mather (1991).

Pigeons and doves

Where and when Woodpigeon is now abundant, having spread into towns and cities in recent decades. Stock Dove is locally common in the countryside (mainly in arable areas) but may breed in mature urban woodlands. Feral Rock Dove (or 'Feral Pigeon') is of course abundant in urban environments, but generally scarce in the countryside. Wild Rock Doves are restricted to rocky coasts of n. and w. Scotland and n., w. and s. Ireland. Collared Dove is common, mainly around habitation, but Turtle Dove is now a very scarce summer visitor from late April to October. It has declined by about 93% since 1970 and is increasingly confined to eastern counties, having disappeared from most of its former range. It is widely predicted to become extinct as a British breeding bird. Collared is associated with suburban gardens and farmyards but Turtle Dove is most likely to be encountered in open, arable farmland or in bushy places.

Woodpigeon *Columba palumbus*

A large pigeon, easily identified by a conspicuous thick white line cutting across the middle of the open wing, and visible at the bend of the wing at rest; it also has a conspicuous white patch on the neck-sides (lacking on juveniles, which are slightly browner than adults and have a dark eye, like Stock Dove). In flight, easily identified by its shape: a small head, full breast, rather swept-back, pointed wings and a fairly long tail; from below, the underwings are grey and the tail is crossed by a pale band. When flushed, it takes off with a loud clatter of wings, often bursting violently from cover. In display, it flies steeply upwards, claps its wings and glides downwards. The song is a well-known, lazy, *coo-COOO coo-coo coo*, peculiarly evocative of balmy summer days.

Stock Dove *Columba oenas*

In winter, often associates with Woodpigeons. In flight, easily identified (with practice) by shape alone: smaller than Woodpigeon and more evenly proportioned, being stocky and compact, with shorter, stiffer, more triangular wings and a shorter tail. They often fly in pairs, one behind the other, even in winter. It lacks white in the plumage but note the broad black border to the end and rear of the upperwing and also the dark grey underwings (silvery-white on Rock Dove and on most grey Feral Pigeons). On the ground, it shows a brightly coloured bill and cere: red with a yellow or whitish tip (*cf.* grey bill and whitish cere of Rock Dove). However, juvenile Stock Dove has a dull greyish or brownish bill. It also has an emerald-green neck patch (lacking on juvenile), a double black bar on the inner greater coverts and tertials (visible at rest). The eye is dark (white on adult Woodpigeon). The male's display flight is straight, with slow, deep wingbeats before clapping its wings over the back and then gliding with the wings slightly above the horizontal. The song is a distinctive deep moaning *ooo-ah*, sometimes given about ten times in accelerating sequence.

Rock Dove *Columba livia*

Wild Rock Doves are attractive, being pale grey above with two thick black bars across the tertials/greater coverts and median coverts, darker grey head and underparts and a white rump. In flight, dark underparts contrast markedly with silvery-white underwings (grey on Stock Dove and Woodpigeon). Grey bill and whitish cere (*cf.* adult Stock). Feral pigeons are highly variable: many resemble their wild ancestors but many are chequered or irregularly patterned. The plumage varies from grey to blackish, white or brown. Adults usually have red or orange eyes (dark on Stock Dove). Shape is important when separating Rock Dove from both Woodpigeon and Stock Dove in flight. It is intermediate between the two and a narrowly pointed, protruding head and swept-back, pointed wings easily separate it from the stockier, more compact Stock Dove. It often glides on V-shaped wings. Its display flight consists of exaggerated slow, deep wingbeats followed by loud wing-claps and a long glide on V-shaped wings. The song is a moaning *oo-oo-oor*. It inhabits cliffs and buildings, and usually ignores trees (although in city centres it often uses thicker branches as a daytime roost).

Collared *Streptopelia decaocto* and Turtle Doves *S. turtur*

Collared is a familiar pale, sandy dove with pale underwings and a long tail, the distal half of the underside being white. In flight it is easily separated from Turtle Dove by proportionately

Woodpigeon in flight shows diagnostic wing-bars.

Stock Dove. Thick black trailing edge to wing.

Juvenile Woodpigeon. Browner-buff than adult and lacks the white neck patch.

Adult Woodpigeon. Long tail, heavy body, small head. Grey-brown above, with white in wing and neck. Note pale eye.

Stock Dove. Short, stocky with shortish tail. Deep blue-grey without patches. Eye dark.

Stock Dove. Dark underwing.

Woodpigeon flies with head held up.

Rock Doves. Narrow-winged, dark head, small white rump and silvery-white underwing.

Rock Dove. Thin black bill, pale grey upperparts, dark head and neck. Note double wing-bar.

Feral Rock Doves. Variable, some grey types shown. Bill thicker than Rock Dove.

Turtle Dove. Note undertail pattern. Collared Dove (left) shows more white.

Collared Doves are pale fawn with rounder wings, longer tail and paler underwing than Turtle Dove. Turtle Dove appears darker with rapid flight.

Collared Dove is plain above, Turtle Dove is patterned.

Turtle Dove

Collared Dove

shorter, more rounded wings that produce a characteristic lolloping flight action. Turtle Dove has dark grey underwings, a contrasting pale belly and a narrower white tip to the tail; it is smaller than Collared Dove, with a shorter, more tapered tail and its swept-back, fairly pointed wings produce a whipping flight action. Differences at rest are obvious, Turtle Dove being easily identified by the rufous tortoise-shell pattern of the upperparts, grey head and vinous breast. Juvenile Turtle Dove is much duller and drabber than the adult, lacking rich rufous upperparts, the grey crown, vinous breast and noticeable neck-patch. Juvenile Collared is also duller than the adult, lacking the half-collar, and it has pale feather fringes on the upperparts. Songs are completely different: Collared has a familiar rather deep cooing *oo-OOO oo*, whereas Turtle gives a soft, purring *turrrr turrrr*. Collared also gives a harsh, nasal 'excitement' call. In display, both species fly up and then glide down on spread wings.

Oriental Turtle Dove *Streptopelia orientalis*

Turtle Doves are extremely rare in winter and any seen at this season should be thoroughly checked for vagrant Oriental Turtle Dove (ten records to 2011). Wintering individuals of both species are likely to turn up in gardens and anyone finding such a bird should endeavour to obtain photographs and seek expert help. There are two races. Nominate *orientalis*, from central Siberia and SE Asia, is significantly larger and heavier than Turtle Dove, with a shorter tail and proportionately shorter, more rounded wings. The latter produce a more ponderous flight, lacking the 'whippy' effect of Turtle Dove. Adults appear darker, purplish-brown on the head and underparts, lacking Turtle Dove's contrasting whitish belly, vent and undertail-coverts, and they have a dark blue-grey rump. Juvenile/first-winters are pinky-brown on the head and breast, with perhaps a tint of purple. The feathers on the upperparts of adults have *rounded* dark grey centres with rufous fringes, forming a strongly scalloped pattern (feather centres narrower and pointed on Turtle Dove). The equivalent juvenile feathers, some of which may persist into winter, are brown with narrower buff fringes. Whitish tips to the median and greater coverts form subtle wing-bars. Other features include: (1) Oriental lacks Turtle Dove's area of bare skin around the eye; (2) the neck-patch consists of four to six narrow black-and-greyish lines (fewer and thicker black-and-white lines on Turtle Dove) and (3) the tip of the tail is greyer. The w. Siberian race *meena* is smaller than *orientalis*, whiter on the vent and undertail-coverts, and has white tips to the tail feathers.

Long-eared and Short-eared Owls

Where and when As a breeder, Long-eared Owl *Asio otus* is widespread but thinly distributed throughout Britain, usually inhabiting coniferous woodland (but sometimes even nesting in hedgerows); in Ireland, it is the commonest owl and, in the absence of competition from Tawny Owl *Strix aluco*, is found more frequently in broadleaved woodland. Short-eared Owl *A. flammeus* breeds in open country, mainly in Scotland, Wales, and n. and e. England. Both species are more widespread in winter, when Short-eared Owls move into lowland areas; variable numbers of Continental immigrants swell the numbers of both species, October being the best time to encounter newly arrived migrants.

Habitat and behaviour Such differences are generally the first clue to identification. Short-eared is frequently seen in broad daylight, particularly on winter afternoons, hunting over rough ground; it usually roosts on the ground in rough vegetation. High-flying Short-eareds (e.g. migrating birds) are very distinctive, flying with the wings 'pushed forward' on deep, flappy wingbeats, but gliding with the wings held in a shallow V. Long-eared is more or less strictly nocturnal, not usually hunting in broad daylight; it typically roosts in conifers, thick hedgerows or bushes, often at some height; when flushed, it will fly off silently through the trees and quickly disappear (often roosts communally in favoured areas). The biggest identification problem arises with migrants, which can appear at coastal migration spots at any time of day. In such circumstances, Long-eared may settle on the ground when trees or bushes are unavailable. Conversely, Short-eareds occasionally roost in trees, even at some height.

Identification at rest Long-eared is a slim, upright owl, usually found in a conifer or thick bush, when often quite approachable. Easily identified by its long 'ear-tufts' and thick vertical dark lines running down the face to the bill; it also has a pale eyebrow and a rather orangey facial shield. The eyes are brilliant orange (best seen when flashed open if startled) and the underparts are heavily streaked and barred down to the belly (some are less heavily streaked). Short-eared is generally paler and less upright, usually seen sitting on the ground or on a fence post. It has short 'ear-tufts' and yellow (or orangey-yellow) eyes which, unlike Long-eared, are surrounded by circular, panda-like black patches; also important is that the underparts streaking is confined mainly to the upper breast.

Identification in flight *General plumage tones* Both species are very variable, but Long-eared is typically darker and browner than the sandier, buffier Short-eared. Because of its darker plumage tones, Long-eared usually appears a more uniform, less contrasting bird. Some Short-eareds are, however, darker than others, while some Long-eareds are particularly pale, almost whitish-buff on the underparts. **1** UPPERWING Both show a pale patch on the upper-primaries. On Long-eared this is a darker rich orangey-buff to rusty colour whereas, on Short-eared, it is paler, sandy-buff to almost white, and therefore more obvious. The upperwing of Short-eared is generally paler than Long-eared's, the upperwing-coverts being more mottled than streaked, giving a less uniform appearance. The paler coloration of Short-eared also means that the dark carpal patches are more obvious than on Long-eared, while the primary and secondary barring also appears stronger and more contrasting (on Short-eared, the primary barring continues onto the secondaries whereas, on Long-eared, barred primaries give way to *finely* barred secondaries). Short-eared has a pale trailing edge to the wing (but on particularly pale Short-eareds this can be surprisingly difficult to see). Long-eared usually lacks a white trailing edge but some individuals can in fact show this feature, but it is never as clear-cut as on Short-eared. **2** UNDERWING Both have pale underwings with a dark carpal patch. Long-eared's primaries are crossed by three to five dark bars whereas on Short-eared the tips of the outer primaries appear more solidly black. **3** TAIL On Long-eared, the upper tail is generally more closely barred on a darker background, producing a more uniform appearance than on Short-eared, which has four or five prominent bars on a paler, sandier background. **4** UNDERPARTS Heavy streaking is confined mainly to the upper breast on Short-eared creating a strong breast-band (although it does extend more finely onto the belly); on Long-eared, heavy streaking extends onto the belly, creating an overall darker, more uniform impression (although on some the streaking is distinctly weaker lower

Underwing tip of **Long-eared Owl** is more finely barred. The body is usually heavily streaked.

Short-eared Owl. More solidly dark on primary tips. Streaking mainly confined to upper breast.

Short-eared Owl. Generally paler and sandier than Long-eared Owl, with paler primary patch. Note white trailing edge to wing (sometimes inconspicuous), coarse blotching on upperparts, and strongly barred tail. Wing-tips less rounded.

Short-eared Owls appear pale-bellie **Long-eared Owls** are rather dark bel

Long-eared Owl. Broader, more rounded wings, primary patch orangier, upperparts patterning less distinct.

Long-eared Owl usually roosts in thick cover above ground; newly arrived migrants may roost in more exposed positions.

Long-eared Owls roost communally in thick bushe Note differing facial expressions and orange eyes.

Short-eared Owl. Yellow eyes surrounded by black. Usually roosts on ground.

258

down). **5 STRUCTURE** Long-eared has shorter and broader wings than Short-eared, with the tips broader and more rounded; it is also proportionately shorter-tailed. Short-eared's wing is narrower and more tapered, emphasised by the fact that it is held more forward than Long-eared's.

Calls Short-eared's most familiar call, given by both sexes, is a hoarse, high-pitched bark, heard mainly on the breeding grounds. It also gives a low hissing screech, rising towards the end. Males have a wing-clapping display flight, as well as a hollow *boo-boo-boo-boo* advertising call (*BWP*). In early spring, male Long-eared gives a low-pitched, flat, dove-like cooing *oo* repeated at intervals of *c.* 2.5 seconds (*BWP*); it is rather quiet, but audible for some distance. It also gives a variety of other calls in the breeding season, including nasal buzzing, wheezing and barking noises. Begging young have a very distinctive, far-carrying, clear, mournful *eee-oo*, like the squeaking of a rusty hinge. This is heard from late May to July and is often the first indication of the presence of breeding birds.

References Davis & Prytherch (1976), Kemp (1982), Robertson (1982).

Common, Alpine and Pallid Swifts

Where and when Common Swift is a familiar and common summer visitor, mainly from late April to late August, with stragglers into October or even November. Alpine Swift is a rare but annual vagrant, mainly in spring, currently averaging 14 records a year (with a record 27 in 2002). First recorded in 1978, Pallid Swift was formerly a gross rarity, but it is currently averaging four records a year (with as many as 16 in 2004). Partly as a result of it being double-brooded, Pallid Swift is prone to periodic influxes during southerly winds in late October and early November. Similarly, there have been very early spring records (late March) with further occurrences right through the spring and summer. Whilst not all early or late swifts will prove to be Pallid, it is clearly worth checking such birds very carefully.

Aberrant Common Swifts *Apus apus*

Before identifying a vagrant swift, it is absolutely essential to eliminate the possibility of an aberrant Common Swift, examples of which are not infrequent. Such birds may show white feathering, those with white on the rump sometimes suggesting the very rare Little *A. affinis*, White-rumped *A. caffer* or Pacific Swifts *A. pacificus*. It is, therefore, essential to take very detailed notes of any swift with a white rump, concentrating in particular on structure, and to look for traces of white feathering elsewhere on the body. The aberrant individual illustrated on p. 260 is loosely based on a bird seen in Dorset in 1983. Aberrant pale, sandy Common Swifts have also been recorded, and these may suggest Pallid (see p. 260).

Alpine Swift *Apus melba*

When faced with a vagrant Alpine Swift, check two things to eliminate an aberrant Common Swift: **1** be sure to accurately evaluate its size (Alpine Swift is about 1½ times the size of a Common Swift) and **2** make sure that the upperparts are pale (they should be similar in tone

Common Swift

Alpine Swift. Very large. White belly and throat (although latter can be hard to see). ▶

Alpine Swift. Pale brown above.

Pallid Swift. Note dark mask, darker 'saddle', dark outer primaries and forewing contrasting with paler mid and hindwing. ▼

Pallid Swift. Paler than Swift, pale fringes to underparts, darker outer primaries, large white throat patch and slightly blunter wing-tips.

Partial albino Swift. Such individuals are not infrequent and can cause confusion with rarer species.

Common Swift

to those of a Sand Martin *Riparia riparia*). However, plumage tone varies according to the light, and Alpine Swifts can sometimes look as dark as Common Swifts, especially when high in the sky. The white belly is always conspicuous, but the white throat can be difficult to see. Long, slim wings produce slower, more deliberate wingbeats and create a rather languid flight action, with more gliding; this is very obvious even at a distance. At other times, its fast, purposeful flight may even suggest Hobby *Falco subbuteo*. Alpine Swift has a distinctive call: a long, slow, dry trill that tapers off at the end; however, this is unlikely to be heard away from its Continental breeding areas.

Pallid Swift *Apus pallidus*

Pallid Swift is a real 'Jekyll and Hyde' species: sometimes easy to identify, sometimes difficult to the point of being impossible. This is because its identification hinges on good views and the accurate assessment of subtle plumage tones. As this is entirely light-dependent, perfect light is required. Paradoxically, bright sunshine is one of the worst conditions, particularly if the bird is high in the sky. Pallid is best viewed in dull, flat light, preferably against a dark background. In such conditions, the following differences (based on Harvey 1981) should be carefully checked. **1** PLUMAGE TONE Pallid has distinctly paler, milkier plumage than Common Swift. **2** WING PATTERN Note particularly the dark outer primaries and leading edge of the underwing contrasting with the rest of the underwing, which is paler. This effect is mirrored on the upperwing, where the greater coverts and secondaries can stand out as being quite noticeably paler, even translucent when seen from below (although Common can show quite silvery upper-secondaries in certain light). **3** BODY COLOUR Its slightly darker body contrasts with the paler underwings, while the darker mantle produces the effect of a slightly darker 'saddle'. **4** FEATHER FRINGES Pale feather fringes on the underparts (particularly on the flanks) may be obvious at close range, forming pale scaling. **5** HEAD The forehead and throat are noticeably white, the throat being larger than on Common Swift; in consequence, the dark eye is more noticeable. **6** WING-TIP SHAPE On Common Swift, the outer primary is longest, producing a sharply pointed wing; on Pallid, the outer two primaries are the same length, producing a slightly blunter wing. This, combined with the slight broader base to the wing, produces somewhat more paddle-shaped wings than Common. Although subtle, this can be surprisingly obvious when looked for; conversely, at any distance they can appear to have pointed wings just like Common. **7** TAIL-FORK Slightly shallower on Pallid. **8** CALL Pallid gives a markedly disyllabic *cheeu-ic* or *cheeur-ic*, slightly lower-pitched than Common, with the second syllable slurred upwards. However, this is unlikely to be heard away from the breeding areas. Common Swifts can in any case also give disyllabic calls, even when feeding away from their breeding sites, but these tend to be not as *strongly* disyllabic as Pallid. Differences in their flight actions are similarly subtle but, whereas Common sweeps back its wings and has strong wingbeats, Pallid perhaps has a slightly more fluttery action with less of a tendency to 'whip back' its wings in normal flapping flight.

Pekinensis Common Swift

A further complication is provided by the paler Siberian race of Common Swift *pekinensis*, which apparently shares many of Pallid's features (large pale throat, pale scaling on the body, and pale and dark contrasts to the upperwing; Fraser *et al.* 2007). Although this form has not been recorded in Britain, it is a potential vagrant. However, as its separation from Pallid has still not been resolved, this subspecies remains something of an unknown quantity. Wing shape may be significant (more pointed, as nominate Common Swift).

References Fraser *et al.* (2007), Harvey (1981).

Red-backed, Daurian, Turkestan, Brown, Woodchat and Masked Shrikes

General characteristics of shrikes Shrikes are conspicuous birds, selecting prominent perches such as telephone wires, fences and the outermost branches of bushes. They survey the surrounding area, watching for prey like a raptor. They pounce on large insects, amphibians, small mammals and birds, often impaling larger victims on thorns. They can, however, be surprisingly secretive and difficult to find (usually because they are simply out of view). Their tails are very mobile, frequently raised, dipped, flicked, swished and wagged. Most species have a variety of harsh, rasping, screeching or chacking calls.

Red-backed Shrike *Lanius collurio*

Where and when Formerly a common summer visitor, Red-backed Shrike became extinct as a regular British breeding bird in 1988. Since then, breeding has been erratic, mostly in Scotland. It remains a scarce late spring and autumn migrant (mid May to June and mid August to October, occasionally into November) mostly on the east coast and in the Northern Isles. It currently averages 200 records a year.

Plumage *Adult male* With its grey head, black facial mask, reddish back and white bases to the outer tail, the adult male is easily identified. *Adult female* Females are less distinctive, being plain rufous-brown or warm brown above, with a greyer crown, nape and rump; the underparts are heavily scaled on a whitish background. The head has whitish lores and supercilium, and a dark brown mask; they are, however, variable, some looking somewhat more 'male-like' than others. *Juvenile/first-winter* In full juvenile plumage, Red-backed Shrikes are rufous-brown above (often with a greyer nape and sometimes crown); the white underparts are lightly scalloped with grey, the upperparts heavily scalloped dark brown and buff (subdued with wear). They show an ill-defined whitish supercilium and dark brown ear-covert patch. Remarkably, their post-juvenile moult begins shortly after fledging, sometimes before their wing and tail feathers are fully grown (*BWP*). This is partial and highly variable, involving the body feathers and some wing-coverts. Consequently, by the time they migrate, their plumage is a variable mixture of scalloped juvenile and plain rufous-brown first-winter upperpart feathering. Like adult females, they often show a strong grey tint to the nape. Note that some females and immatures have a distinctly redder tail, which may suggest Turkestan Shrike (see p. 264).

The 'Isabelline Shrikes'

Where and when 'Isabelline Shrikes' have been almost annual in Britain since the late 1970s (89 records by 2011 with as many as nine in 2011). Two forms occur here: Daurian Shrike and Turkestan Shrike. Despite the fact that it breeds further east, Daurian Shrike is the more frequent (perhaps outnumbering Turkestan by 3:1). Surprisingly, there have been records from late February through to early December, but the peak is from mid September to late November. It is likely that Turkestan will prove to be an earlier autumn vagrant than Isabelline. About one-quarter of records have related to adult or adult-like birds (Slack 2009); oddly, this is a trend seen with other vagrant shrikes.

Taxonomy The taxonomy of the Isabelline Shrike complex seems to have been in a state of continual flux. Formerly regarded as conspecific with Red-backed Shrike, it was finally split from that species in 1980. In recent years there have been moves to split it further, into Turkestan Shrike (sometimes referred to as 'Red-tailed Shrike') which breeds in west-central Asia, and Daurian Shrike (sometimes confusingly referred to as 'Isabelline Shrike') which breeds further east in Mongolia and w. China. Panov (2009) explained that, as a result of their spatial separation and differences in the timing of their respective breeding periods, the two taxa do not normally interbreed in the Tien Shan Mountains, where their ranges come close. He recommended treating Turkestan Shrike as a monotypic species and Daurian Shrike as a polytypic species, containing the subspecies *isabellinus*, *tsaidamensis* and *speculigerus*. However, Pearson *et al.* (2012) demonstrated that the type specimens of *isabellinus* are compatible with populations breeding in Mongolia (i.e. *specu-*

ligerus); thus *speculigerus* is a synonym of *isabellinus* and the name *arenarius* should be used for those breeders in the Tarim Basin. However, the two forms currently remain lumped by the British Ornithologists' Union Records Committee, which has yet to accept *isabellinus* onto the British List. It must be stressed that the identification of the Isabelline Shrike taxa is far from straightforward, not least because hybrids are known to occur between the various forms, so a cautious approach is recommended. Inevitably, the following text cannot possibly cover all of the complications and subtleties, so anyone discovering one of these birds should endeavour to obtain good-quality images and consult detailed texts.

Daurian Shrike *Lanius isabellinus*

Plumage and structure Similar in size and structure to Red-backed, but the tail is marginally longer. Appears uniformly pallid with relatively little contrast between the upperparts and underparts. Most distinctive is a dark orange rump, uppertail-coverts and tail, suggesting a pale oversized female Common Redstart *Phoenicurus phoenicurus*. Its plumage shows soft buff, peach and orange tones, creating a warmer impression than Turkestan Shrike. Also, unlike that species, the sexes are similar. The primary projection is distinctly shorter than Red-backed Shrike's, being about half to two-thirds the tertial length, rather than equal. *Adult male* Plain sandy grey-brown upperparts that are both paler and warmer in tone than Turkestan Shrike. The underparts are strongly washed with soft peachy-buff, becoming warm orangey-buff on the flanks (Turkestan Shrike is much whiter below). There is a strong black facial mask that does not normally extend over the forehead (the pattern of black across the lores is variable). It has a faintly paler supercilium, lacking the prominent white supercilium of male Turkestan Shrike, as well as its rufous coloration on the forehead and crown. Instead, the forehead, crown and ill-defined supercilium are strongly tinged warm peachy-orange. There is a small white patch at the base of the primaries. The bill has a pale pink or greyish base. *Adult female* Similar to male, but has pale lores, often a browner ear-covert patch and, sometimes, subdued scalloping on both the crown and the orangey-buff underparts. It may or may not show a white primary patch. The bill has a pinkish base. *Juvenile/first-winter* Unlike juvenile/first-winter Red-backed, the upperparts are uniform and unmarked, varying slightly from sandy grey-brown through pale fulvous-brown to pale cinnamon-buff. Little contrast with the underparts, which are pale whitish-buff (maybe with an orange wash to the flanks); some are slightly greyer below, others pale peach. Faint scalloping on the sides of the neck and breast, extending down the flanks, is the only patterning on the body plumage (with the possible exception of limited faint scalloping on the crown and back). The pale plumage contrasts strongly with the strikingly orange tail, uppertail-coverts and rump (variously dull orange, cinnamon-orange or reddish-orange); even the undertail is orange (greyish on almost all Red-backed). The head is less strongly patterned than Red-backed with the dark beady eye standing out. The forehead, lores and supercilium are rather pale buff, the ear-covert patch slightly richer (maybe with a slight rufous tinge but some, presumably males, have dark brown or quite blackish ear-coverts). Other points include: (1) a distinct pale wing-panel (pale tertial and inner secondary fringes); (2) lacks white patch at the base of the primaries; (3) paler-based bill (dark tip usually with a pink component to the base) and (4) blackish legs, which contrast with pale underparts (usually blue-grey on Red-backed).

Juvenile/1st-winter Red-backed Shrike. Rufous-brown above with greyer nape.

Juvenile/1st-winter Red-backed Shrike. Rufous-brown, no white in wing.

Juvenile/1st-winter Woodchat Shrike. Whitish rump and white wing patches.

Juvenile/1st-winter Woodchat Shrike of the race *badius*. No white in primaries.

Juvenile/1st-winter Woodchat Shrike. Lacks rufous tones, has pale rump, scapulars and primary bases.

Juvenile/1st-winter Masked Shrike. Monochrome plumage, underparts rather plain.

Juvenile/1st-winter Woodchat Shrike of the race *badius*. Heavy bill, no white in primary bases.

Juvenile/1st-winter Masked Shrike. Delicate and long-tailed, lacks pale rump.

Adult Woodchat Shrike

Adult female Woodchat Shrikes. Birds of the race *badius* (lower bird) have heavier bill and lack white in the base of the primaries.

Adult Woodchat Shrike of the race *badius*. No white in wing.

Turkestan Shrike *Lanius phoenicuroides*

Plumage *Adult male* Although superficially similar to Daurian, adult male is readily separable by its *darker earth-brown mantle and scapulars* (with a subtle greyish tint). These contrast with the crown, rump, uppertail-coverts and tail, which are strongly chestnut or reddish-tinged (tail slightly darker than Daurian). The upperparts show strong contrast with the *silky-white underparts* (sometimes tinged cream or orange-buff on the flanks). Like Daurian, it has a black face mask but, unlike that form, this may extend narrowly above the base of the bill, which is usually completely black (or slightly greyer at the base). Most importantly, the black mask contrasts with *a very obvious white supercilium* (often pure white).

Adult female Red-backed Shrike. Rufous-brown with greyer nape and rump.

Juvenile/1st-winter Red-backed Shrike. Finely scalloped plumage, rufous-brown above, tail brown.

Juvenile Red-backed Shrike. Some have more rufous tail.

Adult female Daurian Shrike. Warm plumage tones above and below, weak 'mask'. Orange tail.

Juvenile/1st-winter Daurian Shrike. Pallid and plain above, apricot below. Brownish wings.

Female shrike with characteristics intermediate between Daurian and Turkestan.

Juvenile/1st-winter Turkestan Shrike. Plain earth-brown above, whitish below, dark wings. Chestnut-orange tail.

Juvenile/1st-winter Brown Shrike. Large-headed, short-winged and long thin tail.

Adult male Daurian Shrike

Adult female Turkestan Shrike. Earth-brown above, whiter below, darker wings. Small white patch at base of primaries.

Adult male Turkestan Shrike. Bold 'mask', earth-brown above, whitish below, white patch in wing.

Juvenile/1st-winter Brown Shrike. Note shorter outer tail feathers.

Adult female Brown Shrike. Like male but with fine scalloping on breast and flanks.

Adult male Brown Shrike. Russet-brown above, yellowy below, long thin tail.

265

It too shows a white patch at the base of the primaries. *Adult female* Similar to Daurian but upperparts are darker, more earth-brown (subtly tinged greyish) and the tail is duller. Also, the white underparts (often tinged buff on the flanks) show obvious scaling. Like the male, the supercilium is whiter and more prominent (lores are also pale). *Juvenile/first-winter* Although similar to Daurian, less pallid, appearing somewhat intermediate between Daurian and Red-backed. The crown, nape, back and scapulars are *a distinctly darker earth-brown*, with a faint grey wash. It usually has a stronger head pattern, with a stronger whitish supercilium and darker ear-coverts, which may be distinctly chestnut-toned, the former scaled with black. Like Daurian, it has strong pale buff fringes to the greater coverts and tertials (perhaps with slightly more contrast between the fringes and the brown feather bases). Also like Daurian, it lacks a white patch at the base of primaries. The tail is less orange than Daurian, being a *darker orange-red* (more similar in tone to Common Nightingale *Luscinia megarhynchos* than Common Redstart). The underparts are usually whiter (light buff when fresh) and scaled dark brown or black (maybe more so than Daurian). Unlike adult male, the bill base is pale pink. Like Daurian, the legs are black (usually blue-grey on Red-backed).

Separation from Red-backed The following should also be noted: (1) some Turkestan Shrikes approach less well-marked Red-backed in mantle colour and in the prominence of their underparts scaling; (2) a proportion of female and immature Red-backed show a decidedly rufous tail; (3) juvenile/first-winter Red-backed shows strong scalloping on the mantle and/or the scapulars and uppertail-coverts, whereas Turkestan Shrike is plainer (maybe some faint scallops remain); (4) *adult* female Red-backed is also plain above (but adults are unlikely to be seen in late autumn). If in doubt, both species can be aged by the fact that juveniles/first-winters have a blackish subterminal bar behind the pale tip of each greater covert, tertial and tail feather (much plainer on adult female).

Brown Shrike *Lanius cristatus*

Where and when Brown Shrike, which replaces Red-backed in Siberia, was first recorded in Britain in 1985, with a further 11 records in 2000–11. It was possibly overlooked in the past.
Structure Similar in size to Red-backed but structural differences are fundamental to its identification. It has a rather large head and, more importantly, a heavier bill, with a distinct upcurve to the lower mandible. Most significant, however, are the relative proportions of the wings and tail. The tail is distinctly long and the primaries short, creating an overall 'short-winged/long-tailed' structure that is somewhat reminiscent of a babbler (Timaliidae). Note in particular that the primary projection is only about one-third to two-thirds the length of the overlying tertials; this is in marked contrast to Red-backed Shrike which has long wings with the primaries and tertials equal in length ('Isabelline' Shrikes have the primary projection two-thirds to three-quarters the tertial length). The overall shape difference is further emphasised by the markedly long and narrow tail, appearing almost wagtail-like at times (the tail feathers are slightly narrower than Red-backed's). Furthermore, its structure is different, being much more graduated, the outer feathers being only about half to two-thirds the length of the visible tail beyond the undertail-coverts (at least 80% on Red-backed and the 'Isabellines'). It must be stressed, however, that the tail does not appear graduated in normal field views as the outer tail feather difference is usually obvious only when viewed from *below*; from above, the tail appears square-ended when closed.

Plumage *Adult male* Darker than the 'Isabellines', being plain russet-brown above, deep fulvous-buff below (throat whiter), with a black mask contrasting with a prominent pure white supercilum; the bill is black, often with a pink base. The tail is also russet-brown but the rump and uppertail-coverts are often more rufous. The race *lucionensis*, which breeds in China, has the crown and mantle more grey-brown (remarkably, an adult in Ireland in 1999 showed characters intermediate between this race and nominate *cristatus*; Slack 2009). *Adult female* Similar to male and sometimes indistinguishable, but the lores are sometimes paler, the mask slightly browner, the breast and flanks scalloped, and the bill base paler (*BWP*). *Juvenile/first-winter* As adult, the upperparts are russet-brown, usually plain, but some retain significant juvenile scalloping on the crown, back and scapulars. The rump and uppertail-coverts are more orangey-brown. The underparts are often strongly suffused with buff, but some are white with buff suffusion confined to the flanks. The breast and flanks are scalloped with brown. There is a diffuse whitish supercilium, a blackish patch through eye and pink on the bill base.

Hybrids

As with many closely related species, the identification of shrikes suffers from problems of hybridisation. There are several known hybrid zones in Asia, between Red-backed and Turkestan Shrikes and between Red-backed and Daurian (race *isabellinus*) while the odd hybrid may occur throughout the range of Turkestan Shrike (Lefranc & Worfolk 1997). One frequent 'colour variety' – known as '*karelini*' – is thought to be a hybrid form between Turkestan and Red-backed, although this view is not universally accepted. Male '*karelini*' have *pale grey* upperparts, but a rufous rump, uppertail-coverts and tail. Hybrids between Red-backed and Woodchat Shrikes have also been recorded in France, some of the offspring then back-crossing with pure Red-backed (Callahan 2012). In October 2008, a possible Brown × Red-backed Shrike was seen in the Isles of Scilly (Vinicombe 2009). As with all hybrids, there is no easy answer to their identification, but the possibility of hybridisation should be considered for any bird that 'doesn't look quite right'. The forensic analysis of good-quality photographs, or even a DNA sample, may be the only way of reaching anything approaching a firm conclusion.

Woodchat Shrike *Lanius senator*

Where and when A rare spring and autumn vagrant, most frequent in s. England. It currently averages 20 records a year, with a peak of 36 in 1997.

Plumage *Adult* Easily identified by its black-and-white plumage and chestnut rear crown. Females are duller than males, with a brown-grey tone to the mantle and a buffier forehead (but some first-summer males may be similar and are apparently impossible to sex). *Juvenile/first-winter* Young Woodchats retain much of their juvenile plumage throughout the autumn. Compared to Red-backed, they lack strong rufous tones and are messily scalloped on a colder, paler, buffy-grey background. Three features suggest their future adult patterning and are diagnostic: (1) a whitish rump; (2) whitish scapulars and median coverts; and (3) a buffy-white patch at the base of the primaries. All three characters vary individually and the whitish scapulars may not be as easy to see in the field as some guides suggest. Some autumn juveniles can show a faint rusty tint to the nape, again an indication of their future adult finery.

Balearic Woodchat

Woodchats that breed on the Balearic Islands, Corsica and Sardinia belong to the race *badius* (nine British records to 2010). Adults can be separated from the nominate by: (1) the lack of a white patch at the base of the primaries and (2) a distinctly heavier bill, with rather rounded upper and lower mandibles. Juveniles can be separated using the same criteria.

Eastern Woodchat Shrike

Another race of Woodchat – *niloticus* – occurs in the Middle East. Although unrecorded in Britain, first-winters have reached Sweden in autumn. The key difference is that, whereas nominate *senator* moults out of juvenile plumage in its winter quarters, juvenile *niloticus* moults mainly on the breeding grounds, *prior* to migrating. Thus, any autumn immature Woodchat showing extensive adult-like plumage could potentially be of this race. Occasionally, some nominate Woodchats also show limited post-juvenile moult, so such individuals should be photographed and expert advice sought; see Rowlands (2010) for more details.

Masked Shrike *Lanius nubicus*

Although there are only two British records (2004 and 2006) this species should be considered when faced with a late autumn juvenile Woodchat. Its plumage is superficially similar to Woodchat, with obvious white in the scapulars, but the key differences are: (1) distinctly greyer, described by Glass *et al.* (2006) as 'cold, grey, white and black, lacking any warm or rufous tones'; (2) darker rump and uppertail-coverts; (3) large white triangular patch at the base of the primaries; (4) breast and belly mostly plain white, the scalloping confined mainly to the flanks. Just as important is structure: it is much more delicate than Woodchat, with a slim bill and a long, slender tail (perched on wires, it can look distinctly wagtail-like).

References Callahan (2012), Campbell (2012), Glass *et al.* (2006), Panov (2009), Pearson *et al.* (2012), Rowlands (2010), Slack (2009), Svensson *et al.* (2009), Vinicombe (2008).

Great Grey, Lesser Grey and Steppe Grey Shrikes

Where and when Great Grey Shrike is a rare autumn and winter visitor, mainly from October to March, with numbers varying from one winter to the next. It currently averages 136 records a year, with a peak of 238 in 1998. It occurs in wild, open country such as moors, heaths and clearings in forestry plantations (although it may occur in surprisingly 'ordinary' farmland). Lesser Grey Shrike is a very rare and declining vagrant from s. and e. Europe (wintering in s. Africa); it averages just two records a year, with most in the Northern Isles and on the east coast. It occurs from May to November with most in late May/early June and September (any grey shrike seen in these peak periods should be carefully checked for Lesser). Steppe Grey Shrike is a very rare vagrant (24 records to 2011); it occurs from mid September to early December, with most in November (also single records in April and June/July). These have been widely scattered but with a basic pattern similar to Lesser Grey.

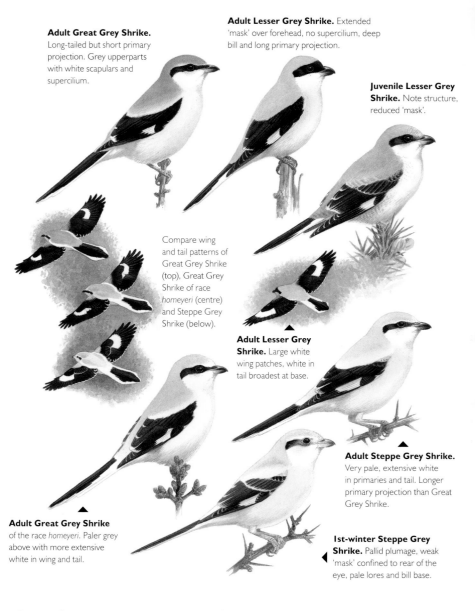

Adult Great Grey Shrike. Long-tailed but short primary projection. Grey upperparts with white scapulars and supercilium.

Adult Lesser Grey Shrike. Extended 'mask' over forehead, no supercilium, deep bill and long primary projection.

Juvenile Lesser Grey Shrike. Note structure, reduced 'mask'.

Compare wing and tail patterns of Great Grey Shrike (top), Great Grey Shrike of race *homeyeri* (centre) and Steppe Grey Shrike (below).

Adult Lesser Grey Shrike. Large white wing patches, white in tail broadest at base.

Adult Steppe Grey Shrike. Very pale, extensive white in primaries and tail. Longer primary projection than Great Grey Shrike.

Adult Great Grey Shrike of the race *homeyeri*. Paler grey above with more extensive white in wing and tail.

1st-winter Steppe Grey Shrike. Pallid plumage, weak 'mask' confined to rear of the eye, pale lores and bill base.

Great Grey *Lanius excubitor* and Lesser Grey Shrikes *L. minor*

Structure Pay particular attention to the length of the exposed primaries beyond the tertials: on Great Grey, the primaries are short, appearing about two-thirds to three-quarters the length of the overlying tertials. Being a long-distance migrant, Lesser Grey has very long primaries, appearing perhaps 1¼–1½ times the tertial length (fresh-plumaged Lessers also show noticeable pale fringes and tips to the primaries). Other differences include Lesser Grey's smaller size, thicker, stubbier bill and proportionately shorter tail (the latter perhaps explaining its tendency to adopt a more upright posture than its longer-tailed relative).

Plumage *General features* Note Lesser Grey's lack of a white supercilium, its indistinct white scapular fringes and, in particular, its larger white patch at the base of the primaries. Both species have white in the outer tail: on Great Grey, this widens at the tip to produce white corners, whereas on Lesser Grey it broadens at the base to produce two basal tail patches (similar to male Red-backed Shrike *L. collurio*). **Adults** Adult Lesser Grey is easily distinguished from Great Grey by a broad area of black over the forecrown; this is extensive on males but may be slightly less extensive and less solid on females (or even absent). Fresh males have a beautiful soft, salmon-pink flush to the breast (greyer on females). *Juveniles/first-winters* Both species' post-juvenile moult commences soon after fledging, so full juvenile plumage is unlikely to be seen in Britain; however, the moult is variable so some autumn migrants may retain at least some juvenile plumage (*BWP*). This shows fine barring and brownish plumage tones, and is most likely to persist on the crown and scapulars. Young Lesser Grey lacks the adult's black forehead, so it could be easily passed off as a Great Grey. Concentrate on structural features, which apply to all ages.

Ageing All grey shrikes can be aged by reference to the greater coverts, which are all black on adults but duller black on first-winters, with buff or whitish tips (although good views are required to confirm this difference, especially as they wear towards spring).

Homeyeri Great Grey Shrike

This race occurs in a band from Bulgaria through Russia and into w. Siberia, immediately south of nominate *excubitor*. It has been claimed in Britain. It differs from the nominate in being paler, with a much larger white primary patch that in flight *extends inwards across the bases of the secondaries towards the base of the wing*. It also has more extensive white at the base of the tail.

Steppe Grey Shrike *Lanius lahtora pallidirostris*

Formerly treated as a race of Great Grey Shrike, this form is now lumped with the 'Southern Grey Shrikes' although, given that these birds are the subject of ongoing research, this arrangement may not persist.

Structure and plumage *First-winter* Similar to Great Grey Shrike but concentrate on the following. **1** PLUMAGE TONE Overall appearance very pallid: pale grey above and whitish below, sometimes with a pinkish-buff hue. Befitting a bird of arid environments, it has distinctly loose and rather fluffy plumage. **2** FACE PATTERN The key feature is the head. Whereas Great Grey has a black mask from the bill across the lores and back through the eye, first-winter Steppe Grey has a small oblong brownish-black patch *behind the eye*. This means that *the lores are obviously pale and whitish*. **3** BILL Note too that the bill base is also very pale whitish-grey (black on Great Grey, sometimes with a grey base). It is also heavier than Great Grey's, with a stronger gonydeal angle. **4** PRIMARY PATCH Much larger and deeper white patch at the base of the outer primaries. **5** PRIMARY PROJECTION Distinctly longer than Great Grey's with an obviously wider gap between the second and third visible primaries (counted back from the wing-tip). **6** TAMENESS Vagrants may be ridiculously tame: one in Lincolnshire in November 2008 perched on birders' heads! *Adult* Black bill and a black eye-patch from the bill back through the eye, making the adult less distinctive than the first-winter (although females may show a grey tint in the lores). Identification therefore needs to be based on the other features outlined above.

Crows: Carrion Crow, Rook, Raven, Jackdaw and Chough

Where and when Rook and Jackdaw occur throughout Britain and Ireland, except in the Scottish Highlands. Carrion Crow is similarly widespread, but is replaced in NW Scotland and Ireland by the Hooded Crow. Raven has been traditionally confined mainly to Scotland, N. and SW England, Wales and much of Ireland. In recent years, however, it has shown a spectacular increase in lowland and even urban areas in the west and has started to spread east (it bred in Kent in 2009). Chough occurs almost exclusively on rocky coasts of Wales (some also inland in the north), the Isle of Man, SW Scotland and Ireland (excluding the east coast). It has recently spread to S. Wales and started to breed again in Cornwall (six pairs in 2010).

Carrion Crow *Corvus corone* and Rook *C. frugilegus*

Adults Seen well, adult Rook is easily separated from Carrion Crow by its bare, whitish face (the crow has a fully feathered face); Rook also has a throat pouch which, when full, appears as a prominent lump.

Shape and structure Particularly useful when separating crows from juvenile and first-year Rooks (see below). Carrion Crow has a rather flat crown and a low forehead, which merge into a blunt-looking bill to produce a wedge-shaped profile (with the bill appearing as a continuation of the head). Perched, Carrion Crow tends to adopt a pot-bellied posture, with the wings and tail held close together and in line with the body at 45 degrees. At rest, Rook has a steep, even vertical, forehead and a peaked crown, creating a distinctly domed head shape. It also has a ragged, 'trousered' effect to its thighs and it often perches with its belly feathers fluffed-out over its legs and feet. A full, rather mobile, rounded tail is often partially spread or pointed vertically downwards. Thus, it often looks angular and 'disjointed' compared to the more compact, hefty and menacing crow. In flight, differences are subtle, but Carrion Crow has slightly broader, more rounded wings and a shorter, squarer tail. Again, Rook appears less compact, with somewhat tapered wings that usually have the primaries slightly angled back. The tail appears distinctly more rounded in flight (the wing and tail shapes may lead to high-flying Rooks being mistaken for Ravens). Carrion Crow's flight is slightly flappier, but differences are slight and depend largely on what the bird is doing.

First-years Juveniles of both species are duller bodied than adults; juvenile crows have a distinct brownish cast to the body plumage prior to their late summer/early autumn moult (as well as a shorter bill, which may initially show extensive dull pink at the base). Their wing feathers remain brownish into their first summer. Note in particular that juvenile Rooks have fully feathered faces and appear crow-like; consequently, they are surprisingly difficult to separate. Only late in their first winter or in spring do they start to lose both the facial feathering and the crow-like wedge of feathering along the top of the bill base. When separating juvenile/first-year Rooks from crows, overall shape is particularly important (see above) and note that their bill is more pointed, less rounded at the tip. At close range, young juvenile Rooks have distinctive blue eyes, but they quickly darken in late summer; also, the bill may

Carrion Crow. Note heavy black bill, flat crown, lack of shaggy thighs and more tidy compact shape.

Adult Rook. Note whitish bill and face, steep forehead, shaggy thighs and less compact shape.

Note difference in calls: Carrion Crow *caw* whereas Rook gives a gruff *arrgh*.

Carrion Crow

Juvenile Rook

Juvenile Rook resembles Carrion Crow until one year old. Note steeper forehead, slimmer, scrawnier shape and feathered thighs. The bill is more slender and pointed. Blue eyes when very young. ▶

Jackdaws. Adult (right) easily identified by small size, pale eye and grey nape. Juvenile (below) has duller eye and lacks the obvious grey nape until autumn moult.

▲

Raven. Huge, similar in size to Buzzard. Note heavy, powerful bill and shaggy throat. Deep croaking calls.

Raven in flight. Huge crow with long wings and diamond-shaped tail. Note shaggy 'beard' and huge black bill.

Rook in flight. Note straighter rear edge to wing, slightly rounded tail and more pointed, narrower wings than Carrion Crow.

Carrion Crow has square-ended tail and bulging rear edge to wing. ▶

Carrion Crows occasionally show silvery-white speckling to wings and tail.

Hooded Crow. Grey body, black wings, tail and head.

Chough. Large, rounded wings. Buoyant, stylish flight. Noisy and acrobatic.

Jackdaws are small with pointed wings and short tail. They appear blunt-headed.

show greyish areas even at an early age. Young Rooks are smaller and skinnier than adults and, by late winter, are heavily abraded about the wings and tail (latter often show protruding feather shafts); the growth of new central tail feathers in late winter will briefly accentuate this graduated look.

White plumage Immature Carrion Crows often show white feathering, especially in the wings, which sometimes show a large white wing-stripe. This pigment loss is thought likely to relate to mineral or vitamin deficiencies as a consequence of poor diet.

Calls An important distinction: Carrion Crow has a characteristic *caw*, whereas Rook has a gruff, strangled *aargh*. Carrion Crow also gives variations on its call, such as a far-carrying ringing *cronnnk* and a hard *k-r-r-r-r* when mobbing a raptor. Both species can make other very peculiar noises, especially when displaying, e.g. Carrion Crow may make loud rattles, reminiscent of a drumming Great Spotted Woodpecker *Dendrocopos major*.

Behaviour Rooks are, in general, more gregarious, but such differences are not absolute as large flocks of crows are not uncommon (especially in spring). Rooks usually breed in colonies (rookeries) whereas crows are solitary nesters.

Hooded Crow *Corvus cornix*

Easily separated by its grey belly and mantle, and its black head, breast, wings and tail. Intermediates between Hooded and Carrion Crows are frequent in the zone of overlap between the two species. Extralimital Hooded Crows (e.g. in s. England) should be studied carefully to ensure that they are pure bred.

Raven *Corvus corax*

Compared to Carrion Crow, Raven is huge (wingspan about a third bigger and similar to Common Buzzard *Buteo buteo*). However, when seen high in the air its enormous size is not always apparent. Shape is significant: Raven is long-winged and (unlike Carrion Crow) the 'hands' are rather tapered and angled back. The tail is strongly graduated and looks diamond-shaped when soaring. However, the tail shape may not be obvious at a distance, particularly when closed. It has a distinctly longer head and bill profile, the huge, deep bill projecting conspicuously from the powerful head, the whole effect often emphasised by the shaggy throat feathering. Beware of misidentifying high-flying Rooks as Ravens, their shape being vaguely similar. On the ground, Raven looks oddly short-legged, while the large head and huge bill can make it look somewhat front-heavy; in direct comparison with Carrion Crow it looks massive.

Calls Very vocal, its loud, croaking *cok cok* and deep *argh* being diagnostic, as is an evocative hollow, drawn out, ringing *kwaark* or *quonk*. It also gives a variety of other calls including a strange 'song' that includes peculiar guttural and bubbling noises. Juveniles may give more subdued notes or a higher-pitched *kar*.

Behaviour Often seen in pairs flying high and purposefully, the female well ahead of the male, the latter often trying to impress her by flipping onto his back with partially closed wings (which enables instant recognition even at great distance). They also indulge in other more spectacular aerial displays, twisting, tumbling or rolling in flight. They may gather in numbers at a good food source and even roost communally (up to 2,000 recorded at a roost on Anglesey). *Juveniles* Like Carrion Crow, juveniles are duller and more matt-coloured than adults (tinged brown on head and body). In flight, ageing is possible at considerable distance, owing to the fact that adults start to moult their primaries as early as mid April, completing this by late summer. Thus, moulting adults in late spring and early summer show conspicuous gaps in their primaries or a distinct step between their short, newly growing inner primaries and their long, unmoulted outers (which often appear quite strongly angled back). At the same time, juveniles have completely fresh fully grown primaries (but the wings are often slightly shorter than adult's).

Jackdaw *Corvus monedula*

Much smaller than Carrion Crow. Unlikely to be confused at rest because of its short bill, grey nape and whitish eye (until early autumn, juveniles have a darker, slightly browner nape and darker eye). In flight, a small, compact, short-winged corvid, with distinctly tapered primaries and an energetic, flapping flight. Like Rook, it is gregarious, and often associates with that species in large mixed flocks. It has a wide vocabulary, but the familiar, excitable *jack jack* is diagnostic. It nests in holes in cliffs and trees, and in chimney pots, on which they often perch. Jackdaws with a horizontal white crescent on the sides of the neck are sometimes seen. Such birds are often identified as 'Nordic Jackdaws', but a prominent white crescent is most typical of the race *soemmerringii* from e. Europe and w. Asia. It may simply be that native British birds occasionally show this feature.

Chough *Pyrrhocorax pyrrhocorax*

Found only on rocky western coastlines (with some inland in N. Wales). Similar in size to Jackdaw. A stunning bird, easily identified by its red legs and long, decurved red bill; juveniles have orange legs and a shorter orange bill. First-years may be browner on wings and tail. Unlike Jackdaw, the wings are long, broad, square and prominently fingered; the primaries may be noticeably kinked back in certain conditions, most markedly when the adjacent inner primaries are missing or partly grown during the summer moult (Grant 1988). Choughs are absolute masters of the air and are a joy to watch as they glide, twist and swoop over their favoured cliffs, often plummeting earthwards with the wings partially closed and swept back close to the body. Loud and evocative *chee-aah* calls are very distinctive.

Reference Grant (1988).

Marsh and Willow Tits

Where and when Both species occur in England and Wales, and in s. Scotland, but are in serious decline. Great care needs to be exercised in their identification, particularly Willow Tit *Poecile montanus*, which is now extremely rare in much of s. England (occurring mainly in the Midlands and n. England). Habitat differences are useful: Marsh Tit *P. palustris* prefers extensive mature deciduous woodland, especially beech and oak, whereas Willow prefers damp woods, carr (fenland overgrown with trees), scrubby habitats and lowland coniferous forest. However, the two species can occur in the same area and, of course, wandering birds present a complication.

General approach Marsh and Willow Tits must be separated by a *combination* of features,

The best way to separate these species is by calls.

Willow Tit. Bull-necked, white of ear-coverts extends some way around nape where the cap narrows. Dull crown, broader 'bib'. Pale fringes to tertials and inner secondaries produce a pale wing panel.

Rounder tail-tip than Marsh Tit.

Bill of Marsh Tit paler at base below nostril.

Marsh Tit. Restricted white ear-coverts, glossy black crown, neat black 'bib', lacks strong pale wing panel. ▶

the most important of which are their calls. In fact, unless they call, they can be very diffi-cult to separate (not until 1900 was it realised that Willow Tit even occurred in Britain). Fortunately, the calls most frequently given are diagnostic, but it must be stressed that, like all tits, both species have a variety of calls. Because of the difficulty in transcribing them, it is recommended that observers familiarise themselves by listening to recordings. Note that there is overlap in most individual plumage and structural features but, despite this, the two species do look different, although their separation becomes much easier with practice.

Calls Willow Tit has an emphatic, loud, full, *deep*, scolding *djur djur djur* or a more nasal *chay chay chay*, sometimes given as *si si chay chay* or *jip jip… jee jee jee jee*. Marsh has a very distinc-tive high-pitched, sneezing call, usually described as *pitch-u*. It is perhaps more accurately transcribed as a double-noted *si-soo, si-swee, swe-oo* or *squee-soo*, often extended into, for example, *swip swip zu zu* or *squit zee zee zee zee*. The important point is that the first part of the call – a rising *si*, *swip*, *squee* or *squit* – has an abrupt, rather explosive quality. Confusion may arise when this first syllable is omitted, leaving only the second part: for example, a *zu zu*, a *zee zee zee* or a *zwee zwee zwee zwee*, which may suggest some transcriptions of Willow Tit's call. However, the *quality* of the calls is totally different, the faster, higher-pitched, rather jaunty calls of Marsh Tit being quite unlike the emphatic slow, deep, sombre tones of Willow Tit. As a final point: beware of the occasional mimicking Great Tit *Parus major*. When feeding, Willow Tits may give soft *see see see* contact calls and very high-pitched *sit* calls. Marsh Tits give a *swit swit* or quite hard *tip tip tip tip* in flight, more in keeping with the explosive quality of their other calls.

Song Marsh has a variable song (one study identified 37 song types!). All are typically tit-like, such as a rapid *swe swe swe swe swe swe swe…*, not unlike a Great Tit but more of a whistle and less ringing; or a disjointed rising and falling *si swoo si swoo si swoo* etc. These songs have a fast delivery: *c.* 5–10 notes per second in bouts of 8–20 notes. Willow has a descending, slow, clear, mournful *sui-swee-swee-swee-swee-swee*, recalling the piping song of Wood Warbler *Phylloscopus sibilatrix*. This is given at a slower rate of about three notes per second in bouts of 2–7 notes. Willow gives various other songs, including a series of thin, high-pitched phrases, and a thin, wistful, descending *si si si soo soo soo*. In late May and June, young Willow Tits utter a descending three- or four-note begging call *jzee jzee jzee*. In contrast, begging juvenile Marsh Tits give a fast, soft, thin, squeaky *si-li-li* or *sid-it*.

Structure and plumage Differences in plumage and structure are subtle and should be used with care. *Head shape and facial pattern* Willow has a larger head and appears bull-necked (with 'less mantle') while the head plumage appears more loosely textured. This bull-necked impression may be emphasised by the facial pattern: a large, swept-back swathe of white curves up behind the eye to the nape, the black of the crown becoming rather narrow on the nape. Note that some Willow Tits are tinged buff across the whole face, sometimes quite obviously so. On Marsh Tit, the white face does not curve up in such a broad swathe behind the eye, giving it a more strictly capped appearance. Also, the front of the face is white but the rear is duskier, sometimes producing a 'two-toned' appearance. *Crown colour* Willow has a dull, matt crown and tends to show a larger throat patch, the black broadening out diffusely to the sides (like a 'bow-tie'). Marsh has a glossy crown and *tends* to show a smaller, neater, 'Hitler moustache'. Remember, however, that the cap gloss is light dependent (see also 'Juveniles' below). *Wing panel* Willow has pale creamy fringes to the secondaries and tertials which, in fresh plumage, produce a distinctive pale wing-panel; however, this may be lacking in worn plumage, particularly in late summer. Marsh Tit is plain-winged, although fresh-plumaged birds may also show a subtle panel (occasionally obvious). *Underparts* Adult Willow *tends* to have buffier underparts, especially on the flanks. *Bill colour* Both species show pale cutting edges to the mandibles, but Marsh Tit also has a pale mark on the upper mandible, on the sides of the bill below the nostril (but this can be difficult to see in the field). *Tail shape* As a consequence of its marginally shorter outer-tail feathers, Willow has a slightly rounder tail tip, visible in good views. *Juveniles* Juvenile plumage is retained from the time of hatching in late May or June until the post-juvenile moult in late September. The following differences from adults should be noted, most of which will impact on the identification process: (1) like Willow Tit, juvenile Marsh Tit has a dull crown; (2) juveniles of both species have a wholly whitish face, juvenile Marsh Tits lacking the adult's grey-brown on the neck-sides and (3) both species are paler below than adults.

Behaviour Both tend to feed at low levels, but Willow Tits can be found feeding very high in pine trees, often on larch cones. Willow Tits excavate their own nest hole from rotten wood, whereas Marsh Tits never initiate a hole from scratch. Note, however, that Willow Tit excavations may be taken over by Marsh Tits; also, Marsh Tits may enlarge their holes, carrying away chippings like Willow Tits.

Borealis Willow Tit

There are a few British records of the northern *borealis* race, which occasionally occurs in northern areas as an irruptive vagrant from Scandinavia. Such birds are larger, paler and greyer than the British race *kleinschmidti*. Willow Tit appears to be more irruptive than the sedentary Marsh Tit, so any Marsh/Willow Tit seen at a coastal migration site is more likely to be Willow. Migrating Willow Tits may be located by an unfamiliar high-pitched *si si si si* or *si-sisit* flight call.

References Broughton (2009), Sharrock & Nightingale (2010).

Skylark, Woodlark and Short-toed Lark

Where and when Skylark breeds fairly commonly throughout Britain and Ireland, with a general retreat from high ground and some southward withdrawal in winter. Woodlark breeds on heathland and forest clearings, mainly in East Anglia, Surrey, Hampshire, Dorset and Devon, some sites (particularly in East Anglia and Devon) being deserted in the winter, when they may occur on nearby stubble, often with Skylarks. A few appear at south and east coast migration sites, especially in April/May and October/November. Short-toed Lark is a regular vagrant, currently averaging 19 records a year, with a peak of 45 in 1996. Most occur in May and September/October at coastal migration sites, particularly in Shetland and Scilly.

Skylark *Alauda arvensis*

General features A familiar bird of open ground, particularly arable farmland. A largish, broad-beamed but rather nondescript buffy-brown lark, with a short crest and quite a thick, pointed bill. The face is rather plain (only a faint eye-stripe behind the eye) with a slight but noticeable supercilium. The general plumage tone varies (becoming paler with wear), but the upperparts are well streaked and the breast shows a finely streaked gorget. Juveniles are noticeably scalloped with buff on the back and scapulars, but acquire adult-like streaking after a complete late summer/early autumn post-juvenile moult.

Flight identification More distinctive in flight, showing a conspicuous white trailing edge to the wing and white outer tail feathers. Relatively slim-winged (for a lark) and evenly proportioned. The flight is flappy, with periodic wing closures, and overhead migrants look rather thrush-like. On the breeding grounds they have a distinctive fluttery flight with the emphasis on the downstroke, so that the wings are bowed downwards; they indulge in aerial chases and hover above long grass and crops before alighting.

Voice Although it may sing from the ground or from a perch, it usually sings from a stationary position high in the sky, with flappy wingbeats and the tail half-spread. The song is a familiar thin, continuous and sustained musical refrain, sometimes containing strongly mimetic phrases. The call is a hard, rippling *chirrup*, *treeip* and so on; flocks also give soft, conversational *see-up* calls.

Pale Skylarks Occasional aberrant, pale sandy-coloured Skylarks may be perplexing.

Woodlark *Lullula arborea*

General features Unlikely to be seen away from traditional southern heathland sites, although small flocks may gather on nearby winter stubble. At their breeding sites, they readily perch in trees (unlike Skylark) but feed on the ground. Distinctly smaller than Skylark but structural differences should instantly attract attention: a stocky lark, with a short crest and noticeably short tail. Although quite thick at the base, the bill is finely pointed. When settled, the most distinctive character is the prominent, creamy-white supercilia which meet on the nape in a V or a pale mottled area (also with a diffuse pale ear-covert surround). The facial pattern is otherwise stronger than Skylark's, with a noticeable dark brown stripe behind the eye and a variable dark brown rear border to orangey-brown or rather richly coloured rusty-brown ear-coverts (with a buff spot in the lower corner). The upperparts are strongly

Juvenile Skylark. Heavily scalloped with poor crest.

Adult Skylark. Large lark with crest and white outer-tail feathers. Note heavy bill and long hindclaw.

Skylark may have crest raised or flattened.

Woodlark. Smaller than Skylark with short, white-tipped tail and black-and-white wingpatch.

Skylark in flight shows white trailing edge to wing.

Woodlark in flight. Note short tail and diagnostic 'wrist' patches.

Juvenile Woodlark. Scalloped above, exhibits the same diagnostic wing and tail patterns as the adult.

Short-toed Lark in flight shows no white in wing.

Short-toed Lark. Ground colour varies, some show chestnut-'capped' effect. Note long tertials hide primary tips.

Short-toed Lark. Pale with heavier bill than Woodlark. Plain-'faced' with little marking on underparts. Black neck patch sometimes obvious, as is dark median-covert bar.

streaked with black on quite a rich rusty-brown background (especially when fresh); the underparts are buffy-white with a clear band of well-defined black streaking across the breast. Thick buffy-white tips to noticeably black primary coverts form a distinctive black-and-buff patch on the edge of the closed wing and this can be noticeable even in flight. The rather short tail shows a whitish tip. Like Skylark, juveniles are scalloped on the back and scapulars but they too acquire adult-like streaking following a complete late summer/early autumn post-juvenile moult.

Flight identification The short tail is particularly obvious, and it looks very round-winged compared to Skylark. The jerky, undulating flight is strangely reminiscent of Lesser Spotted Woodpecker's *Dendrocopos minor*. It lacks Skylark's white trailing edge to the wings; the fringes to the outer tail feathers are very narrow and often sullied with brown, but much more obvious is a thick white tip to the tail, broadest on the outer feathers.

Voice Its beautiful rich and mellow song is delivered in shorter phrases than Skylark, starting slowly but accelerating into a characteristic downward-lilting *lu-lu-lu-lu-lu-lu-lu-lu*; it is often given in a circular song flight. The call is low, soft and musical; transcriptions include: *tuloo-ee, tu-loo, tu-willit*, and *wlee-tloo*.

Short-toed Lark *Calandrella brachydactyla*

General features A small, compact, pale sparrow-like lark that can appear rounded and dumpy or sleek and pipit-like. Most are pale sandy-brown above and whitish below. It lacks a strong crest and often appears rather round-headed; this, combined with the small but rather thick pale bill, noticeable creamy supercilium and dark eye-stripe behind a beady black eye, create a facial expression reminiscent of female House Sparrow *Passer domesticus*; there is usually also a faint brown semi-circular line below the eye. Close inspection usually reveals a small dark patch at the breast-sides (which can be invisible until the bird stretches up); diffuse fine streaking may be present below the dark patches and sometimes also across the upper breast (even in spring). It has creamy fringes to the greater coverts and tertials, and contrasting dark centres to the median coverts, which may produce a dark line across the wing, reminiscent of Tawny Pipit *Anthus campestris*. The legs are noticeably pale, orange or pink, and it moves in erratic bursts with a hesitant, jerky walk. Most autumn vagrants are in complete first-winter plumage, but they may retain some scalloped, dark-centred juvenile scapulars. Note that there is considerable geographical variation in plumage tone: some are distinctly darker and very dingy, being dark buff or even brownish below, some are greyer, while others show reddish tints to the head and, faintly, to the tail. Note that the very long tertials more or less completely cloak the primaries (a diagnostic difference from the extremely rare Lesser Short-toed Lark *C. rufescens*).

Flight identification Slim and evenly proportioned, appearing rather more pipit-like than most larks. The flight is usually fast and dashing, often low over the ground. They typically look pale and sandy in flight, but they lack a white trailing edge to the wings and show only inconspicuous white outer tail feathers.

Call The flight call is usually a hard, clipped *t-t-trr* or *trick trrick*, a more clipped *chr-ip chr-ip* or a more trilling *pirrrick*, all vaguely suggesting a hard, sparrow-like Skylark.

Reference Dennis & Wallace (1975).

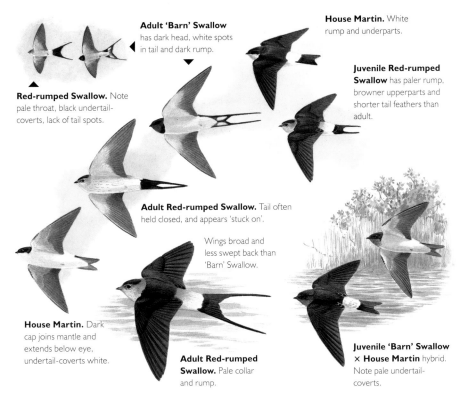

Red-rumped Swallow. Note pale throat, black undertail-coverts, lack of tail spots.

Adult 'Barn' Swallow has dark head, white spots in tail and dark rump.

House Martin. White rump and underparts.

Juvenile Red-rumped Swallow has paler rump, browner upperparts and shorter tail feathers than adult.

Adult Red-rumped Swallow. Tail often held closed, and appears 'stuck on'.

Wings broad and less swept back than 'Barn' Swallow.

House Martin. Dark cap joins mantle and extends below eye, undertail-coverts white.

Adult Red-rumped Swallow. Pale collar and rump.

Juvenile 'Barn' Swallow × House Martin hybrid. Note pale undertail-coverts.

Red-rumped Swallow

Where and when Formerly a great rarity, Red-rumped Swallow *Cecropis daurica* has spread northwards in s. Europe and is now a regular visitor, currently averaging 26 records a year with a peak of at least 61 in 1987. Most occur in spring (mainly April/May) but with records as early as late February. It is less regular in autumn, but occasional late October/early November influxes occur during southerly winds.

Plumage *Adult* Similar in shape to 'Barn' Swallow *Hirundo rustica*, Red-rumped is most easily located by its pale rump. This is narrower and far less eye-catching than the gleaming white rump of House Martin *Delichon urbicum* and is pale buff in colour, often graduating to deep orange-buff towards the upper edge. The undertail-coverts and tail are black and, because of the pale rump, appear 'stuck on' to the rear of the body. Also distinctive is the head pattern: the forehead and face are pale, orange-red, extending back onto the ear-coverts and isolating both the dark eye and the blackish crown, which appears as a 'skull cap'. This isolation is further enhanced by a narrow buff collar around the hindneck, but this can be frustratingly difficult to see. More importantly, it lacks 'Barn' Swallow's red throat and dark blue breast-band; instead, the chin, throat and underparts are orangey-buff, with fine and inconspicuous lines of brown streaking. The underwings are slightly plainer than 'Barn' Swallow's, lacking the strong contrast between the white underwing-coverts and the dark primaries and secondaries. *Juvenile* Late autumn vagrants are most likely to be juveniles. They are similar to adults, but slightly duller with a paler rump and ear-coverts; they can be aged by the narrow buff fringes to the browner wing feathers, those on the tertials being most obvious, especially when perched. Juvenile's tail is slightly shorter than the adult's but is nevertheless surprisingly long.

Structure and flight Both its shape and flight are somewhere intermediate between 'Barn' Swallow and House Martin, having a somewhat stiffer flight action than the former, with the wings less swept back and perhaps more gliding. The tail-streamers are very long but are often held closed together, producing a markedly tapered rear end, oddly reminiscent of a flying Budgerigar *Melopsittacus undulatus*.

Calls British vagrants are generally silent but, if heard, the calls are markedly different from 'Barn' Swallow: quite a distinctive soft, rolling, sparrow-like *shreep, shirip*, a deeper *chrrrp* and so on.

Asian Red-rumped Swallows

A Red-rumped Swallow seen in Orkney and then on Skye in June 2011 showed characters of one of the Asian races *daurica* or *japonica*. These forms differ from European *rufula* as follows: (1) they have a blue-black central hindneck (therefore lacking a complete pale collar) and (2) the dark streaks on the underparts are longer, broader and much more prominent (see Rowlands 2012).

Juvenile House Martins

Although unlikely to be confused with Red-rumped Swallow, it should be noted that autumn juvenile House Martins are distinctly duller than the adults, with some individuals being particularly greyish on the rump, face and flanks, and some even show a hint of a narrow greyish breast-band. They may also show dark scaling on the rump and rear flanks.

Swallow × House Martin hybrids

Another potential pitfall is provided by the occasional occurrence of 'Barn' Swallow × House Martin hybrids, but these are much rarer than Red-rumped Swallows. One such example, a juvenile described by Charlwood (1973), clearly showed intermediate characters, including a large buff rump, a warm buff chin and throat, and a dark breast-band 'less distinct' than that of 'Barn' Swallow.

References Charlwood (1973), Rowlands (2012).

Cetti's Warbler

Where and when First bred in Britain in 1973 but Cetti's Warbler, *Cettia cetti* is now common in wetlands in s. England and s. Wales, although still rare north of the Midlands. Nearly 2,500 singing males were recorded in 2009 but numbers decline significantly following hard winters. It is found mainly in bushy areas at the edges of reedbeds.

Song and calls Even where common, Cetti's can be frustratingly difficult to see and is best located by its very distinctive song and calls. Unlike many secretive birds, it does not come into the open in response to 'pishing'. It sings throughout the year, with the exception of a quiet period in late summer, when moulting. The male's song is unique and unmistakable: a quick, very loud, explosive and positively angry: *You! See you! See you, see you, see you, see you!* Or phonetically: *dji dji djup … di-djup di-djup di-djup di-djup* (rather rhythmic at the end).

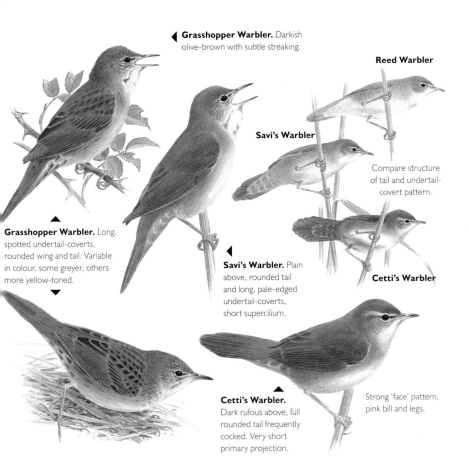

Grasshopper Warbler. Darkish olive-brown with subtle streaking.

Reed Warbler

Savi's Warbler

Compare structure of tail and undertail-covert pattern.

Grasshopper Warbler. Long, spotted undertail-coverts, rounded wing and tail. Variable in colour, some greyer, others more yellow-toned.

Savi's Warbler. Plain above, rounded tail and long, pale-edged undertail-coverts, short supercilium.

Cetti's Warbler

Cetti's Warbler. Dark rufous above, full rounded tail frequently cocked. Very short primary projection.

Strong 'face' pattern, pink bill and legs.

The number of notes and the exact delivery varies individually. Frustratingly, after singing, the bird often flies low and fast through the vegetation to a different part of its territory, thus precluding observation. Its calls are similarly loud and distinctive. Commonest is a loud, abrupt, sharp, dry *stip!* or *plik!* (sometimes repeated) vaguely suggesting the call of Great Spotted Woodpecker *Dendrocopos major*. However, feeding birds often give a quieter, abrupt dry *tp*. Less frequent is a low, hard, dry rattle *t-r-r-r-r-r-r-r-r-r*. Males may also give a loud *swee swee* alarm, very similar to the first two notes of the song. Juveniles have a softer, more unobtrusive version of the call: a more subdued, soft *tip*, given in midsummer when begging for food. This call in particular is reminiscent of some contact calls of Bearded Tits *Panurus biarmicus* as is, to some extent, the more explosive *stip*. However, Bearded Tits' calls typically have a more ringing quality and come from the depths of a reedbed, rather than from scrubbier areas at the edge.

Structure and behaviour Keeps close to or on the ground and is constantly on the move. Most views are of a small dark warbler flying low and fast through the reeds, rarely stopping or moving into view (although it will move higher to sing). In a good view, it is dumpy with an obviously full and rounded tail, which is frequently cocked and sometimes jinked from side to side (unusually, it has only ten tail feathers). The undertail-coverts are shorter than *Acrocephalus* warblers' and show noticeable whitish tips. As befitting a non-migratory species, it has very short rounded wings with a short primary projection (about one-third of the overlying tertial length); they are frequently flicked. Its shape, structure and behaviour are, therefore, completely different from *Acrocephalus* warblers (such as Reed Warbler *A. scirpaceus*, which shares the same habitat). In fact, in many ways, its structure and furtive behaviour are more reminiscent of a giant Wren *Troglodytes troglodytes*. Males are strongly polygynous.

Plumage Very distinctive, being a dark rich reddish-brown above, silvery-white on the throat, but dingy greyish on the neck and breast, with extensive brown on the flanks. It has a fine bill, a narrow white supercilium and a partial white eye-ring (the facial expression is reminiscent of Common Chiffchaff *Phylloscopus collybita*). The legs are pink. Juveniles resemble adults but are duller, with weaker, fluffier plumage, a stronger eye-ring and pinker legs. This plumage is replaced in a late summer body moult.

Locustella warblers: Grasshopper, Savi's and River Warblers (illustrations on p. 283)

Where and when Grasshopper Warbler is a widespread but scarce summer visitor throughout much of Britain and Ireland, frequenting overgrown tangled vegetation on downs and heaths, young forestry plantations and the edges of marshes; it has declined markedly in recent years, particularly in England, where it is now scarce. Savi's Warbler is a very rare summer visitor to selected reedbeds in East Anglia and SE England. Having reached a peak in the late 1970s, its numbers are currently at a low ebb, averaging about seven a year (two confirmed breeding pairs and eight singing males in 2010). It occurs very rarely in other areas on migration, usually being trapped in reedbeds, and is extremely rare in autumn. River Warbler is an e. European bird (40 records in 1961–2011); its appearances here are related to a westward range expansion. Most records relate to singing birds from late May to July, with several long-stayers well inland. There is a second peak in late September/early October, all records at that time coming from Shetland.

General features of *Locustella* warblers All *Locustella* warblers are extremely skulking and difficult to see. They share the same basic shape, being small-headed, rather scrawny with a markedly graduated tail, long undertail-coverts and curved primaries. Unlike *Acrocephalus* warblers, which hop, *Locustella walk* furtively through low vegetation and, when startled, they may scurry off at speed like a mouse. In spring, all species are more likely to be located by song. Note that they tend to sing first thing in the morning and again in late evening, remaining frustratingly quiet during the day. It seems that young birds do not moult out of juvenile plumage until arriving in their winter quarters and so autumn migrants should be aged as 'juveniles' (although some migrant Lanceolated may show some post-juvenile body moult; *BWP*).

Grasshopper Warbler *Locustella naevia*

Song Distinctive, sustained, high-pitched reeling song, recalling an angler's fishing line.
Structure and plumage Olive-brown above, streaked on the crown, mantle and scapulars.

It has a rather bland facial expression with a faint supercilium and a narrow pale eye-ring; the breast is fairly plain, although often lightly streaked, but the undertail-coverts are quite thickly and more obviously streaked. The legs are noticeably pale pink. Adults' underparts vary from buffish-white to pale yellowish, but yellowish is more normal on autumn juveniles. It is easily identified when seen well, but the streaking on the upperparts can be difficult to see, particularly in poor light or at a distance, and this has led to confusion with Savi's Warbler (see below). Except when singing, it is most often seen when flushed from low marshy vegetation or, on migration, low scrubby or even grassy areas. It then flies low and direct before quickly diving back into cover. It appears quite dark brown in flight with a rather long, rounded tail (usually held closed). Sedge Warblers *Acrocephalus schoenobaenus* may behave in a similar manner but show a fairly obvious whitish supercilium and a variable rufous tinge to the rump.

Savi's Warbler *Locustella luscinioides*

Habitat and behaviour Unlike Grasshopper Warbler, Savi's is a bird of reedbeds. In spring, it is invariably located by song but, whereas Grasshopper tends to sing from drier areas within a reedbed (such as bramble bushes), Savi's is most likely to be seen in the wet reedbed itself, climbing to the top of a reed stem to sing.

Song Distinctly *lower-pitched and more buzzing* than Grasshopper Warbler's, which sounds high-pitched, more 'tinny' or 'insect-like' in comparison. Savi's tends to sing in shorter bursts but, like Grasshopper, it may reel for considerable periods once underway; the song is often preceded by accelerating sequences of hard ticking noises, similar to the call, which is a loud, full *tip*, reminiscent of a Robin *Erithacus rubecula*. Note that the songs of both these species may be confused with various bush-crickets (Tettigoniidae): in particular, Roesel's bush-cricket *Metrioptera roeselii* has a song apparently similar to Savi's (see Burton & Johnson 1984, for further details). Being high-pitched, Grasshopper's song becomes difficult to hear as one gets older and there is an adage that, if you can hear a 'Gropper' after the age of 60, it's a Savi's.

Structure Savi's must also to be differentiated from Reed Warbler *Acrocephalus scirpaceus*. In comparison, Savi's is quite a large bulky bird, with a different shape. It has a rather long, spiky, somewhat dagger-like bill (upper mandible black, lower mainly yellowish), a small, rather rounded head and a thin neck, all of which combine to produce rather an emaciated, scrawny appearance in the field. Its most obvious feature, however, is a full, conspicuously rounded tail that is particularly obvious in flight (quite a rich, dark brown colour). Note also Savi's curved primaries (straight on all *Acrocephalus*). When singing, it may adopt quite a distinctive posture with its bill fully open and tilted upwards, its body held parallel to the reed stem but with its tail and full undertail-coverts angled vertically downwards.

Plumage Visual differentiation from Grasshopper Warbler is not difficult *once it is clearly established that there is no streaking* but, as noted above, the streaking on Grasshopper can be surprisingly difficult to detect. Grasshopper, however, has a more buffish basic coloration, rather than the dark brown tone of Savi's. The latter has a very narrow eye-ring and a narrow, pale supercilium, barely visible on some individuals, as well as a slightly darker moustachial stripe at the lower border of the ear-coverts, below the eye. The overall plumage tone is drabber, duller and darker than Reed Warbler. The upperparts are dull brown, lacking olivaceous tones, and the underparts are dingy buffish, but paler on the throat. The cinnamon-buff undertail-coverts are plain, or lightly tipped pale buff. The legs are pinkish or brownish, paler than Reed Warbler's. Remember that its typically furtive *Locustella* behaviour (*walking* up and down branches or reed stems) should also distinguish it from an *Acrocephalus*.

River Warbler *Locustella fluviatilis*

Behaviour Inhabits moist, dense vegetation, such as river floodplains and lowland bogs, but also drier areas than Savi's, such as agricultural margins. More terrestrial, creeping through tangled branches and among leaf litter rather like a chameleon; when startled, it bolts into cover like a mouse.

Song Diagnostic: slower and more rhythmic than Grasshopper's, more like a sewing machine, with the syllables clearly articulated.

Structure and plumage Similar to Savi's, but generally darker, particularly on the underparts, which are dark grey-brown, slightly paler on the belly. Two features must be clearly established when identifying this species: (1) the throat and breast are *noticeably but diffusely streaked* (although some are plainer) and (2) the undertail-coverts have *noticeable pale crescent-shaped tips* to all of the feathers (unmarked or with dull buff tips on Savi's). However, both these features can be frustratingly difficult to confirm in the field. The face shows a narrow pale eye-ring with a very short and faint supercilium extending behind.

Lanceolated *Locustella lanceolata* and Pallas's Grasshopper Warblers *Locustella certhiola*

Both are very rare autumn migrants (September/October) seen almost exclusively in the Northern Isles, with a few records elsewhere, mainly from the coasts of e. Scotland and NE England. Lanceolated currently averages four records a year, Pallas's Grasshopper two. Juvenile Lanceolated can be separated from Grasshopper as follows: (1) smaller, more compact and shorter-tailed, (2) a gorget of pipit-like black streaking across the breast (finely, heavily or diffusely streaked with obvious streaking extending onto the flanks), (3) a more heavily lined back and scapulars, and (4) tertials black-centred, sharply demarcated from the clear-cut brown fringes (centres browner on Grasshopper and fringes more diffuse). Pallas's Grasshopper is *c.* 10% larger than Grasshopper, appearing large and robust with a hefty bill. It can be distinguished by (1) a darker, rather blackish-looking crown, (2) more prominent buff supercilium, (3) upperparts heavily lined black on a rather dark rufous-brown background (it may look dark and oily), (4) a rufous rump and (5) black tail with whitish tips to all but the central two feathers (tips most easily seen in flight).

Reference Burton & Johnson (1984).

Greenish and Arctic Warblers

Where and when Greenish Warbler is a rare but regular vagrant, currently averaging 20 records a year; however, numbers fluctuate considerably and there are occasional influx years, with as many as 47 in 2005. Most occur along the British east coast, with peaks in early June and again in late August and early September; it remains extremely rare in late autumn. An increase in records reflects a series of westward surges of breeding birds into Fenno-Scandia. Arctic Warbler currently averages about eight records a year, with a peak in September; most are seen in eastern counties and, particularly, in Orkney and Shetland. The only 'spring' records have been ten in late June and July, mostly in Shetland (to 2002).

General appearance These two species are similar in size to Willow Warbler *Phylloscopus trochilus* but are grey-green above and whitish below, both showing a strong white supercilium, a dark eye-stripe and, most distinctively, a single narrow white wing-bar on the tips of the greater coverts.

Greenish Warbler *Phylloscopus trochiloides*

Two important confusion species must be eliminated when identifying Greenish Warbler. As well as Arctic Warbler, there has been considerable past confusion with 'Siberian Chiffchaff' *P. collybita tristis* (p. 301). However, confusion with the latter is extremely unlikely in Britain, simply because they occur at different times of the year (*tristis* being a late autumn and winter bird).

Structure Smaller than Willow Warbler *P. trochilus*, with quite a rounded head and shorter wings. Note that the primary projection is shorter than Arctic's, the primaries being about two-thirds the tertial length (Arctic's are almost equal); the primary projection is noticeably shorter than the tail projection beyond the primary tips, so that the tail looks *proportionately* longer than Arctic's. A useful *aide-mémoire* is that Arctic is more similar to Wood Warbler *P. sibilatrix*, both in its larger size and its long-winged/short-tailed structure, whereas Greenish is more like Common Chiffchaff in shape. Greenish's bill is smaller and spikier than Arctic's and is slightly angled upwards from the face (but deeper and broader than that of Chiffchaff and rather more 'parallel' when viewed from the side).

Plumage Autumn adults are very rare in Britain, so the following details relate to first-winter unless otherwise stated. The upperparts are green with a distinctive grey cast, but the exact shade may vary according to the light; the underparts are silky-white but, at *close range* diffuse yellow or greyish hues may be discernible on the breast and belly. There are narrow bright green fringes to the primaries, secondaries and tertials (and tail feathers). Four important features stand out. **1 WING-BAR** Narrow but noticeable, sharply defined and clear-cut, formed by white or yellowish-white tips to the outer four to six greater coverts. It very rarely shows a trace of a second bar on the median coverts. **2 HEAD PATTERN** Also obvious is a prominent, clear-cut, long whitish or yellowish supercilium, broadest behind the eye but becoming duller as it fades into the nape; it is usually straight, but in certain postures there may be a slight upward-kink at the rear (although less pronounced than shown by Arctic). Note that, unlike Arctic, the supericilia often (but not always) meet above the bill. The supercilium is emphasised by a strong greenish eye-stripe but this is weaker than on Arctic, particularly on the lores, and it does not usually reach the bill (it usually does so on Arctic). **3 BILL** Dark along the culmen, but the cutting edges and lower mandible (except the tip) are yellow or pinkish-orange. **4 LEGS** Darker than Arctic's, generally looking medium-brown (in close views brownish-grey at the front but pinkish-grey at the back and sides) but note that, in bright sunlight or at certain angles, they can look quite pale, even acquiring a yellowish tint (prompting confusion with Arctic). For further differences from Arctic, see that species. **Voice** One of the best differences from Arctic. Greenish gives a distinctly disyllabic, soft, rather musical *tswe-ut*, reminiscent of the soft conversational call of Pied Wagtail *Motacilla alba*, but can be louder and more explosive when agitated. Spring birds often sing: a fast, high-pitched, chattering musical jangle: *chi-chi-cheu chi chi-chi-chi-chi-chi-chi-chip*; the last notes forming a trill that is reminiscent of Wren *Troglodytes troglodytes*.

Behaviour Reminiscent of Yellow-browed Warbler *P. inornatus*: extremely active, moving quickly, leaping around and almost tumbling through the foliage. It constantly flicks its wings and occasionally flicks its tail in a nervous downward-dipping (unlike the more consistent dipping of Common Chiffchaff).

Late summer/autumn adults Autumn adults are very rare in Britain. Between June and early August (when they moult) they may lose the fresh olive tone to the upperparts and become relatively dull grey-brown; the wing-bar may abrade away or become irregular. Such individuals can be identified by a combination of head pattern, bare-part colour, structure and call. Note that adult Greenish undertakes only a body moult in late summer, so in autumn it has worn primaries and tail feathers.

Two-barred Greenish Warbler
Phylloscopus (trochiloides) plumbeitarsus

There have been four British records (to 2006) of this eastern counterpart of Greenish Warbler. Its taxonomic position has been the subject of debate, but in Britain it is currently treated as a race of Greenish. Unlike Greenish, records have been in late autumn (late September to late October; Slack 2009). The main difference from Greenish is that it shows a long and *broad* wing-bar on the greater coverts and a strong second bar on the median coverts. Any 'Greenish' showing this combination should be very carefully scrutinised, photographed and a full description taken. Note that, unlike Greenish (but like Arctic) its supercilia do not usually meet above the bill, and it usually has a stronger loral line and more mottled ear-coverts. In addition, it may utter a *trisyllabic*, soft, sparrow-like call: *cheeuwee*.

Green Warbler *Phylloscopus nitidus*

Any Greenish showing strong yellow tones to the face, underparts and wing-bars could be a Green Warbler (one record from Scilly in late September/early October 1983); immediately seek expert advice, try to obtain photographs and consult detailed texts.

Arctic Warbler *Phylloscopus borealis*

Structure Intermediate in size between Willow and Wood Warblers; indeed, its overall shape and long-winged/short-tailed appearance can recall Wood Warbler. It is larger, chunkier and larger-headed than Greenish, with a longer primary projection: the primaries are almost equal

Western Bonelli's Warbler. Crown and upperparts greyish, bland facial expression with pale lores. Back half of bird (wings, rump, tail) greener, underparts silky-white.

Greenish Warbler. Note thin supercilia which meet over the bill. Dark eye-stripe, small wing-bar and short primary projection.

Eastern Bonelli's Warbler. Not safely separated from Western without calls, but averages greyer.

Arctic Warbler. Long supercilium and blotchy ear-coverts. Neat wing-bar and long primary projection, sturdy bill and pale legs.

Dusky Warbler. Brown above and dusky beneath. Strong cream supercilium and dark eye-stripe.

Radde's Warbler. Olive-green above, yellowish below with (usually) orange or yellow undertail-coverts. Heavy bill, pale legs.

to the tertial length (about two-thirds on Greenish). Also note that the primary projection is about equal to the tail projection beyond the primary tips (primaries noticeably shorter on Greenish). Consequently, because of its long wing length it may look proportionately short-tailed (again recalling Wood Warbler). The bill is substantial and hefty for a *Phylloscopus* and is noticeably broad at the base.

Plumage *First-winter* The overall plumage tone is similar to Greenish, but the upper-parts may appear richer green (although the tone varies, some having a slight greyish cast, dependent on the light, surrounding foliage etc.). The following differences should, in combination, separate it from Greenish. **1** HEAD PATTERN The supercilium is longer, passing beyond the ear-coverts and virtually reaching the nape, where it often kinks upwards, but the uptilt is dependent on posture (Greenish can also show a slight up-kink when the head is sunk into the shoulders). On Greenish, the supercilia *often* meet over bill, but they stop short on Arctic. The lores often have a dark smudge on Greenish, whereas Arctic has a more definite dark line across the lores, unbroken from the bill to the eye. Arctic usually shows dark mottling on the ear-coverts, whereas Greenish has rather plain, pale, unmarked ear-coverts (Stoddart 2009). **2** WING-BARS There is frequently a suggestion of a second wing-bar on the median coverts (rare on Greenish). The lower bar often appears 'broken' (individual spots on the outer webs at the tips of the feathers). **3** BILL Noticeably orange-yellow, but darker along the culmen and towards the tip of the lower mandible. **4** LEGS Paler than Greenish, recalling Willow Warbler, looking orange, orange-yellow or even pinkish-yellow in some lights; at closer range they may look yellow with a browner wash, but yellowest at the rear. **5** CALL If heard, *the* best difference: a hard, metallic *dzik* (recalling Dipper *Cinclus cinclus*) and quite unlike Greenish. Although extremely unlikely to be heard in Britain, Arctic's song is completely different from Greenish's: a slow, dry, downward-dribbling trill, sometimes changing pitch between deliveries; it may be preceded by musical *tui tui tui tui tui tui* notes. *Adults late summer/autumn* Like Greenish, worn adult Arctics can lose one or both wing-bars. They too have a body moult only in late summer, so any autumn adults should show worn remiges and rectrices. However, apart from late June/July 'spring' overshoots, adults are unlikely to occur here.

References Slack (2009), Stoddart (2009).

Dusky and Radde's Warblers (illustrations on p. 289)

Where and when Both are rare late autumn migrants to coastal migration spots, mainly in late September/October, but Dusky, in particular, continue to turn up through November with some wintering (remaining into April, sometimes inland). A few May records of Dusky are thought to involve wintering birds en route back to Siberia. They occur in similar numbers: Radde's currently averages 13 a year, with a record 31 in 2000, and Dusky averages 15, with a maximum of 26 in 2001.

Structure and behaviour Both feed close to or on the ground, rather like a Dunnock *Prunella modularis*. Dusky is similar to Common Chiffchaff *Phylloscopus collybita* but is dumpier and sturdier with a proportionately shorter-looking tail (with rounded corners) and a stronger, more wedge-shaped and rather pointed, spiky bill. Its primaries are short and bunched. Compared to Dusky, Radde's is a larger, robust, thickset, big-headed and bull-necked warbler, with sturdier legs and a noticeably thicker, more tit-like bill. It shares Dusky's short, bunched primaries and rounded corners to the tail. Unlike Chiffchaff, neither species tail-dips, but the tail is constantly flicked open as well as twitched and jinked upwards. Radde's may even cock its tail or jink the whole rear end from side to side. Both species continually and nervously flick open their wings, like a Dunnock.

Dusky Warbler *Phylloscopus fuscatus*

Much browner than Radde's, the upperparts being dull brown, with slightly warmer fringes to the wing feathers. The exact plumage tone varies individually and according to the light. The underparts are dull whitish-buff, whiter on the throat, but darker on the flanks and undertail-coverts, the latter showing a hint of orange in good light. The facial pattern is rather 'sharp', with a long, pointed buff supercilium above a thick brown eye-stripe that is stronger and clearer than Chiffchaff's. A narrow whitish eye-ring is not obvious (often more prominent below the eye). There is a noticeable pale base to the lower mandible and the legs average darker than Radde's: brownish through horn to pale yellow and in, good light, frequently orange (often bright), sometimes with yellower feet. Prior to spring migration, Dusky undergoes a body moult (plus tertials and central tail feathers) and so usually looks scruffy in late March/early April.

Radde's Warbler *Phylloscopus schwarzii*

Unlike the rather drab 'brown-and-buff' Dusky, autumn Radde's is a more colourful green-and-yellow. Some are brighter than others, but the upperparts are typically olive-green and the underparts dull yellow, varying from buffy-yellow to bright pale yellow (some are whiter on the throat). Like Dusky, the undertail-coverts may be more deeply coloured, varying from dark yellow through mustard to dark orange. Radde's shares Dusky's long, clear-cut supercilium and eye-stripe, but these are pale yellow and olive-green respectively. Whereas Dusky's legs often look quite dark, Radde's always look pale, varying with the light from pale horn through pale yellow and orange-pink to bright orange.

Calls Dusky is very vocal, the call being a soft, dry *twik* rather like the sound of two small pebbles being knocked together (easily imitated by tongue-clicking). This may also sound like a quiet *tic* or a subdued *tuc*, but sometimes more of a *tchk* or *tchak* (perhaps similar to Blackcap *Sylvia atricapilla*). Radde's is usually less vocal than Dusky, the call being quiet and unobtrusive: a low, soft, rather liquid *tlip*, *qulip* or *cluk*, or a more disyllabic *quillup*, with something of a clucking quality (vaguely suggesting the subdued clucking of a Blackbird *Turdus merula*).

Yellow-browed, Hume's and Pallas's Warblers

(illustrations on p. 294)

Where and when Formerly an official rarity, Yellow-browed Warbler has increased spectacularly in recent decades, having apparently established a regular wintering range in the Western Palearctic. It currently averages 570 records a year, with a remarkable 1,445 in 2005 (and a huge arrival in 2013). They occur from late September, with a peak in October, trailing off in November with some remaining to winter, usually in milder areas. Wintering birds occur in habitats favoured by wintering Common Chiffchaffs *Phylloscopus collybita*, particularly around water and frequently at sewage works. Formerly treated as a southern race of Yellow-browed, Hume's Warbler was split in 1997. First recorded in 1966, it has been annual since 1989 (121 records by 2011, with a record 28 in 2003). It occurs later than most Yellow-browed and is extremely unlikely before mid October; most occur in late October and November, with a few remaining to winter (until late April or even May). With only three British records prior to 1958, Pallas's Warbler increased in the 1970s to a current average of 96 a year, with as many as 313 in 2003. Like Hume's, it is a late October/November migrant, with occasional winter and early spring records.

General approach These three species are being increasingly recorded inland and a good way of finding them is by checking late autumn and winter tit flocks. Knowing their calls is a useful short cut.

Yellow-browed Warbler *Phylloscopus inornatus*

Intermediate in size between Goldcrest *Regulus regulus* and Common Chiffchaff, but proportionately larger-headed with a long and conspicuous yellowish supercilium, and clear-cut green eye-stripe. Most distinctive are two broad yellowish wing-bars, as well as contrasting whitish fringes to the tertials. Many show a diffuse paler area on the central crown. When it descends to lower levels, it shows rather bright green upperparts which, together with the silky-white underparts, create a very 'clean' impression. In October, the species is most frequently found feeding in the tops of willows, where a typical view is of its very white underparts and shortish tail. It can be frustratingly difficult to follow as it moves swiftly through the foliage, often quickly flicking open its wings.

Call Best located by its slightly variable loud, clear, high-pitched, upslurred, penetrating call: *swee-oo-ee*, *suweet*, *stu-eet* and so forth (but sometimes a more monosyllabic *sweet*). It is vaguely reminiscent of the call of Coal Tit *Periparus ater* but much clearer and more incisive. Feeding birds have bursts of calling, especially towards dusk on still evenings, but they may then remain silent for frustratingly long periods.

Hume's Warbler *Phylloscopus humei*

Any late autumn 'Yellow-browed' should be carefully checked for Hume's. Although similar to Yellow-browed, it can be separated as follows. **1** PLUMAGE TONE Distinctly duller, the upperparts being dull green, suffused grey. The duller underparts vary from dusky greyish-white to silky-white, with perhaps a yellow tint to the breast-sides and flanks. **2** HEAD PATTERN Weaker and more subdued than Yellow-browed, with a dull and diffuse greyish crown-stripe (most prominent at the rear) and duller supercilium, either dull buff or whitish, but perhaps suffused with pale yellow. **3** WING-BARS Can be broad and prominent but the upper bar can be almost non-existent, while the lower may be fainter than Yellow-browed's. They vary from dull buff (lacking any yellow tones) to white suffused with pale yellow. The tertial, secondary and primary fringes are lime-green to pale yellow, this colour contrasting with the dullness of the rest of the upperparts. The ground colour of the wing is also duller, greyer and less black than Yellow-browed. **4** BARE PARTS Darker: whereas Yellow-browed has a predominantly pale orange lower mandible and legs, the bill of Hume's is darker, with dull orange confined to the base of the lower mandible. Its legs are a duller brownish-orange, dark brownish or blackish (with orangey feet). **5** MOULT By late March, wintering birds undergo a body moult prior to migrating, appearing rather scruffy. **6** CALLS Fundamentally different from Yellow-browed's. Rather than that species' clear, penetrating, rather Coal Tit-like calls (see above), Hume's calls are much less distinctive. Like Yellow-browed, there is variation in the calls (which is difficult to describe) but the usual call is a markedly disyllabic, high-pitched *tseeu-wit* (with the emphasis on the first syllable). Reminiscent of the conversational call of Pied Wagtail *Motacilla alba*, loud and penetrating when close, but 'thinner' at a distance. Other transcriptions include an upslurred *du-weet* or *speeip* and a flat, monosyllabic *weet*. The calls are more reminiscent of Common Chiffchaff than those of Yellow-browed, but they possess a harder first syllable and are more emphatic. Hume's may also call in flight: a single *dsip*. Song may be heard in late winter, prior to spring migration: a *swee-oo swee-oo* (perhaps reminiscent of certain Coal Tit songs) followed by a very thin, penetrating, fading *zweeee*. Yellow-browed's song resembles a musical rendition of it calls.

Pallas's Warbler *Phylloscopus proregulus*

Resembles Yellow-browed in that it is a clean green and silky-white warbler, with two prominent wing-bars. Like Yellow-browed, it is often viewed from below as it feeds in the canopy, but is Goldcrest-sized and rather short-tailed, with a distinctly large-headed, bull-necked appearance. The most obvious distinguishing features are the long, broad yellow supercilium (yellower than Yellow-browed) and narrow but clear-cut yellow crown-stripe. Most distinctive, however, is a square, pale yellow rump patch but this can be surprisingly difficult to see, mainly because Pallas's are often viewed from below. Usually, it is most easily seen when the bird hovers, which it does frequently, or when it zaps from one twig to another in its relentless search for food. When feeding, Pallas's are often difficult to follow as they move erratically through the foliage with frequent changes of direction, sometimes tumbling through the branches. They continually quickly flick their wings when feeding and also make rather distinctive quick, swooping flycatching sallies, down and then up into the canopy.

Calls Much less distinctive than Yellow-browed's and do not draw attention to the bird in the same way. They are softer, quieter and squeakier, variably described as *djueet*, *che-weet*, *choo-eet*, *joo-it* and so on, with an upward inflection. It may recall a squeaky toy; it thus lacks Yellow-browed's loud, clear, incisive penetration.

References Madge & Quinn (1997), Svensson (1992).

Goldcrest and Firecrest

Where and when Firecrest *Regulus ignicapilla* is a rare but increasing breeding species, with a record 800 singing males in 2010, mostly south of a line from the Severn to the Wash. Also an early spring (March/April) and autumn migrant (September to November) with many remaining to winter; it is commonest in the south, becoming progressively rarer further north (especially rare in Scotland). Numbers vary, but large influxes sometimes occur in the milder south and south-west. They prefer a mixture of deciduous and coniferous trees and bushes, and in winter are often found near flowing water.

Structure and plumage The yardstick is the familiar and abundant Goldcrest *R. regulus*. Poor illustrations may suggest that Firecrest is simply a 'Goldcrest with a supercilum', but it is very much more than, being a stunning bird and one of our most attractive passerines. Although similar in size to Goldcrest, it is somewhat sturdier and more robust, an effect emphasised by

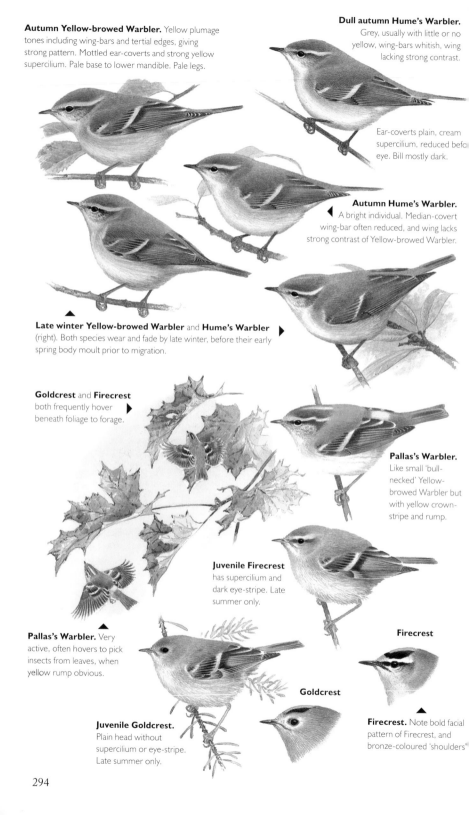

Autumn Yellow-browed Warbler. Yellow plumage tones including wing-bars and tertial edges, giving strong pattern. Mottled ear-coverts and strong yellow supercilium. Pale base to lower mandible. Pale legs.

Dull autumn Hume's Warbler. Grey, usually with little or no yellow, wing-bars whitish, wing lacking strong contrast.

Ear-coverts plain, cream supercilium, reduced befo eye. Bill mostly dark.

Autumn Hume's Warbler. A bright individual. Median-covert wing-bar often reduced, and wing lacks strong contrast of Yellow-browed Warbler.

Late winter Yellow-browed Warbler and **Hume's Warbler** (right). Both species wear and fade by late winter, before their early spring body moult prior to migration.

Goldcrest and **Firecrest** both frequently hover beneath foliage to forage.

Pallas's Warbler. Like small 'bull-necked' Yellow-browed Warbler but with yellow crown-stripe and rump.

Juvenile Firecrest has supercilium and dark eye-stripe. Late summer only.

Firecrest

Pallas's Warbler. Very active, often hovers to pick insects from leaves, when yellow rump obvious.

Goldcrest

Juvenile Goldcrest. Plain head without supercilium or eye-stripe. Late summer only.

Firecrest. Note bold facial pattern of Firecrest, and bronze-coloured 'shoulders' its striking facial pattern: very broad, rather wedge-shaped white supercilium, black eye-stripe (strongest before the eye), black lateral crown-stripe and yellow or orange crown-stripe, a curved swathe of white below the eye and a narrow but quite noticeable black moustachial stripe. All this combines to create a rather fierce appearance, leading to its epithet 'a Goldcrest with attitude'. In addition, the upperparts are brighter green, the underparts whiter and the polishing touch is a characteristic greenish-yellow or 'bronze' patch on the shoulder.

Calls Although its calls resemble the thin, high-pitched, *see see see see see* of Goldcrest, with practice the two species can be readily separated. Firecrest's calls are distinctly thicker, more emphatic and 'zittier', with 'z's rather than 's's. The important point is that the call *rises* up the scale and slightly speeds up: *zi zi zi-zi-zit* (imagine the call written with an upcurved arrow above it). When feeding, Firecrests also give a single *sit* or *zit* that is stronger, deeper and more emphatic than the single notes given by Goldcrest. This is often strung out into a *sit..sit..sit.. sit*, which is often given in flight as it moves from bush to bush. Goldcrests can occasionally give 'zittier' calls or even rising calls that may recall Firecrest, but their quality is thinner and higher-pitched so they do not persistently sound like Firecrests.

Song Goldcrest's familiar but somewhat variable song is a very high-pitched, pulsating, rhythmic *si-si-so-si-si-so-si-si-so-si-si-so* usually with a variable garbled final flourish. Firecrest's is a gently rising and accelerating *si-si-si-si-si-si-si-si-si-si-si* lacking Goldcrest's distinctive terminal flourish.

Sexing If a Firecrest has a *completely* orange crown-stripe, it is a male; if it has a completely yellow crown-stripe it should be a female. Intermediate individuals could be either sex: some females have a tinge of pale orange in the middle, but some first-year males are similar (Svensson 1992). *Juveniles* This plumage can be seen in the breeding areas in late summer. A body moult from to July to September means that it is lost prior to autumn migration (and before the vaguely similar Yellow-browed *Phylloscopus inornatus* and Pallas's Warblers *P. proregulus* arrive). Juvenile Firecrests lack the adult's distinctive black, white and orange/yellow head-stripes; instead, they show a greyish crown and eye-stripe (darkest on the lores when front-on) and a short, weak yellowish supercilium and subocular crescent; thus, the pattern weakly mirrors that of the adult's. Similarly, late summer juvenile Goldcrests lack the adult's black and orange/ yellow crown-stripes and instead have an oddly plain and rather greyish head, apart from a diffuse whitish eye-ring. Juveniles of both species have obviously weak and fluffy body plumage.

Wood Warbler and Western and Eastern Bonelli's Warblers

Where and when Wood Warbler is a summer visitor, associated with mature deciduous woodland, particularly sessile oak woods of w. and n. Britain (rare in Ireland). Unaccountably rare on migration, it being particularly infrequent at coastal migration sites. Western Bonelli's Warbler is a rare vagrant, mainly in August to October, but also occasionally in spring; it currently averages about four records a year, mostly in s. England. Eastern Bonelli's is a real rarity, with just five accepted records (to 2011).

Autumn Wood Warbler. Well-defined eye-stripe, pale ear-coverts, silky-white underparts.

Spring Wood Warbler. Strong 'face' pattern, brighter yellow chin and throat. Green upperparts with pale-fringed tertials.

Long-winged and short-tailed impression.

Wood Warbler (top) is silky-white below, with long wings and short tail compared to **Willow Warbler** (below).

Autumn Willow Warbler. Strong yellow suffusion below.

Wood Warbler *Phylloscopus sibilatrix*

Structure Compared to Willow *P. trochilus*, a sleek, attenuated, streamlined bird, always looking long-winged and proportionately short-tailed (primary projection is equal to tertial length). From below, the tail is splayed (with a slight cleft) and there is only a short tail projection beyond the long undertail-coverts.

Plumage Once known, a very distinctive *Phylloscopus*, but beginners often misidentify brightly coloured Willow Warblers, particularly in autumn when they are much yellower. Out-of-context Wood Warblers at, for example, a coastal migration site, may confuse even experienced observers. Wood Warbler is, however, very much a 'super *Phyllosc*': it is very bright green above with obvious narrow green fringes to the primaries, secondaries and tertials; the underparts are a clear, silky-white, and the chin and throat are primrose-yellow, merging into the white underparts (it does not normally show the *clear-cut*, intensely yellow throat illustrated in most field guides). The head pattern is strong, with a clear-cut yellow supercilium and well-defined green eye-stripe.

Behaviour A typical view is from below, with the bird high in the tree canopy: the silky-white underparts, long wings and long undertail-coverts are then obvious. It is rather slow and deliberate in its movements. When feeding, it often falls and glides through the canopy with its wings half-open. It does not tail-dip. Its wings and tail are quivered during the trill at the end of the song.

Song Diagnostic: a beautiful fast, dribbling trill, often preceded by clear, piping *pew pew pew pew pew* notes, highly evocative of midsummer mature deciduous woodland.

Call Quite unlike Willow Warbler: a loud, clear, far-carrying, mournful *pew* or *duu*, similar to the introduction to the song (in late summer, this is often the only indication of their presence).

Western Bonelli's Warbler *Phylloscopus bonelli*   (illustrations on p. 289)

Structure Closely related to Wood Warbler, and the song is similar, but in shape and structure it is more similar to Willow Warbler (often looks rather round-headed).

Plumage Superficially rather nondescript, but distinctly pale grey-brown or greenish-brown above and silky-white below. A large black eye (with a narrow whitish eye-ring) stands out prominently within a bland face, which lacks a strong eye-stripe (the lores may be completely pale) and there is only a faint pale supercilium (the head pattern and jizz can recall a *Hippolais*). Most likely to attract attention are prominent but narrow lime-green fringes to the primaries, secondaries and tertials, forming a rich green panel on the closed wing, contrasting with dark-centred tertials. The rump is tinged pale yellow, but this can be very difficult to see in the field (best seen when hovering) and is in any case duller on first-winters. The spiky bill has an orangey cutting edge and lower mandible; the legs are dark blackish-brown.

Call A loud, sweet, upslurred *poo–weet*, similar to Willow but with a consonant at the beginning and rising more sharply.

Song A short trill, like the start of Wood Warbler's song, but slower, fuller, more liquid and all on one note (but sometimes changing pitch between deliveries).

Pitfalls There are at least three recorded cases of peculiar Wood Warbler-like birds that have strongly resembled Bonelli's. Like Bonelli's, they were very grey above and white below (with little or no yellow) and strong green fringes to the wing and tail feathers. However, they had a stronger eye-stripe and supercilium, typical of Wood Warbler. Their structure (particularly very long wings) and call indicated that they were aberrant Wood Warblers, rather than Wood × Bonelli's hybrids. Such oddities, although very rare, need to be borne in mind when identifying vagrant Bonelli's. Other confusion species are Siberian Chiffchaff (race *tristis*; p. 301) and Booted Warbler *Iduna caligata* (p. 310). The latter has been confused with Bonelli's on several occasions (even in the hand, when their separation is complicated by the fact that they have very similar wing formulae).

Eastern Bonelli's Warbler *Phylloscopus orientalis*   (illustrations on p. 289)

Although Eastern averages slightly greyer than Western, with a slightly stronger facial pattern (and perhaps whiter underwing-coverts and axillaries) the only way that they can be separated is by call. Eastern gives a dry *twick* or *tswick* (sometimes likened to the call of a Crossbill *Loxia curvirostra*). Sometimes, a string of calls are given in succession, some on a slightly different pitch from the others.

Willow Warbler and the chiffchaffs

Where and when Willow Warbler is a common summer visitor throughout Britain and Ireland, although it has recently declined in the south. It occurs from late March to September, with stragglers into October; but does not usually occur in winter. As its name suggests, Common Chiffchaff is also common, although it is absent from much of the Scottish Highlands. It occurs from mid March to October with small numbers wintering, mostly in southern areas. Browner or greyer Scandinavian Chiffchaffs (race *abietinus*) can occur on migration, although their precise status is unclear. Siberian Chiffchaffs (race *tristis*) occur in small numbers in late autumn and winter. Iberian Chiffchaff is a rare overshooting vagrant from Iberia and SW France. First recorded in 1972, it is now almost annual (28 records to 2011). All have been in spring (early April to late June).

Willow Warbler *Phylloscopus trochilus* and Common Chiffchaff *P. collybita*

General features Field identification of these two common species, which can at first appear very similar, is much easier with practice. Apart from their songs (see below) the following features are most useful. **1** SHAPE Although subtle, Willow is rather more attenuated, sleeker and long-winged compared to the rounder-headed, rounder-bodied, 'podgy' Chiffchaff. **2** PRIMARY PROJECTION Pay particular attention to the extension of the primaries beyond the overlying tertials (although this can be frustratingly difficult to see on a rapidly moving bird). Willow is longer-winged than Chiffchaff and the primary projection is about three-quarters of, or equal to, the length of the tertials. On Chiffchaff, the wings are shorter and the exposed primaries are only about one-third to half the tertial length. **3** TAIL LENGTH Partly as a consequence of its shorter primaries, Chiffchaff's tail appears proportionately longer than Willow's. Furthermore, Chiffchaff tends to hold its tail more tightly closed. In contrast, Willow appears *proportionately* longer-winged and shorter-tailed, and tends to hold the tail less tightly closed, revealing more of a cleft tip. **4** TAIL-DIPPING A significant behavioural trait when feeding is Chiffchaff's habitual downward tail-dipping. Willow usually holds its tail still, although it will also tail-dip, but not as obviously or as *persistently* as Chiffchaff (usually the odd desultory dip, especially after alighting). **5** LEG AND BILL COLOURS Willow has paler, orangey legs and often has an obvious orange bill base; Chiffchaff has blackish legs and a darker-looking bill (usually with little orange evident). However, leg colour differences are not absolute as, although Chiffchaffs do not show orange legs, some Willows can show blackish legs (although they usually still have dark orange feet). **6** FLIGHT The idea of separating them in flight may seem fanciful, but, because of its longer wings, Willow has a more dashing, flycatcher-like flight compared to the weaker, more tit-like flight of the more pot-bellied Chiffchaff. The differences are of course subtle, but become more apparent with practice.

Plumage Differences are both subtle and variable but, in autumn, a basic understanding of ageing and moult times helps considerably. **1** FACIAL PATTERN Willow has a more 'severe' facial expression, with a stronger eye-stripe and a longer, more definite yellowish supercilium, and it is this combination that grabs the attention. Chiffchaff has a weaker, more subdued head pattern but, particularly in autumn and winter, it has a well-marked and noticeable whitish eye-ring, which is the most obvious facial feature. Willow does not show an obvious eye-ring. **2** WING-PANEL Willow *tends* to show a better-marked green panel on the tertials/secondaries; Chiffchaff looks uniform and plain-winged. The following differences are related to season. Note that the two species are quite easy to separate in autumn. **3** SPRING ADULTS Plumage differences are most subtle in spring and the two species can then be difficult to separate. However, Willow is typically paler and more washed-out than Chiffchaff, with greener upperparts and a primrose-yellow tone to the supercilium, throat and upper-breast. Chiffchaff is darker, more olive-green above and olive-yellow below. **4** JUVENILES (LATE SUMMER) Compared to summer adults, juvenile Willows are noticeably primrose-yellow on the supercilium and underparts, and very green on the upperparts; both the bill and the legs are very orangey compared to Chiffchaff. They also show weak, rather fluffy plumage. Juvenile Chiffchaffs have very fine, wispy plumage and often look scruffy; they are greener above than adults, lightly streaked yellow below (yellowest on the breast-sides) and slightly but distinctly greyer on the head, with a prominent dull yellow eye-ring. There is extensive dull orange at the bill base. Juvenile Chiffchaff's calls are monosyllabic (see below). **5 FIRST-WINTERS AND WINTER ADULTS** In late summer and early autumn, the two species are easily separated, even at a distance, a consequence of their different breeding and moult strategies. Unlike the normally double-brooded Chiffchaff, Willows are single-brooded and their breeding season ends earlier. This enables them to migrate earlier than Chiffchaffs, mainly from July to early September as opposed to mid September to October. Because of this, Willows moult earlier and more quickly, the adults undertaking a complete moult and the juveniles a body moult, from June onwards. Consequently, by the time they migrate, mainly in late July and August, Willows of all ages are in fresh, immaculate plumage which is very bright: distinctly green above and evenly yellow on the supercilium and underparts (first-winters average yellower than adults). Because Chiffchaffs migrate later (mainly September/October) they have more time to moult and so look quite scruffy throughout August and into September, when Willows are very yellow, sleek and immaculate. Consequently, any 'Willow/Chiff' at this time showing heavy moult will be a Chiffchaff, such birds being very distinctive by virtue of their sheer scruffiness. Once they have completed their moult, usually by mid September, adult and first-winter Chiffchaffs appear identical. Both ages are much less yellow than autumn Willows, being rather olive above and buff or *dull* yellow below, with a noticeable pale eye-ring and weaker supercilium (although some are brighter and yellower than others). When their plumage is very fresh, they may show a subdued wing-bar across the tips of the greater coverts. On typical individuals the only strong colour is provided by the bright yellow axillaries and underwing-coverts, which may protrude from the wing bend to produce what appears to be a yellow rim on the breast-sides.

Song Diagnostic: Willow has a familiar, pleasant, soft, descending refrain, a subdued version of which is often given on spring migration. Chiffchaff's song repeats its name (usually preceded by a peculiar nasal wheezing). There are several recorded instances of Chiffchaffs finishing their song with the end of a Willow Warbler's; this may be due to individual idiosyncrasy, but probable hybridisation has been recorded. Not infrequently, Chiffchaffs may also give aberrant songs, such as a repeated *Chif-chiff-chiff-chiff...* (i.e. with no 'chaff'), or more complicated songs, such as *sit-soo-sit sit-soo-sit twee-twee-twee-twee*. Note that Iberian Chiffchaff has a different song (see p. 301).

Call The commonest calls are very similar. Willow gives a soft, disyllabic *hoo-eet*, which tends to be slightly more disyllabic than the equivalent call of Chiffchaff. Young juvenile Willow has a squeakier, fuller, more monosyllabic call than its parents. Chiffchaffs may be extremely vocal, calling much more loudly and more persistently than Willow. Although Chiffchaffs also call *hoo-eet*, they also give a variety of other far more distinctive calls and it has even been suggested that, in recent years, they have started to change their vocabulary. Autumn birds, perhaps, mainly juveniles/first-winters, give a very distinctive, loud, down-slurred musical *swee-oo* (variations including *spee-u, swee-ut, pee-up* and so on). This call is now being increasingly heard from adults in spring (oddly, in some years more than others). Begging juvenile Chiffchaffs give a soft, monosyllabic *suut* or *sweet*.

Other races Scandinavian Chiffchaffs (race *abietinus*) are similar to the nominate but tend to be more colourless, averaging greyer or browner above and whiter below, with a better-defined supercilium. Calls include a thin, piping *pee-u* or occasionally *pee-it*. Despite these

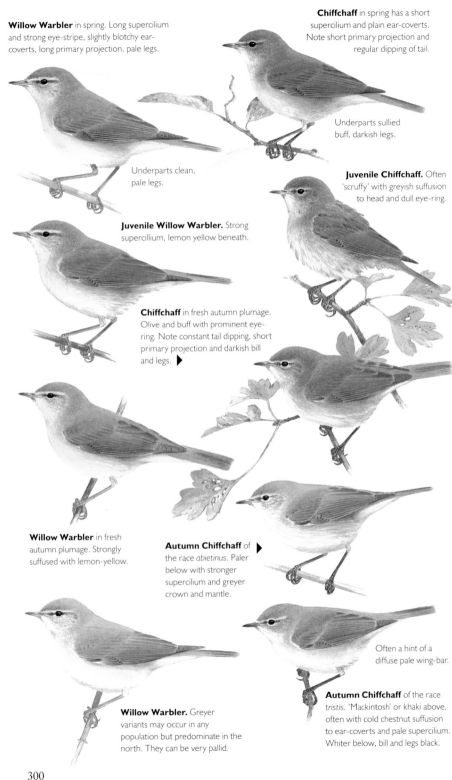

Willow Warbler in spring. Long supercilium and strong eye-stripe, slightly blotchy ear-coverts, long primary projection, pale legs.

Chiffchaff in spring has a short supercilium and plain ear-coverts. Note short primary projection and regular dipping of tail.

Underparts sullied buff, darkish legs.

Underparts clean, pale legs.

Juvenile Chiffchaff. Often 'scruffy' with greyish suffusion to head and dull eye-ring.

Juvenile Willow Warbler. Strong supercillium, lemon yellow beneath.

Chiffchaff in fresh autumn plumage. Olive and buff with prominent eye-ring. Note constant tail dipping, short primary projection and darkish bill and legs. ▶

Willow Warbler in fresh autumn plumage. Strongly suffused with lemon-yellow.

Autumn Chiffchaff of ▶ the race *abietinus*. Paler below with stronger supercilium and greyer crown and mantle.

Often a hint of a diffuse pale wing-bar.

Willow Warbler. Greyer variants may occur in any population but predominate in the north. They can be very pallid.

Autumn Chiffchaff of the race *tristis*. 'Mackintosh' or khaki above, often with cold chestnut suffusion to ear-coverts and pale supercilium. Whiter below, bill and legs black.

differences, their identification in Britain is not recommended. Classic individuals of the race *tristis* or Siberian Chiffchaff are usually identifiable and are dealt with below. Northern Willow Warblers (race *acredula*, from n. Scandinavia eastwards) may be particularly pale and washed-out, with mainly whitish underparts and an insipid grey-green above. Extreme individuals may be very distinctive but there is much intergradation between *trochilus* and *acredula*. Such birds are usually seen in late autumn but washed-out, rather greyish Willow Warblers are also seen at south coast migration sites in May and are thought likely to be en route to n. Scotland.

Siberian Chiffchaff *Phylloscopus collybita tristis* and '*fulvescens*'

Chiffchaffs show a reduction in yellow and green coloration towards the east of their range. The Siberian race *tristis* could be confused with both Greenish *P. trochiloides* and Bonelli's Warblers *P. bonelli* and *orientalis* but, since *tristis* is likely to be found in this country only in the period late October to March, in reality, confusion with these three species is unlikely. Siberian Chiffchaff resembles the nominate race except that it is virtually devoid of yellow or green. The upperpart colour is difficult to describe, 'buffy-brown' being perhaps the most common description (but also 'cold beige', 'khaki' or 'mackintosh'). The underparts are cold whitish, often appearing very white at a distance; however, there may be a beige suffusion across the breast and flanks, while the undertail-coverts may look cream. It can show some yellowish at the bend of the wing (protruding yellow axillaries) and narrow green fringes to the tertials, secondaries, primaries and tail feathers. It has a distinct pale whitish-buff supercilium (narrow but strong to the end of the ear-coverts) and a faint whitish eye-ring; the ear-coverts may show a warmer rusty-buff tint. The beady black eye stands out against the pale plumage. Also distinctive are slight but noticeable pale tips to the greater coverts, often forming a diffuse pale wing-bar. Bare-part colours are as nominate *collybita* but note in particular that the black bill and legs contrast strongly with the pale plumage.

Call Particularly distinctive: an often loud, plaintive, piping, monosyllabic *weet*, *peep* or *sip* recalling a lost chick or, vaguely, a Bullfinch *Pyrrhula pyrrhula*.

Song Migrants may sing, even in autumn, and this is quite distinctive: a fast, thin musical, rhythmic and undulating but somewhat random *chee-weewee chee-weewee chee-weewee* etc. This distinctive song may be the best indication that *tristis* should be treated as a separate species.

Moult In spring, Siberian Chiffchaffs have a body moult before they migrate, so March and early April birds will often look scruffy and dishevelled, certainly in comparison to the newly arriving, immaculate *collybita*.

Complications A great deal has been written in recent times about what constitutes an acceptable Siberian Chiffchaff. The w. Siberian form is sometimes treated as a separate race '*fulvescens*', with nominate *tristis* being confined to e. Siberia. Another complication is that, at the western edge of its range, there is an intergrade zone between '*fulvescens*' and Scandinavian *abietinus*, the intermediate forms sometimes being referred to as '*raphaeus*' (see Dean *et al.* 2010). However, Morova *et al.* (2009) suggested that the zone of active intergradation in one study area in the s. Ural Mountains is only 10km wide, the implication being that the intergrade problem may have been over-stated. The main difference between '*fulvescens*' and 'classic *tristis*' is that the former can show a greenish tone to the upperparts, but the obvious question is: how do you separate '*fulvescens*' from hybrids/intergrades? To circumvent this conundrum, the current convention is to accept only 'classic *tristis*', with any non-classic individuals simply categorised as 'Eastern Chiffchaffs'. However, since Siberian Chiffchaff's plumage may vary more than is generally acknowledged, even in the core of its breeding range, this may be tantamount to throwing out the baby (*'fulvescens'*) with the bath water. Interestingly, recent DNA analysis of intermediate-looking birds in the Netherlands suggests that this may indeed be the case (De Knijff *et al.* 2012). Another question concerns the status of *abietinus* in Britain, there having been suggestions that it may not actually be as frequent as is widely believed (at least in late autumn and winter). A more prosaic problem concerns the identification of potential *tristis* from photographs, the vagaries of photographic reproduction making it impossible to accurately judge plumage tones (it is far easier in the field). Paradoxically, a final stumbling block is accurately describing the colours seen in the field. Many observers are imprecise with their colour descriptions, often describing *tristis* as 'grey' above, even though this may not, strictly speaking, be the case. Despite these semantics, we should not lose sight of the fact that, when seen well, 'classic *tristis*' is actually a very distinctive and recognisable bird.

Iberian Chiffchaff *Phylloscopus ibericus*

All British records have been identified by song, the species frequently establishing territory and singing for long periods. The song is distinctive, starting rather like a Common Chiffchaff but ending with a fast downward stutter: *chup chup chup weet weet chi-chi-chi-chi-jup* or *chip chip chip cheep dee-dee-dee-dee-dee*. However, less distinctive and half-hearted versions are sometimes given, and these may be much more reminiscent of Common Chiffchaff: *chit chit chit chit chid-it chit* or *chiff chiff chiff chiff ch-chiff*. Its call is a downward-inflected, plaintive *seeu... seeu*. In appearance, it may suggest a cross between Willow Warbler and Common Chiffchaff. Although most similar to Common Chiffchaff, it usually shows the following subtle differences. **1** BILL Fine, pointed and slightly spikier. **2** FACE Quite a strong primrose wash to the well-defined supercilium, subdued eye-ring and throat. **3** UPPERPARTS Quite green. **4** BELLY Probably whiter than the average Common Chiffchaff. **5** PRIMARY PROJECTION Marginally longer. Like Willow Warbler, the tail is dipped infrequently. Common and Iberian Chiffchaffs sometimes hybridise, so apparent intermediates can occur. Consequently, only 'classic' individuals should be positively identified. Would-be finders of an Iberian Chiffchaff should endeavour to obtain sound recordings and alert other observers.

References Dean *et al.* (2010), De Knijff *et al.* (2012), Morova *et al.* (2009).

Aquatic Warbler

Where and when A very rare autumn passage migrant from e. Europe, mainly to s. England (mostly Sussex to Scilly, particularly in the SW Peninsula). Usually found in reedbeds or rank vegetation, particularly sedges, but at migration points it may resort to crops or bushes. Although as many as 102 have been recorded in a single year (1976) it currently averages just 16 records a year. Most occur in August and early September, extreme dates ranging from late July to early November. This globally threatened species is Europe's rarest migratory passerine and its decline is clearly reflected by the trend in British records.

Juvenile Aquatic Warbler. Note rounder head, paler, thicker bill, pale lores and broad down-turned supercilium. ▶

◀ **Aquatic Warbler** (left) and **Sedge Warbler** (right). Note heavily streaked Aquatic Warbler with spiky tail. Sedge Warbler shows warm brown rump.

Juvenile Sedge Warbler. ▶ Note dark lores and 'sharper' expression than Aquatic Warbler.

Juvenile Sedge Warbler. Breast spotted not streaked.

Juvenile Sedge Warbler ▼ shows pale median crown-stripe.

◀ **Juvenile Aquatic Warbler.** Pale lores, bronze forehead.

Adult Aquatic Warbler. Worn autumn birds are much duller than juvenile and may show dark lores. Streaked flanks. ▲

◀ **Adult Sedge Warbler** in autumn. Worn plumage, plain rump.

Juvenile Aquatic Warbler. Shape and colour distinctive, as are pale 'tramlines' on mantle. Note clean pink legs. ▲

Identification Although similar to Sedge Warbler *Acrocephalus schoenobaenus*, it must be stressed that Aquatic *A. paludicola* is distinctive in its own right. For the rarity minded, a useful *aide-mémoire* is that Aquatic bears an uncanny resemblance to a miniature, thin-billed Bobolink *Dolichonyx oryzivorus*. The major pitfall is pale, buffy juvenile Sedge Warblers, which can also show an obvious crown-stripe. Aquatic should be identified with caution, and not claimed until all the main differences have been noted.

General appearance *Juvenile* Note that juvenile Aquatics and juvenile Sedge do not undertake a body moult until arriving on their winter quarters; consequently, young autumn migrants of both species should be referred to as 'juveniles' (not 'first-winters'). **1** PLUMAGE TONE Juvenile Sedge is distinctly paler and buffer than the adult, but Aquatic is yellower than that species. **2** UPPERPARTS Heavily striped black and yellowish-buff (giving a 'tiger-striped' appearance), with two rather broad, creamy 'tramlines' or 'braces' towards the sides of the mantle; the rump is noticeably streaked black on a buffish background, in contrast to the plain (or almost plain) chestnut-tinged rump of Sedge. Even in a brief flight view, Aquatic looks a mass of streaks. **3** TAIL FEATHERS Quite sharply pointed, giving Aquatic a spiky-tailed appearance (surprisingly different from Sedge, which has more rounded tail feathers); the central pair protrudes beyond the rest, contributing to a more graduated tail shape. **4** HEAD PATTERN Given a reasonable view of a perched Aquatic, concentrate on the head. Surprisingly, the crown-stripe is not always obvious, particularly if the head is seen side-on, but note the pale, unmarked lores (Sedge has a dark line from the bill to the eye, producing a more 'severe' facial expression); the prominent supercilium is yellow-ish-buff and rather downcurved (not as wedge-shaped as Sedge), while the black eye-stripe is rather stronger and broader behind the eye. The crown-stripe is seen best when the bird bends its head: the crown is very dark, almost black (browner on Sedge), contrasting strongly with a thin, sharply defined buff crown-stripe, which broadens above the bill into a small bronze patch. On well-marked juvenile Sedge, the crown-stripe is broader, messier and less sharply defined, and, unlike Aquatic, there may be an intrusion of dark streaking, blurring the demarcation between the crown-stripe and the crown itself. **5** BILL Slightly shorter and thicker-based than Sedge's, with a pale lower mandible. The head may look more rounded. These features, combined with the more curved supercilium and pale lores, create a softer, more 'open' expression than Sedge. **6** OTHER DIFFERENCES Aquatic has distinctly paler and brighter fleshy pink or grey-pink legs (darker and browner on Sedge). Subtle differences in overall shape may be perceived in prolonged views, Aquatic often looking rather more rakish, with a proportionately smaller head, longer neck and, perhaps, a slimmer body; it may show a narrow gorget of very faint black breast streaking and thin black flank streaks (juvenile Sedge has a variable pectoral band of diffuse spots). ***Adult*** Adult Aquatics are much rarer here than juveniles. Like adult Sedge, they have only a body moult prior to the autumn migration and this is variable in extent, some showing little moult, others extensive (*BWP*). However, the primaries, secondaries and tail feathers are not moulted until they reach their winter quarters. Consequently, autumn adults may be abraded, with worn primaries and tail. As they often show little body moult, they may lose much of the pale streaking on their upperparts, appearing peculiarly dark above. They are less yellow than juvenile, the breast is finely streaked and the face may show a better developed loral line (more similar to Sedge).

Call Aquatic and Sedge Warblers may be quite vocal on migration, particularly in the early morning and late evening. Sedge gives a soft *tchek* or a hard *trrr* or *trr-r-r* (often repeated) reminiscent of a 'hard' Wren *Troglodytes troglodytes*. Aquatic utters a *tucc*, significantly deeper than that of Sedge (Rumsey 1984) and a rolling *trrrr*, lower-pitched than Sedge.

'Pishing' When faced with a potential Aquatic, the most immediate problem may be that of obtaining a good view. Flushing it into the open is often counterproductive. Since Sedge and Aquatic readily respond to 'pishing', it is advisable to retreat about 20m and try this technique: providing conditions are calm, this will usually bring the bird into the open.

References Porter (1983), Rumsey (1984).

Unstreaked *Acrocephalus* warblers

Reed *Acrocephalus scirpaceus* and Marsh Warblers *A. palustris*

Where and when Reed Warbler is a common summer visitor from early April to October (rarely November) to reedbeds in England and Wales, but largely absent from Scotland and Ireland. Marsh Warbler is a summer visitor from late May to October, but it is currently close to extinction as a British breeder (four to nine pairs in 2010). Migrants may also appear at coastal sites in late spring (late May/June) and autumn (mainly September/October) currently averaging 40 a year, mostly in the Northern Isles and the east coast.

Habitat When breeding, this may be the first clue to the bird's identity. In summer, Reed Warbler inhabits reedbeds, whereas Marsh breeds in rank vegetation, such as nettle beds, rosebay willowherb and osier beds. It should be stressed, however, that Reed frequently occurs in drier habitats, sometimes breeding in dense shrubbery (usually near reeds) and on migration is often found well away from water.

Song The most reliable means of separation. Reed has a well-known somewhat grumpy sounding song, including characteristic, repetitive *chara-chara-chara* or *crik-crik-crik* phrases, usually from the depths of a reedbed. The song can, however, become quite excited, and it may even mimic other species, although this is rarely persistent and is generally infrequent and unconvincing. Marsh, on the other hand, is a superb songster with a very varied song (no two bursts are quite the same); a strong mimic, it is often difficult at first to be sure whether it is a Marsh Warbler starting to sing or the species that it is mimicking. The song is generally sweet and warbling, with chattering, grating and sweet, high-pitched sounds intermixed; although many phrases are unmistakably Acrocephaline, they are thinner and faster than those of Reed. The song phrases are long and frequently contain mimicry (the extent of which varies individually): songs or calls of other species include Great Tit *Parus major*, Skylark *Alauda arvensis*, Swallow *Hirundo rustica*, Common Whitethroat *Sylvia communis*, Starling *Sturnus vulgaris*, Blackbird *Turdus merula*, Song Thrush *T. philomelos*, Yellow Wagtail *Motacilla flava*, Chaffinch *Fringilla coelebs*, Greenfinch *Chloris chloris*, Goldfinch *C. carduelis*, Bullfinch *Pyrrhula pyrrhula* and even African species copied in their winter quarters. When singing, Marsh tends to be less secretive than Reed, regularly climbing to the tops of plants or bushes, and singing with the bill wide open, revealing a brilliant orange gape. Reed tends to sing well down in the vegetation, opening its bill only a few millimetres (fully open only when 'carried away').

Calls Reed's commonest call is a rather soft *chrer*, sometimes extended into a slightly harder *cheurrr*, a rolled *chr-r-r-r* or *ch-r*. It must be stressed that Reed Warblers do not give *tac* or *chack* calls. Begging juvenile Reeds give a soft *joo joo*, like a squeaky toy, a familiar sound in late summer reedbeds. Differences from Marsh do not seem to be absolute, but Marsh gives thinner, often harder and more emphatic calls, such as an abrupt, clipped *dik*, *stit* or *chit*, or a thin, rather low-pitched, rolled, hard, churring *trrrrr*.

Plumage and structure Separation of Reed and Marsh Warblers is extremely difficult, and they represent one of the most difficult 'species pairs' on the British List. Nevertheless, to the experienced eye, Marsh *does* look different, although it is often difficult to claim a non-singing individual as anything other than a 'probable'. The following differences are most significant, but it should be stressed that Marsh Warbler should be identified by a *combination* of most of

Spring Reed Warbler. Warm plumage tones, especially rump, dusky flanks. Shorter primary projection than Marsh Warbler, with primaries slightly bunched at tip.

1st-winter Reed Warbler. Warm rufous wash to plumage, especially rump. White throat conspicuous. Buff undertail-coverts.

▲ **1st-winter Reed Warbler** of the race *fuscus*. Buff-brown above, very pale below.

▲ **Spring Marsh Warbler.** Note olive plumage, no warm rump, long, evenly spaced primary projection. Straw-coloured legs, and short pale claws.

▲ **1st-winter Marsh Warbler.** Browner than adult, stout bill.

◄ **Spring Blyth's Reed Warbler.** Uniform tertials, upperparts greyish-brown; no rump contrast.

▲ **1st-winter Paddyfield Warbler.** Rather pallid, strong head pattern and long-tailed appearance. Short primary projection and dark-centred tertials.

1st-winter Blyth's Reed Warbler. Warmer than adult, particularly in the wing. Primary projection short. Silky-white undertail-coverts. 'Tacking' calls like Blackcap.

these characters. **1** PLUMAGE TONE Marsh is slightly, sometimes noticeably, paler than Reed, looking generally more pallid with absolutely no rufous tones: the rump especially is olive grey-brown (quite rufous on fresh Reed). The upperparts are paler brown, even sandy-brown, and they often show a faint greenish or greyish tint. The underparts are creamy or yellowish-cream, not as buff as Reed, and lacking rufous on the flanks. The throat may look noticeably white, especially when singing. Juvenile and first-winter Reed Warblers are richly coloured with rufous tones above (especially on the rump) and are very buff on the flanks; prior to their late summer body moult, adult Reeds are a colder grey-brown above and whiter below than young birds, with the buff on the undertail-coverts less intense. **2** WINGS Pay particular attention to the remiges: Marsh Warbler's primaries generally look slightly longer, narrower and less bunched than Reed's, this leading to a slightly longer-winged impression. More significantly, Marsh shows narrow pale fringes to the tertials and secondaries, while whitish crescent-shaped tips to the seven or eight well-spaced primaries may stand out from the darker background (and may provoke thoughts of Icterine Warbler *Hippolais icterina*). In comparison, most Reed show plainer, more uniform remiges, although faint pale feather fringes are apparent when fresh. Note that, on late summer and autumn adults of both species, the primaries are noticeably worn and the paler fringes may disappear (adults do not moult their primaries until arrival in the winter quarters). **3** LEGS Those of adult Reed are dark and may appear almost black by midsummer. The legs of adult Marsh are slightly paler and browner, but close in shade to Reed, so differences are not always especially striking. The legs of juvenile and first-winter Marsh are paler than Reed, often noticeably orangey, yellowish-brown or straw-coloured, with yellower feet; the legs of similarly aged Reed are a darker, brownish-grey (also with yellower feet). **4** STRUCTURE Differences in shape are very subtle, but Marsh has a slightly shorter and distinctly heavier and broader bill than Reed, and this no doubt contributes to a somewhat *Hippolais*-like (or even *Sylvia*-like) look to the head, with a rather spiky rear crown and, particularly when singing, a more obvious 'jowl' (a somewhat bulging throat). **5** ALULA Marsh has more obviously dark alula, contrasting with paler wing-coverts, but, frustratingly, the alula is often concealed by the flank feathers. **6** OUTER-TAIL FEATHERS In flight, Marsh may show a faint *hint* of paler edges to the outer-tail feathers. For differences from *Hippolais* and *Iduna* warblers, see p. 312 (some species of which can appear very similar to *Acrocephalus*).

'Caspian' Reed Warbler

A difficult problem in late autumn is posed by the eastern race of Reed Warbler *fuscus*, sometimes called 'Caspian Reed Warbler', which has been claimed in Britain. These birds are paler and longer-winged than nominate *scirpaceus*, which grades into *fuscus* around the Caspian Sea. The existence of such birds explains why cautious observers usually attach the epithet 'probable' when reporting suspected late autumn Marsh Warblers.

Blyth's Reed Warbler *Acrocephalus dumetorum*

Where and when It has a similar pattern of occurrence to Marsh, with spring records in May and early June, and a larger autumn peak in late September and October. It currently averages about eight records per year, but it is being increasingly identified here, with as many as 16 in 2007. Most occur in the Northern Isles, on the east coast and as far west as Scilly.

The followings details relate to first-winter plumage compared to first-winter Reed Warbler. On both species, ageing can be checked in autumn by reference to the primaries: worn on adult, fresh on first-winter.

Structure Slightly but distinctly smaller and slighter than Reed and less robust; at times, it is almost reminiscent of a *Phylloscopus*. The body shape is rather more rounded, less 'cigar-shaped' and often more upright. The bill is weaker, finely-pointed and distinctly spiky. The legs are also fine and rather *Phylloscopus*-like.

Plumage tones Different from first-winter Reed, being mid brown above, with a slight olive tint at times (greyer in some lights) and lacking any rufous tones to the rump (only faintly richer in tone). The underparts are predominantly white (faintly buffer on the flanks). Note that the chin and throat are very white and, in particular, the undertail-coverts are a pure silky-white (lacking the strong buff tones of first-winter Reed).

Head The supercilium often bulges before the eye; both it and the eye-ring are creamy-buff, paler and more obvious than on Reed.

Wings Plain and short-winged compared to Reed and the primary projection is shorter (about two-thirds the tertial length, compared with four-fifths on Reed). The wings are noticeably plain, with the tertials, secondaries and primaries finely fringed with brown, the centres of the tertials being hardly any darker than the fringes (more contrasting on Reed).

Leg colour Brown or dull grey, with a faint yellow tint to the feet (legs browner on Reed).

Behaviour Does not inhabit reeds, but is found mainly in trees and bushes, autumn vagrants sometimes favouring ivy. It is usually very quick, lively and often positively mercurial, rarely still and moving smoothly but rapidly through the branches. It may hold its tail slightly above the horizontal and occasionally partly cocked. Reed may also raise its tail and twitch it open or even up and down when nervous, but this is done with the whole body pivoted forward so, strictly speaking, the tail is not cocked.

Call Quite different from Reed, the usual call being similar to Blackcap *Sylvia atricapilla*: an abrupt *tchk*, *chk* or *tchak*, or a *tac tac tac*: sometimes soft, sometimes hard. It also gives a low, hard buzzing *zzzzr zzzzr* or *churrr*, apparently when defending territory, which it will do even on migration. Occasionally gives a *trrr*, more like Reed.

Paddyfield Warbler *Acrocephalus agricola*

Where and when Paddyfield Warbler has a similar pattern of occurrence to Blyth's Reed, but it is proving to be much rarer, with a current average of four records a year.

General appearance A pale *Acrocephalus* with a mobile and longer-looking tail than Reed, but its length is accentuated by the shorter undertail-coverts, shorter primaries (about two-thirds of tertial length) and by its habit of frequently twitching its tail in all directions (often twitched open). In shape, the tail appears square when closed but obviously rounded when spread. In flight, it looks very pale and quite long-tailed.

Plumage The upperparts are noticeably pale: sandy-brown with obvious pale fringes to the tertials, secondaries and primaries. Very white throat (quite sharply demarcated from its grey-brown ear-coverts) and pale buffy-white underparts, appearing much whiter below than Reed. Most distinctive is the head pattern: a distinct narrow cream supercilium, flaring behind the eye and sometimes turning up at the rear. This is bordered above by a diffuse dark lateral crown-stripe and below by a fairly distinct dark eye-stripe. As a consequence of its short wings, long tail and pale plumage, its overall appearance is oddly reminiscent of a small babbler (Timaliidae).

Bare parts The bill is narrower and more pointed than Reed's; horn-coloured with a dark culmen. The legs are dark horn.

Call A throaty, squeaky *chji* (recalling a soft Sedge Warbler *A. schoenobaenus*) and a chattering *ch-d-d-d*.

Iduna warblers: Eastern, Booted, Sykes's and Olivaceous Warblers

Where and when Eastern Olivaceous Warbler is a great rarity (17 records to 2011) with a wide scatter, both geographically and temporally (late May to early July and mid August to mid October, with a peak in September). Booted Warbler has proved to be fairly regular in recent years (127 records to 2011). Most occur in the Northern Isles and on the east coast (with another cluster in Scilly) from late August to early November, with a peak in late September. There have also been five June records. Prior to its split from Booted in 2002, there had been just four records of Sykes's Warbler, but there have been ten since (to 2010, with four in that year). Occurrence patterns are similar to those of Booted, from mid August to early November, with peaks in late August and October (plus one in early July).

Eastern Olivaceous Warbler *Iduna pallida*

Structure Structurally rather similar to Melodious Warbler *Hippolais polyglotta*, with a similarly short primary projection (about two-thirds the tertial length). Eastern Olivaceous is generally smaller than Western, some being noticeably small, prompting confusion with Booted Warbler.

Plumage Eastern is rather like a Melodious Warbler totally lacking in green and yellow: the upperparts are brownish-grey, the underparts strikingly pale, greyish-white; the wings are relatively plain, although it has narrow pale fringes to the remiges (strongest on the tertials and secondaries) which can give the effect of a faint panel at times, rather like that shown by some Melodious. Eastern Olivaceous also needs to be separated from Reed *Acrocephalus scirpaceus* and Marsh Warblers *A. palustris*, but its shape is more like *Hippolais*, often with a slightly ragged effect to the rear crown and a long, noticeably orange bill; it also lacks warm buff plumage tones. Like Melodious and Icterine *H. icterina*, Eastern Olivaceous has a bland face with plain, pale lores. A faint supercilium may be slightly emphasised by a vague *hint* of darkening at the edge of the crown. Pay particular attention to the tail: it looks very square with noticeably pale outer edges.

Behaviour A most useful and significant behavioural trait is that *it continually and deliberately dips its tail downwards when feeding* (rather like Common Chiffchaff *Phylloscopus collybita*; Reed and Marsh do not do this). *Acrocephalus* warblers generally have longer undertail-coverts than *Iduna*, but note that some Eastern Olivaceous can also show fairly long undertail-coverts. It is a lively bird, and far more arboreal than either Reed or Marsh.

Voice More vocal than Melodious and Icterine, giving a slightly sneezing *stt stt*, suggesting a soft Blackcap *Sylvia atricapilla* (easily imitated by tongue-clicking). Its song is a repetitive warbling and grating, continually and regularly rising and falling in pitch.

1st-winter Booted Warbler. Rather *Phylloscopus*-like, strong facial pattern with dark lateral crown-stripe, pale fringes to wing feathers and short primary projection.

Dark tip to lower mandible.

1st-winter Sykes's Warbler. Like Booted Warbler but longer bill, longer tail and primary projection, subdued facial pattern, wings plainer.

Juvenile Icterine Warbler. Note pale lores and orange bill, long primary projection (equal to tertial length), pale wing-panel, grey legs.

Juvenile Melodious Warbler. Shortish primary projection and plain wings. Note pale lores and orange bill, legs tend to be browner than Icterine Warbler's.

1st-winter Eastern Olivaceous Warbler lacks green in plumage, may show pale wing-panel. 'Pumps' tail rhythmically with *tak* calls.

Booted Warbler *Iduna caligata*

Quite different from the preceding species. Although confusion with Eastern Olivaceous Warbler has occurred, Booted should not really cause difficulties with that species.

Size and structure Smaller than Eastern Olivaceous, being closer in size to Common Chiffchaff, but slightly larger and bulkier with a rather domed head shape. A small warbler with, in many ways, the character of a *Phylloscopus*. However, the bill is longer and sturdier than Chiffchaff's (and rather pointed), and appears slightly angled upwards from the face; it is pale orangey-flesh or horn, with a darker culmen and tip.

Plumage A strikingly pale warbler, the upperparts being pale brown (frequently described as the colour of milky tea) and the underparts are also pale, off-white. Unlike Eastern Olivaceous, it has a strong facial pattern, again rendering it more *Phylloscopus*-like in character. The lores are fairly plain but a narrow dark eye-stripe begins just before the eye, tapering behind (note that the gape line may give the *impression* of darker lores). A noticeable creamy supercilium extends behind the rather beady eye, but fades at the rear (although at times it looks rather wedge-shaped). Another distinctive feature is a very faint, but quite definite, narrow dark lateral crown-stripe, extending from the bill and sometimes broadening to form a rather wedge-shaped line above the supercilium. Other features include: broad, slightly ill-defined pale creamy fringes to the greater coverts, tertials and secondaries, on the latter forming a fairly noticeable warmer-toned panel; a short primary projection (about half to two-thirds the tertial length) with a distinct 'tertial' step' between the primaries and tertials; the rump and rather square tail are slightly darker and warmer than the upperparts; it has very narrow pale edges to the outer-tail feathers (but these can be very difficult to detect in the field and are best seen in flight); and pale grey or greyish-horn legs.

Behaviour Tends to frequent low bushes and weedy fields. Less lively than Chiffchaff, and does not habitually fly-catch or hover. Unlike both Eastern Olivaceous Warbler and Chiffchaff, there is no persistent tail-dipping.

Calls See Sykes's Warbler (below).

Sykes's Warbler *Iduna rama*

Very similar to Booted Warbler and their separation is difficult (see Lidster 2009). Like Booted, it is a bird of low, weedy vegetation and bushes (it apparently prefers taller bushes, although of course a vagrant may not have the choice). It is, therefore, much less arboreal than Eastern Olivaceous. The most reliable differences are structural. **1** BILL Compared to Booted, it has a distinctly longer, finer and rather spiky bill. **2** HEAD SHAPE It tends to show a flatter forehead and forecrown, with a slight peak at the rear and a somewhat ragged effect to the feathering on the rear crown (therefore, the head is less domed than Booted). **3** TAIL Longer than Booted's, with variable tail movements (including cocking and spreading). **4** LEGS Longer-legged. All these structural differences combine to give it something of the jizz of an *Acrocephalus* warbler, rather than the more *Phylloscopus*-like look of Booted. **5** HEAD PATTERN Rather subdued facial pattern; although similar in length to Booted's, the supercilium is duller and weaker, petering out behind the eye. It too shows a narrow dark lateral crown-stripe; this can be thin and clear-cut or weak and diffuse, depending on plumage sleeking. A number of other features also need to be checked, but it must be stressed that some of these are very subtle and difficult to establish. **6** PLUMAGE Being a bird of arid environments, it has pallid, weak-looking plumage (late autumn vagrants of both species may look rather worn). It is perhaps perceptibly greyer above than Booted and whiter below. **7** WINGS In good light, browner than the rest of the upperparts and somewhat plainer and more concolorous than Booted's: the tertials, secondaries and primaries are a dull, rather sandy-brown, showing relatively little contrast with the paler grey-brown fringes. It therefore lacks Booted's stronger contrast between the clear-cut pale buff fringes and the darker grey-brown feather centres. **8** TAIL FEATHER TIPS Apparently shows more prominent white tips to the penultimate outer-tail feathers (T5). **9** CALLS Perhaps more vocal than Booted and its calls are subtly different. Recordings by Jännes (2003) indicate that Sykes's has an abrupt, clipped, tongue-clicking *cht cht*, reminiscent of Sedge Warbler *A. schoenobaenus*, whereas Booted's call is more reminiscent of Reed Warbler *A. scirpaceus*: a rather soft *chr chr...* (with something of a rolling 'r' sound) as well as a soft: *cht cht cht ...* Anyone finding a Booted or Sykes's Warbler should obtain expert assistance and good-quality photographs (the identity of one Booted on Scilly was confirmed by DNA).

Western Olivaceous *Iduna opaca* and Olive-tree Warblers *I. olivetorum*

For the sake of completeness, Western Olivaceous (no British records but a potential vagrant) is larger than Eastern with pale brown upperparts, lacking contrasting fringes to the secondaries, and it does not tail-dip. Most importantly, it has a long, broad bill with straight or slightly *convex* sides when viewed from below (slightly *concave* on Eastern; Svensson *et al.* 2009). Olive-tree Warbler has occurred once (Shetland in August 2006; Harrop *et al.* 2008, on which some of following is based). Larger than Eastern Olivaceous, recalling a large colourless Icterine Warbler *Hippolais icterina*. Pale grey above and white below, with a whitish supercilium, a long, hefty, orangey bill, blue-grey legs, a prominent whitish wing-panel (fringes to the tertials, secondaries and greater coverts) and long wings (like Icterine, the primaries are similar in length to the overlying tertials). It has whitish outer-tail feathers and tail-dips like Eastern Olivaceous (as well as waving its tail in an oval movement). It can be slow-moving and heavy, but at other times mobile and impetuous. Very deep calls are reminiscent of Great Reed Warbler *Acrocephalus arundinaceus*.

References Harrop *et al.* (2008), Jännes (2003), Lidster (2009), Svensson *et al.* (2009)

Hippolais warblers: Melodious and Icterine Warblers

(illustrations on p. 310)

Where and when Melodious *Hippolais polyglotta* and Icterine Warblers *H. icterina* are rare but regular migrants, Melodious currently averaging 21 records a year, Icterine 72 (but both are currently declining). Coming from the east, Icterine is most regular in the Northern Isles and on the British east coast, but also annual further west at well-watched migration sites such as the Isles of Scilly. Coming from the south, Melodious has a more southerly occurrence pattern and is most regular from Dorset west to Scilly and at the Irish Sea bird observatories; it is very rare on the British east coast and elsewhere. Both occur in autumn (mainly August to October) but small numbers of Icterine also occur in spring (mainly late May and early June); Melodious is very rare at this season. Icterine has recently bred in Scotland (on four occasions up to 2009) and inland singing Melodious have also been recorded in England (but very rarely).

Separation from other warblers Before separating them from each other, Melodious and Icterine must be separated from other large, sturdy warblers (Garden *Sylvia borin*, Reed *Acrocephalus scirpaceus* and Marsh *A. palustris*) as well as from Willow Warbler *Phylloscopus trochilus*. Their separation from the first three is straightforward: Melodious and Icterine are essentially green and yellow (rather than brown and buff) and have long, pale, predominantly orange bills. Garden is a podgy, featureless warbler, olive-brown above and pale buff below, with a rather stubby *blackish* bill, a large, dark eye and rather plain-faced expression. As on Icterine, its primary projection is long (approximately equal to the tertial length) but the wings are often held slightly drooped, revealing the rump. Unlike Melodious and Icterine, it often feeds on berries in autumn. Reed and Marsh are more similar to *Hippolais*, but both have a stronger facial pattern, with a distinct eye-stripe *before* the eye (which is brown, not black) and a pale supercilium. They also show a rounder tail and longer undertail-coverts (which afford a more tapered shape to the rear end) and both are browner than Melodious and Icterine (Reed often shows rich buff and rufous tones, particularly fresh first-winters). Both *Acrocephalus* are *far more vocal*, but also more furtive, usually (but not always) feeding in lower vegetation. Willow Warbler is easily separated by its small size, restless, flitting behaviour, strong facial pattern, weak bill, *hu-eet* call and, in autumn, by its yellower plumage.

Plumage tone Virtually all autumn *Hippolais* warblers seen in Britain and Ireland are juveniles (as opposed to first-winters). Unlike most warblers, their post-juvenile moult takes place entirely in the winter quarters so the following details refer to that age unless otherwise stated. Icterine varies slightly, but is generally a pale washed-out grey-green above (some are slightly more olive), this paleness being noticeable even in flight (it looks much paler than Willow Warbler). In the field, the underparts are suffused with pale yellow but the strongest yellow is often confined to a pale primrose area on the chin and throat. Melodious is similar, but generally rather more olive-green above and more uniformly dull yellow below, richest on the throat and upper breast. Spring adults of both species tend to look brighter and more uniformly yellow below.

Head Both have a long, hefty, wide-based bill that is noticeably orange (apart from a darker culmen). The bill's size is emphasised by the typical *Hippolais* head shape: a rather sloping forehead and peaked crown. Unlike *Acrocephalus* and *Phylloscopus* warblers, including Willow, both have a bland, open-faced expression, lacking any dark between the eye and the bill, and having only a faint supercilium and eye-ring.

Wing structure To separate them from each other, it is imperative to concentrate on the structure of the closed wing. On Icterine, the length of the exposed primaries is approximately equal to the length of the overlying tertials; on Melodious, the primaries are shorter and rather bunched, and the projection is only about half the tertial length. The long-winged appearance of Icterine is usually very apparent in the field, even in flight, which is fast and rather flycatcher-like compared to the more fluttery flight of the shorter-winged Melodious.

Wing-panel Another good field character is Icterine's wing-panel. This is usually conspicuous and is formed by broad pale fringes to the secondaries (these are usually whitish in autumn, but may be yellower on spring adults). Melodious often looks completely plain-winged, but some individuals show a slight pale panel (particularly in spring) although the fringes are narrower and less well marked than on Icterine and rarely stand out as prominently (although their obviousness is slightly dependent on the angle of view).

Subsidiary features Other differences are subtle, but Icterine tends to show more obvious narrow pale edges to the outer-tail feathers in flight, and it has bluer legs than Melodious.

Behaviour and character The two species tend to have different behaviours. Melodious is a more rounded-looking bird with a tendency to skulk; it is more lethargic and is a rather slow and methodical feeder (often remaining in the same clump of bushes for considerable periods). By contrast, Icterine has a rather long, slim appearance and is more lively and impetuous, shooting off over open ground when disturbed, with a dashing, flycatcher-like flight, swerving suddenly into the back of a bush when alighting; it often perches in full view on the tops of bushes, unlike the more reticent Melodious.

Voice Except when singing, neither is very vocal so, if faced with a persistently calling *Hippolais*, check that it is not a Reed or Marsh Warbler. Icterine has a short, hard soft *tuc* and a soft scolding alarm; Melodious has a low, sparrow-like chatter *st-st-st-st-st-st* (which seems to be given mainly in the presence of other Melodious). Icterine's song is a hesitant, disjointed, high-pitched, sweet *Acrocephalus*-like refrain, perhaps recalling Marsh Warbler, but containing discordant whistling, chacking, grating and squealing, as well as mellower, more thrush-like phrases; each sequence is often repeated before moving on to the next. Melodious starts hesitantly, but proceeds into a softer, gentler, smoother but much faster and more sustained musical rambling (also distinctly Acrocephaline) that lacks Icterine's 'rough edges'.

Sylvia warblers: Common and Lesser Whitethroats and Subalpine, Garden, Moltoni's and Barred Warblers

Common Whitethroat *Sylvia communis*

Where and when As its name suggests, Common Whitethroat is a common summer visitor (mid April to early October) to most of Britain and Ireland except the Scottish Highlands and the Northern Isles. In 1968/69 it suffered a population crash of *c.* 75%, the result of a severe drought in the Sahel zone, south of the Sahara, where it winters. Since then, numbers have significantly increased again.

Plumage A common and familiar bird that should act as a yardstick when identifying rarer members of the genus. Easily identified: adult male in summer has a grey head with a prominent white throat, brownish back and pinkish-white underparts. Females have a brownish head and buffer underparts (first-summer males are also browner on the head). Most significant are conspicuous rufous fringes to the greater coverts, tertials and secondaries, distinguishing both sexes from all other British warblers (except the vagrant Spectacled *S. conspicillata*, see p. 318). Both sexes have a narrow white eye-ring that may be conspicuous (again suggesting Spectacled). *Juvenile* Has rather weak and fluffy plumage, rather a dull grey-brown head, back and scapulars and buffish breast sides. Their eyes are darker than adults, they may retain a yellow gape line and the white outer-tail feathers are sullied with brown. There is a body moult prior to autumn migration when all ages are brown on the head.

Habitat, structure and behaviour Habitat is a good first clue to its identity: low bushes (particularly brambles), hedgerows, scrub and low vegetation. A characteristic view is of a predominantly brown, rather long-winged, long-tailed warbler, flying jerkily into cover ahead of the observer, flicking its rather long, square tail and revealing narrow but conspicuous white outer-tail feathers. Unlike Lesser Whitethroat, it is a lively and excitable bird with an extrovert character. When breeding, it is very responsive to human approach, often scolding the observer with its body tilted forward and its tail jerked, jinked and twisted; when aggravated, it raises its crown feathers to produce a slight crested effect and fluffs out its throat feathers to produce a 'jowled' look.

Calls Also distinctive: a low, deep, scolding *churr* or *churrit*, but also a hoarse, angry scolding *vid vid vid vid vid* (number of notes varies); occasionally, also a less distinctive *tack tack tack...* similar to Blackcap *S. atricapilla* but not quite as 'chacking'. Begging juveniles give a more abrupt, softer *cht* or *cht–t–t*, sometimes tagged on to the churring call.

Song Virtually defies description, but basically a short, fast 3–4-second burst, sweet, throaty and scratchy all at the same time: *dji–do, ji–do ji–doji–do ji–do*. This is very distinctive once learnt, but it varies considerably in both length and content. It is often given from a prominent perch, even telegraph wires, or in the air during a jerky, 'dancing' song flight. It then rises to a height of *c.* 10m with its crown feathers raised and tail spread, descending in a series of jerky swoops with exaggerated wingbeats.

Lesser Whitethroat *Sylvia curruca*

Where and when Lesser Whitethroat is also fairly common (mid April to October) but is restricted mainly to England and Wales; in recent years, very small numbers have also been seen in winter, often at birdfeeders.

Structure and plumage A slightly smaller, shorter-tailed, more compact, bird than Common Whitethroat. Once seen, however, it is easily identified by its uniformly grey-brown upperparts, which lack Common Whitethroat's rusty tertial and secondary fringes; instead the wings appear plain and uniform, with just narrow brown or grey-brown feather fringes. The head is grey, often with darker lores and ear-coverts, and it can show a narrow white eye-ring; the underparts appear clean silky-white, contrasting strongly with the distinctly grey-brown upperparts. Generally regarded as one of our smartest and most attractive warblers. Also of note is that both the bill and the legs are dark (the latter grey in good light), unlike Common's pale pinkish or yellowish legs. In flight, Lesser appears a more compact bird, lacking Common's rather long, 'mobile' tail. *Juvenile* Has rather weak and fluffy plumage, a dark eye and a weak white eye-ring. After a late summer body moult, ageing is difficult, but first-winter retains a dark eye and has the outer-tail feathers sullied with brown.

Habitat and behaviour Frequents taller, thicker bushes and hedges than Common Whitethroat, such as hawthorn and blackthorn. As it dives into cover, it appears essentially grey above with narrow white outer-tail feathers. Unlike Common Whitethroat, Lesser is far less excitable and is distinctly introvert in character, usually keeping to the depths of bushes and refusing to come into full view, even when singing. In late summer, however, they become much easier to see, often feeding in willows and joining mixed tit and warbler flocks.

Calls An *abrupt*, clipped, dry, tutting *cht* or *st*, thinner and more abrupt than that of Blackcap, and more similar to the call of Sedge Warbler *Acrocephalus schoenobaenus*; the call tends to be given singly, unlike the often-repeated chacking of Blackcap. Once learnt, it often reveals that the species is far more numerous than originally anticipated. Also a scolding *chr–r–r–r*.

Song A distinctive one-note 'rattle', perhaps more accurately described as a relatively short but far-carrying one-noted rhythmic burst: *j–j–j–j–j–j–j–jut*, with a deep, rather hollow, almost ringing quality, and given with great gusto. At close range it is often preceded by subdued, disjointed warbling and chattering.

Other races In late autumn and winter, extralimital races of Lesser Whitethroats may occur. **1** Individuals resembling Siberian birds (formerly a separate race *blythi* but now sometimes regarded as synonymous with nominate *curruca*). **2** The central Asian race *halimodendri*. **3** The central and east Asian race *minula*. Such records are usually from garden bird feeders. The identification of these forms is both complicated and controversial, not least because the taxonomy of the entire Lesser Whitethroat complex is currently subject to re-examination.

Spring Lesser Whitethroat. Grey-brown above with darker 'mask' and silky-white underparts. Note white outer-tail feathers and grey legs.

Autumn Lesser Whitethroat. Pale grey and brown with variable pale over eye.

Spring Common Whitethroat. Larger than Lesser Whitethroat with longer tail. Rufous wings, brown rump and pink legs.

1st-winter Common Whitethroat. Very rufous-orange in wing.

Adult female Subalpine Warbler. Small, grey-brown above. Note reddish orbital-ring.

Pink-brown legs.

Common (above) and **Lesser Whitethroat** (below) in autumn. Common is essentially rufous, the Lesser is grey. Both have white outer-tail feathers.

1st-winter Subalpine Warbler. Sandy-brown with grey rump and neck-sides. Prominent pale eye-ring.

Garden Warbler. Large and stocky. Long-winged and short-tailed. 'Plain-faced' with grey neck-sides. Bill and legs grey.

1st-winter Barred Warbler. A large pale grey warbler with pale-fringed wing feathers and a hint of barring to the body.

Spring male Subalpine Warblers: Western (above) and Eastern (below). Note strength of white moustachial stripe and colour and extent of underparts.

316

On finding such a bird, the best approach would be to consult detailed texts, obtain photographs and sound record it (some forms have distinctive calls).

Eastern and Western Subalpine Warblers *Sylvia cantillans* and *inornata*

Where and when Eastern and Western Subalpine Warblers are rare vagrants from the Mediterranean (mainly April to June and again in autumn, mostly October, with one winter record, they average about 18 records a year (as many as 37 in 1995). There would appear to be differences in the occurrence patterns of the two species in that most late March and April records relate to overshooting Westerns, which often turn up in the South West, whereas later spring records are more likely to involve Easterns, particularly those seen in the Northern Isles and down the east coast.

Plumage *Males spring* Both species are easily identified by their bluish-grey upperparts and orange-brown or pinky-brown underparts, which contrast with narrow white moustachial stripes; also obvious is a red orbital-ring (and orange eye). First-summer males have a much duller eye and orbital-ring, and also show very brown and often very worn primaries, secondaries and tail feathers. Western, which breeds in France and Iberia, has pale cinnamon-orange or chestnut-orange underparts, this coloration extending further down the body sides than on Eastern, leaving only a small and rather ill-defined pale belly. Eastern, which breeds from Italy east to western Turkey, has darker, predominantly brick-red or chestnut-brown underparts, extending only slightly beyond the breast, being sharply demarcated from the more extensive and whiter belly. *Female* Females and first-winters of both species are much less distinctive and structure is then important in their separation from other *Sylvia* warblers. They appear small, 'innocent looking', delicate and dinky (even smaller and daintier than Lesser Whitethroat) with a small, dark bill, rounded head, rather plump body, medium-length tail and short wings. Pale brownish-grey above, with a clear line of demarcation between the ear-coverts and the throat. Whitish-cream below, but with a variable pinkish-buff suffusion on the breast-sides and flanks. Such birds are easily separated from Lesser Whitethroat by their prominent buffish-white eye-ring (faint on Lesser Whitethroat, or appears as two separate white crescents), their lack of a dark mask (particularly on the lores) and *pale yellow-brown* legs (dark grey on Lesser Whitethroat). Female Eastern tends to be more pinky-orange on the breast but older adult females of both races are more male-like, with stronger and more colourful plumage tones. *First-winter* A late summer post-juvenile body moult means that autumn vagrants to Britain will be in first-winter plumage, which resembles fresh adult female. First-winters can be aged by the lack of pure white in the outer-tail feathers, which are instead sullied brown. First-winter males acquire a bluish tint to the upperparts and salmon-pink or chestnut underpart coloration, the shade and extent of which being dependant on the species, Eastern having the deep pinky-red confined to the throat. First-winter males also show the white moustachial stripes, but they retain a dull brown eye and a dull pinkish or yellowish orbital ring. First-winter females may appear rather a pale sandy-brown above.

Calls Soft, weak and rather higher-pitched than other *Sylvia*: a soft *tuc*, *st-t* and a *chur*. It becomes angrier when agitated, sometimes giving a rapid staccato *st-da-st-da-st-da...* or *st-st-st-st-st-st*, not as hard as Sardinian *S. melanocephala*.

Song A fast, sustained warbling and chattering, rather sweet and high-pitched for a *Sylvia* and mostly lacking harsher notes.

Moltoni's Warbler *Sylvia moltoni*

A third species occurs in the Balearics, Corsica, Sardinia and parts of north-western Italy. There are currently two putative records (Norfolk in 2007 and Shetland in 2009). It resembles Western but lacks any orange hue, instead having pale to medium salmon-pink underparts, pinker than Western, but similar in extent. Best distinguished by call, which is a markedly different hard *trrrr*, strongly recalling Wren *Troglodytes troglodytes* but trailing off at the end. It also has a different moult strategy in that first-winter apparently has a complete moult in its winter quarters, whereas Western has only a body moult. This means that any first-summer Western-type Subalpine with fresh remiges should, in theory, be Moltoni's.

Confusion with other species

With their rather featureless appearance, 'colourless' first-winter Subalpines have, in the past, often been misidentified as rarer *Sylvia* warblers. The following extreme rarities could cause confusion. **1** SPECTACLED WARBLER *S. conspicillata* Only seven British records (to 2011). There should be no problem in separating Subalpine and Spectacled, but confusion has arisen probably because some old field guide illustrations show them as being far more similar than they really are. Spectacled is a small, energetic warbler, strongly reminiscent of a diminutive, brightly coloured Common Whitethroat. Like Whitethroat, it has bright rusty fringes to its tertials and secondaries *at all ages* but these are much broader than on Whitethroat (although Subalpine has narrow buffish secondary fringes, they never give the impression of Spectacled's clear-cut rusty panel). The rest of the plumage is similar to, but brighter than, that of Whitethroat, but note Spectacled's generally more prominent eye-ring and, on spring males, a very black area on the lores (the latter useful in separating them from male Whitethroats). Spectacled has a different character from both Subalpine and Whitethroat, preferring low, ground-loving vegetation, where it moves in a quick, nervous manner, frequently shooting off across open ground with a weak, tit-like flight. One of the best features is a characteristic grating call *trrrrr*, reminiscent of a Wren *Troglodytes troglodytes* and very distinctive once learnt (but Moltoni's Warbler has a similar call). **2** SARDINIAN WARBLER *S. melanocephala* A fairly regular vagrant, with 77 British records to 2011. A larger, bulkier, much darker and duller warbler than Subalpine. Male has distinctive black head and a red eye and orbital-ring. Female has a dark grey head, darkish grey-brown back and very dull underparts, being particularly dingy and brownish on the flanks (quite unlike Subalpine). The eye and orbital-ring are also red (but dark on juvenile). The typical call is a rhythmic, almost machine gun-like *st-t-t-t-t*, but it also gives a more Whitethroat-like *chur* and other grating noises.

Garden Warbler *Sylvia borin*

Where and when A locally common summer visitor (mid April to October) but rare in n. Scotland and largely absent from Ireland.

Plumage and structure Because it is a rather featureless bird, Garden Warbler often causes problems for beginners and even more experienced observers. Paradoxically, its lack of features becomes a feature in itself. It is worth spending time familiarising yourself with Garden Warbler, so that when you eventually come across something rarer, you will be in a better position to identify it. Firstly, Garden Warbler is a bird of bushy places, and does not usually occur in mature woodland (except at edges or if there is a bushy understorey). Large, bulky, but unobtrusive, often appearing rather podgy and rounded, with long primaries that are often drooped. Thick, rather stubby dark bill and dark grey legs. A large, round eye stands out strongly in a plain face (only a hint of a supercilium and eye-ring), giving it a benign expression. Its plumage is a warm olive-brown, lacking strong characters, but note that many show a subtle warm grey patch on the neck-sides. Like Blackcap, Garden feeds extensively on soft berries in autumn, a habit not readily associated with *Acrocephalus*, *Iduna* or *Hippolais* warblers.

Song In spring and summer, best located by song. Many people struggle to separate this from Blackcap *S. atricapilla* but it is actually very different: if you cannot separate them, it may well be that you are not hearing Garden Warblers at all. Whereas Blackcap's song is dominated by high-pitched discordant whistles, Garden Warbler's remarkable song is fast and extremely complicated, with beautiful deep, mellow phrases, the quality of which recalls Blackbird *Turdus merula*. Some phrases can also recall the deep sound of burbling water in a brook. A useful *aide-mémoire* is that Garden Warbler sounds like someone trying to talk too fast, whereas Blackcap sounds like a madman trying to whistle. Note, however, that Blackcap will often introduce its song with a weaker, more subdued preamble, more similar to the song of Garden Warbler, so caution is required. Also, newly arrived spring Blackcaps often indulge in a peculiar and complicated low, fast, scratchy subsong, although they eventually break into their high-pitched manic whistling. Blackcaps may also mimic other birds, such as Song Thrush *Turdus philomelos*, Reed Warbler *Acrocephalus scirpaceus* and Common Whitethroat.

Calls Garden Warbler's call is also very different from Blackcap's: a repeated deep, emphatic but soft *vit vit vit vit vit vit vit…* (number of notes varies), which is quite similar to the equivalent call of Common Whitethroat. Occasionally it also gives a throaty, slightly rasping *eeup* or *eeip*. Blackcap's normal call is a characteristic abrupt chacking *chet* or *teck*, often given persistently. Territorial Blackcaps may also give a bewildering variety of other odd calls, including nasal buzzing and a rather screaming *eeea eeea eeea*. Begging juveniles give an unobtrusive but distinctive, rather flat *eeh* or a slightly disyllabic *eeut*, almost like a soft toy trumpet.

Barred Warbler *Sylvia nisoria*

Where and when Barred Warbler is an autumn visitor from mid August to late October, mainly in Shetland and on the British east coast, with smaller numbers on the south coast and further west; surprisingly, it is virtually unknown in spring. It currently averages 170 records a year, with a record 297 in 2002.

Plumage and structure Adult Barred Warblers (i.e. with bars) are extremely rare in Britain, the overwhelming majority of records relating to first-winters in autumn, to which the following details refer. A large, hefty warbler, with a slightly ragged effect to the head, and a rounded back. It often shows a pronounced 'jowl', strong, sturdy grey legs and a full tail. Quite slow and heavy in its movements, and its general character and appearance vaguely recall a shrike *Lanius*. In flight, can suggest a 'full-tailed' Spotted Flycatcher *Muscicapa striata*. The head and upperparts are pale grey and the face is plain, except for a slightly darker line through the eye and slight eye-ring. Unlike adults, the eye is brown. Very white below, with a buffier wash on the flanks. Features assisting the identification include the following. **1** Most noticeable is a narrow double whitish wing-bar, produced by white tips to the median coverts and pale grey fringes to the greater coverts. **2** Dark bases to the median coverts highlight the pale wing-bar. **3** Less conspicuous pale fringes and tips to the tertials, and whitish fringes and tips to the primaries. **4** Unlike Garden Warbler, it has narrow white outer-tail feathers. **5** Bill strong and pointed, dark with a pale grey base to the lower mandible. **6** Often has variable but subdued barring on the flanks and undertail-coverts, and pale fringes to the uppertail-coverts. **Call** A loud, hard, rattling *trrrrrrrrr*, also a loud chacking, very chat-like and very different from Blackcap and other common *Sylvia* warblers.

Rose-coloured Starling

Where and when Rose-coloured Starling *Sturnus roseus* is a rare vagrant from SE Europe and SW Asia, averaging 34 records a year. The species is, however, irruptive and in some years there are significant late spring and summer influxes of adults, with a remarkable 195 in 2002. There is a further wave in autumn (mainly juveniles in September/October). Both adults and juveniles may overwinter, often in gardens.

Plumage *Adult* Easily identified by its pink mantle and belly, which contrast with the black head, breast, wings and tail; it also has elongated rear crown feathers and a pink bill. In winter plumage, much of the pink is obscured by buffish feather tips, which eventually wear off; the bill can also be duller. First-summers tend to be a buffier shade of pink. Occasional 'washed-out' partially leucistic adult 'Common' Starlings *S. vulgaris* have shown patterns very similar to Rose-coloured: look for the latter's shorter, pale pink bill. *Juvenile* A pleasant, bland-faced, innocent-looking bird, with rather loose, fluffy plumage. Easily identified by its very pale, sandy plumage (rump somewhat paler in flight). The wings are darker brown with the feathers noticeably fringed pale buff. To confirm the identification, it is essential to check the bill: shorter and stubbier than Starling's, pale to bright yellow at the base, grading towards pink, dull reddish or brownish at the tip; any purple at the base is usually blackberry staining (juvenile Starling's bill is uniformly blackish). It is important to remember that juvenile Rose-coloured Starlings occur in Britain *after* juvenile Starlings have gained most of their adult-like first-winter plumage. This is because juvenile Starlings moult into adult plumage from June to October, whereas juvenile Rose-coloured moults mostly from October to February, *after* arriving in their winter quarters. In consequence, any pale juvenile starling seen in Britain before the end of August will undoubtedly prove to be an aberrant Common Starling (check bill colour to clinch the identification). Moulting juvenile Rose-coloured in winter appears scruffy, a consequence of the patchy pink-and-black adult feathering starting to appear. The new 'pink' feathers on such birds are rather buff or brown in tone, and they may also show white 'peppering' in their new black head plumage.

Juvenile Starling. Dark grey-brown, black bill and dark legs.

Juvenile Starlings are mostly moulted into adult plumage by late autumn when unmoulted juvenile Rose-coloured Starlings arrive.

Juvenile Rose-coloured Starling begins to moult after migration, in late autumn.

Wintering juvenile, partially moulted in December.

Juvenile Rose-coloured Starling. Paler than juvenile Starling, especially the rump and underparts. Wings darker. Note blunt yellow bill and pale legs. Beware leucistic Starlings.

1st-winter Rose-coloured Starling. Fully moulted by February.

Ring Ouzel and Blackbird

Where and when Ring Ouzel *Turdus torquatus* is very much the upland counterpart of the familiar and abundant Blackbird *T. merula*. It occurs in summer on mountains and moors in Scotland, n. England, Wales and (more sparsely) SW England and Ireland. On migration, it can be found at coastal sites and on high ground inland, mainly from mid March to late April and late August to early November; late autumn migrants (and very occasional winterers) may associate with Fieldfares *T. pilaris* and Redwings *T. iliacus*.

General features Male Ring Ouzel in spring is easily identified by the conspicuous large white crescent on its breast. Also distinctive are the narrow white fringes to the wing feathers which, at any distance, produce an overall whitish impression to the wings, even in flight. Structural differences from Blackbird are also important, Ring Ouzel being longer-winged, flatter-backed, sleeker and more streamlined. These differences are obvious both on the ground and in flight, when Ring Ouzel is much more similar to Fieldfare in shape. It also has longer legs and a thicker, more wedge-shaped bill than Blackbird.

Flight Befitting a summer migrant, Ring Ouzel is longer-winged than Blackbird; at rest, the exposed primaries are about equal to the overlying tertial length (about two-thirds on

Spring male Ring Ouzel. Note lemon bill, silvery wings, Fieldfare-like shape.

Male Blackbird

Adult male Blackbird. Jet-black throughout. Large yellow eye-ring.

Male Ring Ouzel shows long silvery wings.

1st-winter male Blackbirds have black bills and browner wings than adults. Some have pale feather tips recalling Ring Ouzel. There is no pale fringing to the wing feathers. The primaries are short.

Female Ring Ouzel is duller than male, browner with less distinct gorget.

Female Blackbird. Some have very pale throats. Partial albino Blackbirds are not uncommon. If in doubt check wing for pale fringing and primary length..

1st-winter female Ring Ouzel shows almost no gorget. Note wing fringing and long primaries.

Juvenile Blackbird is very chestnut above and heavily flecked and speckled.

Juvenile Ring Ouzel. Note pale neck patch and contrasting cream and chocolate-brown plumage.

Blackbird). This gives Ring Ouzel a much stronger, more continuously flapping flight (it may even suggest Starling *Sturnus vulgaris* when viewed back-on). It is normally very shy compared to Blackbirds and flushed birds invariably fly off into the distance, usually at some height.

Calls A loud, deep, hollow chacking call, rather like two pebbles being banged together: *tchak-tchak-tchak-tchak-tchak.*

Song The atmospheric song is usually a simple series of high-pitched but mellow, vaguely Blackbird-like *chree chree chree* notes, followed by lower, shorter, more garbled and rather guttural notes.

Plumage *Adults* Although spring males are distinctive, many Ring Ouzels are difficult to sex, a few adult females resembling males. However, most females are browner than males (lacking the rufous tones sometimes shown by female Blackbird) and they usually have a more obscure breast-band (often with heavy brown scalloping) plus fine whitish scalloping across the entire body. In autumn, ageing and sexing is complicated by the fact that some males have their white breast crescent slightly obscured by pale brown feather tips, while first-winter females can virtually lack the crescent altogether (or show just a ghost of it). On such birds, their whiter throat and white belly scaling may be more obvious (as well as the whitish fringes to the wing feathers). *Juvenile* Darker than juvenile Blackbird, lacking the latter's rufous tones. It completely lacks a breast crescent and most distinctive is broad chocolate and cream barring on the entire underparts; it also has a pale throat and, usually, a noticeable whitish patch on the neck-sides. Juvenile plumage is lost in a July–September body moult, prior to autumn migration.

Aberrant Blackbirds Occasionally show white feathering in their plumage and, when this is present on the breast, it can cause confusion with Ring Ouzel. Confusion could also arise with first-winter male Blackbirds of the so-called 'stockamsel' type (from Germany and Poland) which have a dull bill and eye-ring, browner wings, a paler chin and heavy pale fringing to the underparts feathers. Because of such oddities, it is imperative that the entire suite of structural and plumage differences is noted when identifying a Ring Ouzel, especially those seen in atypical locations or at unusual times of the year.

Song and Mistle Thrushes

Where and when Song Thrush is a common resident throughout Britain and Ireland. It is the most familiar thrush, most commonly found in woodland, but also in farmland and gardens. Mistle Thrush is less common, but still fairly numerous. It usually occurs in open countryside, parks and large gardens.

Song Thrush *Turdus philomelos*

Behaviour More retiring and less demonstrative than Mistle Thrush, being rather skittish and usually very wary of humans, preferring to keep within easy reach of cover.

Structure and plumage Much smaller, more compact and more evenly proportioned than Mistle Thrush, with warmer brown upperparts and rather rich buff underparts, particularly

Juvenile Song Thrush. Heavily speckled and rather yellowish.

Flying thrushes: note underwing colours.

Redwing

Song Thrush

Fieldfare

Mistle Thrush

Adult Song Thrush. Olive-brown above, uniform wings and tail. Breast flushed yellow-orange.

Adult Mistle Thrush. Large, pale, grey-brown, patterned wings. Appears 'open-faced'. Note white tail corners, obvious in flight (far right).

Juvenile Mistle Thrush. Very pa

on the breast and flanks; the underparts show heavy, rather arrow-shaped spotting. A buff eye-ring and subdued fore-supercilium are usually quite noticeable. Juvenile has buff streaks on the mantle and scapulars but, after the late summer/early autumn post-juvenile moult, ageing is difficult (note that both adults and first-winters show two buff wing-bars).

Flight identification Similar to Redwing *T. iliacus* but the most obvious difference is the latter's noticeable whitish supercilium and the rather densely streaked breast and flanks. Song Thrush has orangey-yellow underwing-coverts, whereas Redwing's are reddish (although this difference may be very difficult to see in flight). Structurally, Redwing is more compact, shorter-tailed and has a consistent habit of only partially closing its wings between bouts of flapping. Other behavioural differences are also useful: when flushed, Song Thrush usually flies low and dives back into cover, whereas the more skittish Redwings (which often occur in flocks) fly high, usually landing in the treetops or flying into the distance.

Song and calls The song is loud, clear and penetrating, consisting of a varied series of repeated phrases, delivered with great gusto, often from a prominent perch. It starts to sing very early (often by November) and continues well into July, after most other birds have finished. The usual call is an unobtrusive *tic*, often given by overhead migrants; also a dry chattering alarm, *dj-dj-dj-djup*, when flushed, or a longer, more anxious *dji-dji dji djip djip djip djip*. Redwing has a distinctive flight call: a thin, drawn-out *zeeeep*, a familiar sound from nocturnal migrants on autumn nights. At rest, Redwing also gives an abrupt but subdued, slightly muffled *ick*, while various alarm calls include a loud, deep rattling *chrrrrk*, a hard *trr trr* or an abrupt *dip djip*, more similar to Song Thrush. In spring, flocks of Redwings may give a disjointed whistling subsong.

Mistle Thrush *Turdus viscivorus*

Behaviour Often occurs in pairs, small parties, or even small flocks. Less retiring than Song Thrush and often found well away from cover. Bold and upright when feeding on the ground, often moving with a strong, bounding hop. Feeds on berries in autumn and winter, often vigorously defending its food supply (particularly mistletoe).

Structure and plumage Mistle Thrush is a bold, 'necky', upright thrush. Much greyer above than Song Thrush, with noticeable whitish-buff fringes to all of the wing feathers (those on the greater coverts may form a prominent pale mid-wing panel). At a distance has rather a bland, greyish face with a prominent dark eye, but closer views reveal a complicated pattern with a subdued vertical dark line below the eye, hollow pale centre to the ear-coverts, subdued dark border to the rear, as well as a distinctive whitish ear-covert surround (often reduced to an ill-defined pale crescent). The underparts are rich orangey-buff when fresh (whiter when worn) and much more boldly *spotted* than Song Thrush, *right down to the flanks and belly*. The spots often coalesce on the breast-sides to form a fairly solid dark brown patch. Until their late summer/autumn body moult, juveniles appear very pale, with whitish spotting, streaking and dark flecking on the upperparts.

Flight identification Obviously larger, longer-winged and longer-tailed than Song Thrush, with a slower, less hurried but more undulating flight, often at treetop height (often landing in the treetops). White underwing-coverts are obvious, while buffish-white tips to the outer-tail feathers may be particularly noticeable on take-off or landing. When seen overhead, easily confused with Fieldfare *T. pilaris* but note the latter's (1) dark undertail, (2) heavily streaked throat and upper breast on a peach or orangey background, and (3) its diagnostic muffled *chack chack chack chack chack* call; Fieldfare's flight tends to be more direct.

Song and calls Its far-carrying, wistful song is given in a series of disjointed and rather hesitant phrases, mellow and ethereal in quality. The tone is more similar to Blackbird *T. merula* than the clearer, repeated phrases of Song Thrush. Like Song Thrush, it is often heard as early as mid November, continuing until late spring. The most distinctive call is a loud, angry, grating *trrr-rr-rr-rr-rr-rr*, often likened to the sound of an old-fashioned football rattle; quite persistent when alarmed.

Common and Thrush Nightingales

Where and when Common Nightingale *Luscinia megarhynchos* breeds in s. England, mostly south of a line from the Severn to the Wash. Numbers have seriously declined in recent decades and it is now increasingly confined to the south-east. Thrush Nightingale *L. luscinia* is remarkably rare in Britain, although it is almost annual in spring. It currently averages five records a year, mainly in May and early June, with most in the Northern Isles and on the east coast. Rarer in autumn, with a scatter of records from mid August to late October.

Songs and calls In spring, the best chance of locating either species is by song. That of Common Nightingale is stunning, with deep, guttural noises, a rapid chugging *jug jug jug*, high-pitched 'fluted' whistles and a distinctive recurring *lu lu lu lu*. Although similar, the song of Thrush Nightingale is more powerful, but slower, more monotonous and somewhat reminiscent of a 'mellow' Song Thrush *Turdus philomelos*, being higher-pitched than Common Nightingale. Like that species, the individual phrases are often repeated. It also has harder notes, including 'tongue-clicking rattles' and some grating phrases that recall Sedge Warbler *Acrocephalus schoenobaenus*. The *lu lu lu lu* crescendo of Common Nightingale is given less commonly, but it has a distinctive far-carrying deep *choc choc choc choc*, similar to the *jug jug jug* of Common, albeit louder (Beaman & Madge 1998). Common Nightingale can also be located by some distinctive calls, notably a deep *tuc* and very distinctive low, deep, frog-like *trrrrrrr* (and variations of both) as well as a loud, very penetrating *wheet* or *peep*. All three calls are given on the breeding grounds and the first two also from migrants. Thrush Nightingale has similar calls but they are more clipped and higher-pitched, while Common's penetrating *wheet* is replaced by a monosyllabic *whit* (Cramp 1988).

Character and plumage Thrush Nightingale resembles Common Nightingale in that it is an alert, sturdy bird with plain plumage and a large eye, accentuated by a narrow pale eye-ring. It too skulks close to or on the ground, where it is rather upright as it hops along on rather long, pale pink legs. Its rounded tail is often slowly raised up and down, wavered and partly cocked. Whereas Common Nightingale has bright rufous-coloured upperparts with a slightly more cinnamon-rufous tail, Thrush Nightingale is altogether darker, duller and more earth-brown, showing less contrast between the upperparts and the tail. The latter is slightly more rufous than the rest of the upperparts but lacks the obvious orangey tones of Common Nightingale. This difference in tail colour is readily apparent in flight, even as the bird dives into cover. The other area on which to concentrate is the underparts. Common Nightingale is smoothly buffy-white below, often with a faintly darker breast, and a velvety grey tint to the faint supercilium and neck-sides. Thrush Nightingale has a noticeably darker breast (similar in shade to the upperparts) that is delicately and faintly mottled. The mottling can be hard to see but should be apparent in a good view. It also has browner flanks. Often there is a more distinct lateral throat-stripe, a somewhat thicker, stubbier-looking bill and many show an obvious yellow gape line, rather like that on recent fledglings (although many Common Nightingales also show this to some extent). A minor difference is that Thrush Nightingale has a shorter first primary than Common, not extending beyond the primary coverts at rest (on Thrush Nightingale it falls 1–10mm *short* of the tips of the primary coverts, on Common 1–5mm *beyond*; Svensson 1992). This is, of course, of little use as a field character but it may be visible on good-quality photographs (and is an essential feature to check in the hand).

Ageing In autumn, first-winters can be separated from adults by the presence of pale spots at the tips of the greater coverts and tertials, and sometimes on the uppertail-coverts. On adults, these feathers are plain brown. In spring, many first-years retain traces of these marks and such birds also have more worn tail feathers (Svensson 1992).

Pitfalls Some Common Nightingales can also show fairly prominent lateral throat-stripes as well as a grey wash to the breast, which can manifest itself as apparent mottling (Rogers *et al.* 2005). This is particularly true in autumn, when pale tips to the grey-tinged breast feathers can give the appearance of a diffusely spotted breast. A slight breast-band can also be seen on spring birds, with very faint mottling on the breast-sides. Also, some Thrush Nightingales are warmer brown above than others, with occasional individuals being very problematic (Beaman & Madge 1998). Consequently, caution must be exercised when identifying a potential Thrush Nightingale, especially in autumn.

Thrush Nightingale. Note dark brown plumage with grey or olive tones. Dark lateral throat-stripe contrasting with paler throat and submoustachial stripe, dappled grey breast; dark tips to undertail-coverts often evident.

Nightingale. Warm rufous, especially on rump and tail. Pale lores and vague supercilium, non-contrasting throat, underparts creamy. 1st primary longer than primary coverts.

Evenly spaced primary tips. Note 1st primary very short, not reaching tips of primary coverts.

Some birds ambiguous. This Nightingale is greyer than usual, though the tail and rump are still rufous. In long, careful observation the length of the 1st primary can be judged.

Golzii Common Nightingale

Cramp (1988) recognised three races of Common Nightingale: nominate *megarhynchos* is replaced in the Caucasus, e. Turkey and the Middle East by *africana*. Further east in central Asia is the race *golzii* (formerly *hafizi*, but the name *golzii* has been shown to have priority: Dickinson 2008). Differences between the three races appear to be clinal, but *golzii* differs markedly from nominate *megarhynchos* in being paler overall – greyer above, sandy-buff on the breast, and white on the belly and flanks – with whitish lores and a rather distinct fore-supercilium. The fringes to the flight feathers are sandy-grey, and pale tips to the median and greater coverts form two pale wing-bars; the axillaries and underwing-coverts are cream (*africana* is somewhat intermediate). There are three British records of *golzii* (all in October). Photographs of one in the Isles of Scilly in 1987 show a surprisingly distinctive bird which, as well as being colder sandy-grey above and whiter below, was larger and longer-tailed than the nominate, with noticeable pale tips to the greater coverts and uppertail-coverts (Bradshaw *et al.* 2004, Rogers *et al.* 2004).

References Beaman & Madge (1998), Bradshaw *et al.* (2004), Cramp (1988), Dickinson (2008), Rogers *et al.* (2004), Svensson (1992).

Common and Black Redstarts

Where and when Common Redstart *Phoenicurus phoenicurus* is a summer visitor (late March–October) to mature deciduous woodland; it is commonest in N. and W. Britain (largely absent from Ireland). More widespread on migration, mainly April/May and July–October. Black Redstart *P. ochruros* is a rare breeder, mainly in English cities (especially London), often on industrial buildings. Commoner on migration (mainly March/April and October/November) in coastal areas of England, Wales and Ireland (most numerous in SW England), with small numbers in winter.

Habitat Often the first clue to identity. Common Redstart is a bird of trees and woods and, even on migration, is rarely found far from cover. Black Redstart prefers open ground, perching prominently on rocks and buildings; on migration, it often occurs in ploughed fields or on sheltered beaches.

Behaviour Both are sleek and energetic, swooping down to the ground to prey on insects; on landing, the tail is often 'shimmied' revealing the orange rump, uppertail-coverts and outer-tail feathers.

Plumage *Females/first-winters* Both species have bright orange tails (brighter and more intense on Common Redstart) but are easily separated by plumage tone. Female and first-winter male Common Redstarts are grey-brown above and buffy-brown below, often with a strong peach tint to the breast and flanks; there are narrow buff fringes to the tertials and a faint dark malar stripe; the eye is large, with a faint pale eye-ring. Older, or worn late summer females, may acquire male-like characters, such as white mottling on the forehead, a slightly darker throat, greyer upperparts and a stronger orange tone to the underparts. Female and first-winter Black Redstarts are plain slate-grey at all times (never showing pale buff or peach tones) but they vary in plumage tone to some extent (beware of slightly paler individuals); most males do not acquire black on the face and throat until their first spring or later, so first-winter males are impossible to separate from females. *Males* Highly variable in autumn. Both species have only one complete moult each year (in late summer) so the males' breeding plumage is acquired by abrasion, pale feather tips gradually wearing off to reveal the full plumage. By late autumn, these tips often wear sufficiently on adult males of both species to reveal traces of black on the throat (and, on Common Redstart, white on the forehead and orange on the underparts). First-winter male Common Redstarts have broader pale feather fringes than adults so are, on average, more subdued. In contrast, most first-winter male Black Redstarts remain grey throughout their first year of life (and sing in this plumage the following spring). Note that, unlike adult males, those that do acquire black on the throat lack a white wing-panel and retain brown juvenile wings until the following late summer moult. *Juveniles* Easily separated. Black is *plain* slate-grey, similar to female. Juvenile Common is spotted and similar to juvenile Robin *Erithacus rubecula*, but is generally paler, buffer below and rather yellowish on the undertail-coverts; most importantly, it has an obviously orange rump and tail, the latter being habitually quivered, even when still very short. Juvenile Common Redstarts can disperse in July and appear well away from their breeding grounds, but true autumn migrants lose all traces of juvenile plumage in the late summer body moult, prior to migration. Juvenile Nightingale *L. megarhynchos* is also spotted, but has a duller, rounded, orange-chestnut tail (unlike Common Redstart, it lacks darker central feathers). It is larger

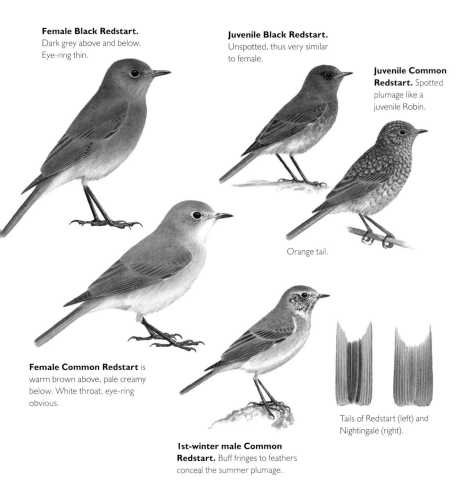

Female Black Redstart. Dark grey above and below. Eye-ring thin.

Juvenile Black Redstart. Unspotted, thus very similar to female.

Juvenile Common Redstart. Spotted plumage like a juvenile Robin.

Orange tail.

Female Common Redstart is warm brown above, pale creamy below. White throat, eye-ring obvious.

1st-winter male Common Redstart. Buff fringes to feathers conceal the summer plumage.

Tails of Redstart (left) and Nightingale (right).

and very furtive, hopping on the ground and rarely straying from cover; it also habitually raises and lowers its tail. Nightingale gives deep, almost frog-like *tuc* calls and variations (including a deep, guttural *turrrrr*).

Calls Common Redstart is very vocal both on its breeding grounds and sometimes on migration; commonest is a far-carrying *hoo-eet*, resembling Willow Warbler *Phylloscopus trochilus*, but louder, clearer, fuller and more emphatic. On the breeding grounds it may be given incessantly. They also give an abrupt, dry *tp*, *tup* or *ddp* alarm, particularly when flushed; in quality 'drier' and 'fuller' than Robin's familiar *tic*. This call is sometimes preceded by the *hoo-eet*. In contrast, Black Redstart is astonishingly silent on migration, but territorial birds in winter may give a loud, thin, whistled *weet*, often repeated.

Song Common Redstart has a very distinctive song: a loud rich, hollow, ringing *wee joo joo jooo ji-ji-ji-ji-jit* or, alternatively, *ji jur-jur-jurt jid-it* (the middle section 'hollow', the latter fast). Black Redstart's song is completely different: a variable short, fast, sweet, rising warble, *swoo swee-see-see-see-see*, often with a strange sound at the end, sometimes likened to a handful of ball bearings being shaken together.

Eastern Black Redstarts Adult males of the nominate *phoenicuroides* from Central Asia have more extensive reddish on the belly and underwing-coverts, with abrupt demarcation with the blackish throat/upper breast. A minority of first-winters males also show this pattern and such birds have occasionally been seen in Britain in late autumn. Such individuals should be photographed and expert advice sought (wing structure may be important in their identification).

European and Eastern Stonechats and Whinchat

Where and when European Stonechat breeds on heaths and rough ground, often along clifftops, mainly in w. Britain and Ireland (being absent from much of e. and central England). It is usually associated with gorse, heather and bracken. Following winter dispersal and migration the species' upland breeding sites are deserted and it is more widespread in lowland areas, frequenting all manner of rough ground. Numbers may crash after hard winters. Whinchat is a summer visitor, breeding on rough ground, particularly with bracken, mainly in uplands of Scotland, N. and SW England and Wales (scarcer in Ireland); it is more widespread on migration in April/May and July–October. Vagrant Eastern Stonechats occur at coastal sites, mainly in n. and e. Britain, in late September/October (a few also in winter); they currently average about five records a year (with as many as 32 in 1991).

European Stonechat *Saxicola rubicola* and Whinchat *S. rubetra*

Structure In shape, European Stonechat looks more rounded than Whinchat, with shorter wings (the primary projection is only about half the length of the overlying tertials). Whinchat is flatter-crowned and more angular-headed (an effect emphasised by its supercilium), flatter-backed, noticeably shorter-tailed and longer-winged (primary projection about three-quarters that of the overlying tertials).

Plumage *Males* Stonechat is easily identified in spring and summer by its black head and white neck patches. This plumage, however, is acquired not by moult, but by wear. In autumn and winter they are streaked above, their distinctive summer plumage being largely obscured by orangey-brown feather fringes, which gradually wear off. Male Whinchat is easily identified by its bold white supercilium, blackish crown and ear-coverts, and prominent white patches at the base of the tail. Some are much brighter and more striking (they wear darker and older males are apparently brighter than younger ones). They also show a noticeable white patch on the primary coverts and a less obvious one on the inner greater coverts. In autumn, they are difficult to age and sex as males, females and first-winters look similar (see below). Unlike Stonechat, males acquire summer plumage by a winter body moult. *Females* Stonechat is identified by its brown head *lacking an obvious supercilium*, darker upperparts, orange underparts, *lack of white tail patches* and larger white patches on the inner wing-coverts (obvious in flight); throat colour varies from brown to whitish. Note that, by late summer, some female Stonechats wear so dark that they resemble males. Female Whinchat has a *prominent, broad, whitish supercilium* from the bill, curving back towards the nape. Both the upper- and underparts are paler and buffer than those of Stonechat, being less orange in tone. Most obviously, female Whinchat also has *large white patches at the base of the tail*. Female Whinchat lacks Stonechat's *extensive* white on the inner wing-coverts. **Juveniles** Stonechat is generally very dark, streaked or speckled with buff on the head, breast and upperparts, the pattern recalling that of juvenile Robin *Erithacus rubecula*; it has two broad buff wing-bars and a dirty greyish breast. Juvenile Whinchat is paler than Stonechat, with buff shaft-streaks on the head and upperparts, and dark mottling on the breast; unlike Stonechat, it has a noticeable supercilium, but this is duller and less well defined than on the adults. Both species lose their juvenile plumage following a late summer body moult, *prior* to autumn migration. **Autumn** Ages and sexes are similar at this season, when all Whinchats are very buff and resemble females. They have noticeable but thin white crescent-shaped tips to the black mantle and scapular feathers, forming a distinctive scalloped appearance; they also show a pale buff panel on the tertials and secondaries (*cf.* Eastern Stonechat below). The white patches at the base of the tail can be dull buff and inconspicuous.

Behaviour Both species perch prominently atop vegetation, nervously flicking open their wings and tail (the latter often slowly raised and lowered). Stonechat's short wings produce a rather whirring flight, whereas the longer-winged Whinchat has a stronger, more purposeful flight. On autumn migration, Whinchat is rather more catholic in its choice of habitat, often occurring in cabbage and kale fields.

Calls and song Calls similar. Stonechat has a hard, emphatic *chit chit* or *swit, chit chit*. Whinchat has a softer, less emphatic *stit stit* often preceded by a soft, plaintive *hew*. Songs of both species are unmemorable. Stonechat's is a high-pitched warble, consisting of short, repetitive phrases that in quality suggest Dunnock *Prunella modularis*. Whinchat's is variable, but comprises short, rather feeble whistled phrases with some scratchy notes admixed. This may suggest a *Sylvia* warbler, with longer phrases vaguely suggesting Blackcap *S. atricapilla*; the song can also include some mimicry.

Eastern Stonechat *Saxicola maurus*

Taxonomic background The race of Stonechat breeding in Britain is *hibernans* (also in Brittany, coastal w. Iberia and possibly SW Norway). The Continental race *rubicola* breeds in mainland Europe east to Iran, as well as in NW Africa. The two forms in Siberia, *maurus* in the west and *stejnegeri* in the east, plus *hemprichii* and *variegatus* (see below) have now been split as a separate species: Eastern Stonechat. Note that Svensson (1984) could not distinguish *maurus* and *stejnegeri* in museum skins and Hellström & Wærn (2011) considered it inadvisable to separate vagrants (although a *stejnegeri* in Dorset in 2012 was identified by DNA analysis). All references below to European Stonechat refer to British and European *hibernans* and *rubicola*, which may be synonymous.

Autumn Eastern is very distinctive and, in a superficial view, is perhaps as likely to be passed off as Whinchat as European Stonechat. Eastern is typically Stonechat-like in shape, although the wings are slightly longer (with the primary projection about two-thirds the overlying tertial length). It is easily separated from European Stonechat by its pale, Whinchat-like plumage (being basically far less colour saturated). Buffy-brown above, streaked darker (but looking relatively plain at a distance) and pale buff below, with a distinctly whiter throat that contrasts with a variable but relatively faint orangey-buff breast. By late autumn, some first-winter males gradually acquire black on the head and throat, and this is usually obvious on later individuals (presumably acquired through abrasion, but see 'Spring males' below). Like Whinchat, it has a prominent buff tertial/secondary panel. When identifying Eastern Stonechat, concentrate on the rump (usually most obvious in flight): this is unstreaked, ginger or orange at first but wears to white later in autumn and shows as a discrete patch, sometimes having a 'wrap-around' effect (European Stonechat has a chestnut rump, usually streaked darker). Males can be sexed by their jet-black underwing-coverts and axillaries (pale greyish on females). To eliminate Whinchat, concentrate on structure (see above) and the following. **1 SUPERCILIUM** Eastern Stonechat has a relatively faint supercilium and looks rather plain-faced at a distance (Whinchat always has a prominent supercilium). **2 TAIL** Eastern has a black tail with narrow whitish edges and tip (Whinchat has prominent white tail patches, but occasional individuals have dull buff patches; see above). **3 MANTLE** Although streaked, Eastern Stonechat appears relatively plain (Whinchat is strongly patterned above and often delicately scalloped with white in autumn).

Spring male Unlike European Stonechat, first-winter Eastern appears to have a variable, but at least partial, pre-breeding body moult into summer plumage. Some first-summer males in spring are readily identifiable, appearing similar to autumn birds, being pale but with a black face and throat. Adult males vary from fairly straightforward to very difficult. 'Classic' spring birds are similar to European Stonechat but usually show: (1) more white on the inner wing-coverts (sometimes forming a large panel); (2) often very large neck patches that can form a collar; and (3) paler underparts with the orange or pink wash on the breast more restricted and better demarcated from the white flanks and belly. Also, the underwing-coverts and axillaries are jet black. Note, however, that some spring continental European Stonechats (*rubicola*), which may appear in southern coastal areas, can look similarly striking in the field, but have grey axillaries and duller underwing-coverts.

Caspian Stonechat *Saxicola maurus hemprichii*

There are three British records (to 2012) of this distinctive race of Eastern Stonechat, which occurs on the steppes west and north-west of the Caspian Sea, south to the e. Caucasus. It resembles Eastern Stonechat but, like Whinchat, has obvious white patches at the base of the tail (taking up at least one-third to half the tail length). Such birds should be photographed and expert advice sought.

References Hellström & Wærn (2011), Robertson (1977), Svensson (1984), Svensson *et al.* (2012).

Autumn female European Stonechat. Darker than Whinchat with no supercilium. No white in base of tail. Primary projection short.

Autumn Whinchat. Paler than Stonechat with strong supercilium, long primary projection and scalloped effect on upperparts. Note white in tail base.

Eastern Stonechat shows a pale rump in flight.

Note longer primary projection of Eastern Stonechat.

Autumn female Eastern Stonechat. Paler than European Stonechat with stronger supercilium and whitish rump. Note pale wing-panel.

Female European Stonechat appears very dark in flight.

1st-winter Whinchat. Note tail pattern. Adults show white on primary coverts and inner greater coverts.

Juvenile European Stonechat. Darker than Whinchat, breast more orange.

Juvenile Whinchat. Paler than juvenile Stonechat, underparts especially so. Note supercilium and white tail base. Compare primary projection.

House and Tree Sparrows

Where and when Although the familiar and once abundant House Sparrow *Passer domesticus* has declined enormously in recent years, it is still locally common and can be found almost anywhere, usually around habitation. Tree Sparrow *P. montanus* is now very scarce, having declined by an astonishing 93% between 1970 and 2008, a result of changes in agricultural practices. Although still widespread, it is now most likely to be found in arable areas in the Midlands and E. and SE England, with small pockets elsewhere. It is generally absent from n. Scotland and pastoral areas of SW England, Wales and W. Ireland.

Summer male House Sparrow. Extensive 'bib', plain ear-coverts and grey crown and rump. In winter the head pattern is subdued and the bill is pale at base.

Winter male House Sparrow

Tree Sparrow. Neat 'bib', white collar. ▼

Summer Tree Sparrow. ▶ Sexes alike. Chestnut crown, white collar, black 'cheek' spot. Rump brown, underparts buffy-brown.

Winter male House Sparrow. Diffuse 'bib'.

Juvenile Tree Sparrow. Crown olive-tinged.

Tree Sparrow × House Sparrow hybrid shows mixed characteristics.

Identification Female House Sparrow is rather plain grey-brown (tinged olive) and streaked black above, with an unobtrusive pale supercilium. Juveniles are similar but generally paler and fluffier, with brighter yellow at the bill base and, when young, a yellow gape line. Male House Sparrow and both sexes of Tree Sparrow are more similar, but their separation is straightforward: note Tree's brown crown, black ear-covert spot, white collar and buffer plumage tones. Whereas male House Sparrow has a large black bib that extends to the upper breast, Tree Sparrows of both sexes have a small bib confined to the chin and throat. Note, however, that in early winter, male House Sparrow is duller and also has a small bib (see 'Moult'). Juvenile Tree Sparrows resemble adults but, like juvenile House Sparrows, are rather fluffy with yellow at the bill base and a prominent yellow gape line when young; their head markings are also slightly subdued compared to the adults. Both species fly fast and direct, and give muffled chirps in flight.

Calls The harder, more abrupt flight notes of Tree Sparrow are distinctive once learnt, as are its hard, dry 'tecking' and chattering noises at rest.

Moult Adults and juveniles of both species have one complete moult a year, in late summer and autumn. As a consequence, freshly moulted male House Sparrows have subdued colouring; most obvious is that the black bib is obscured by grey feather fringes, and is largely confined to the chin and throat. As winter progresses, the grey fringes gradually wear to reveal the full extent of the black bib on the chin, throat and upper breast. Moulting males in autumn (particularly juveniles) may show an odd combination of female-like characters (particularly a pale supercilium) and male characters (grey on the crown and black on the lores and throat).

Hybrids Rare. That illustrated opposite is based on one that showed characters obviously intermediate between the two.

Spanish Sparrow *Passer hispaniolensis*

A very rare vagrant (nine records to 2012, many if not all of which have probably been ship-assisted). Resembles House Sparrow but males have a chestnut-brown crown, very white face and extensively black underparts with thick streaking extending onto the flanks. The mantle is also more heavily striped black, buff and white. In fresh plumage (autumn and early winter) much of the plumage is obscured by buff feather fringes that gradually wear off (although its basic pattern is still obvious). Its 'song' is high-pitched and rather musical compared to House Sparrow: *twoo-oo-lit... twoo-oo-lit....* Females are very similar to female House Sparrows but have a slightly larger bill and they are plainer-headed with diffuse black streaking on the breast and flanks (obvious on some, faint on others).

Grey, Yellow, Eastern Yellow and Citrine Wagtails

Where and when Grey Wagtail is typically associated with upland streams, breeding throughout much of Britain and Ireland except across a large swathe of e. England from Humberside to Essex, where it is thin on the ground. More widespread in winter, when there is a general retreat from high ground, a southward withdrawal and some emigration (often appears in city centres at this season, where it may even breed). Yellow Wagtail is a summer visitor from late March to October, breeding mainly in England (generally absent from the south-west), e. Wales and s. Scotland (but virtually absent from Ireland); it has seriously declined in recent years. It occurs in marshy areas and on agricultural land, with a tendency to nest in pea fields. Several continental races also occur (see below) as well as Eastern Yellow Wagtails from e. Siberia (this group is now increasingly considered as a separate species). Citrine Wagtail is a rare vagrant with a few in spring (April–June) but most autumn (mid August–October, with a peak in September). It has increased considerably since the mid 1990s, reflecting a westward spread into e. Europe; it currently averages 12 records a year (with as many as 21 in 2008).

Grey Wagtail *Motacilla cinerea*

Structure and plumage Easily identified: lively and restless with a very long tail (much longer than Yellow Wagtail's) which is energetically wagged, particularly when alighting. Unlike Yellow, the head and mantle are grey (with a very slight green tint), it has a narrow white supercilium and yellowish green rump. On juveniles and winter females, the underparts vary from pale yellow to washed-out peachy-yellow, the only intense yellow being on the undertail-coverts. Males tend to be yellower on the breast and, in summer, have a black chin and throat (as can some summer females). The legs are pinkish or pale brown (blackish in Yellow).

In flight A slim, streamlined wagtail with a long, quivering tail and a broad white wing-bar.

Calls A loud, abrupt, penetrating, metallic *tzip* or *tzizip*, sometimes prolonged into an almost musical *tiss-is-is-is-is-is*. Generally rather solitary, frequenting streams, rivers, rocky shores, farmyards and rooftops; occasionally in grassy fields (but it does not habitually feed among cattle). They are strongly territorial with squabbling frequently occurring between close-feeding birds.

Yellow Wagtail *Motacilla flava*

Structure and plumage A more evenly proportioned, shorter-tailed wagtail, green above and yellow below; habitually associates with cattle. Spring males are yellow and green on the head; females are duller green with a yellowish supercilium; all have a thin double white wing-bar and narrow white tertial fringes. Juveniles are duller and buffer (some, particularly females, being very colourless). They have a messy blackish lateral throat-stripe and blackish necklace on the lower throat (juvenile plumage is lost after the post-juvenile moult in late summer/early autumn, but migrants often retain traces of this plumage).

Call A very distinctive *swee-up* or *sweep*, useful when locating overhead migrants. Often occurs in small flocks and, in autumn, large numbers may roost in reedbeds.

Racial identification The racial complexity of Yellow Wagtails is notorious. Apart from spring males, most birders do not bother to racially identify them. Many of the extralimital races claimed in this country (e.g. Sykes's Wagtail *beema*) may just be intergrades or local variations. Such birds may appear intermediate between Yellow and Blue-headed (often with lavender-coloured heads) and are often referred to as 'Channel Wagtails' as they tend to breed in coastal SE England, where the two races frequently come into contact. The following is a guide to the various head patterns (races listed in approximate order of frequency).

British Yellow Wagtail *flavissima*

Our native breeding form, male easily identified by its green-and-yellow head. Even females show strong yellow tones to the supercilium and throat.

Blue-headed Wagtail *flava*

The common form breeding on the near Continent. Males are greyish-blue on the head with a white supercilium and often a paler area in the centre of the ear-coverts (or 'subocular patch'). Females can have slight bluish tones to the head and a whitish supercilium and throat, lacking the strong yellow tones of female *flavissima*.

Grey-headed Wagtail *thunbergi*

Breeds across n. Europe and n. Siberia. A rare late spring passage migrant on the east coast and very rare elsewhere. The male's head is dark blue-grey with generally darker, blackish-grey lores. Most importantly, it lacks a supercilium, although some may show a narrow, faint, short one behind the eye. Some also show a necklace of dark spots on the upper breast. Females are not always safely separable from *flava* but they typically have a narrower, less clear-cut supercilum (sometimes lacking) and slightly darker and more 'solid' ear-coverts.

Black-headed Wagtail *feldegg*

This distinctive form breeds in SE Europe, Turkey and central Asia. Currently 17 late spring records (to 2011). In fresh plumage, males have a glossy black head that extends well down the nape to merge with the green upper mantle. The throat is yellow but in the easternmost part of the range a greater proportion show a white submoustachial stripe. Some males have a black head with a white supercilium; currently, such birds are not considered acceptable but there is uncertainty as to whether they are intergrades or pure first-summer male Black-headed. Females are often readily identifiable by their blackish heads, like washed-out males. Most show greyer upperparts and whiter underparts than other European races. Often has a more buzzy call: *tzeeup*. When identifying *feldegg*, great care is needed to eliminate grey-headed *thunbergi*, which can look dark-headed in certain lights.

Spanish Yellow Wagtail *iberiae*

Breeds in Iberia and NW Africa. Very similar to *flava*, but males have a large white throat that contrasts sharply with the bright yellow breast. In this respect it is similar to its close Mediterranean neighbour, Ashy-headed Wagtail *cinereocapilla*. Compared to *flava*, *iberiae* is slightly darker and greyer on the head, usually lacks a white eye-ring below the eye and a pale subocular patch. The supercilium is often lacking before the eye, but this is variable. Females are similar to female *flava* but tend to show a narrower, less distinct supercilium and lack a pale subocular patch; most significantly, they too show a large, contrasting pure white throat. Rather harsh call is similar to Black-headed.

Ashy-headed Wagtail *cinereocapilla*

Breeds in Italy, Sicily and Sardinia. Very similar to *iberiae* and may be best combined with it (call is also similar). Like that form, males have a large white throat, but most lack a supercilium or show just a faint narrow stripe above and behind the eye.

1st-winter Grey Wagtail. Grey above, lemon-yellow below, strongest where it surrounds the base of the very long tail.

1st-winter Yellow Wagtail. Note ▶ white chin, buffy-yellow breast, olive rump, yellowish undertail-coverts.

Grey Wagtail in flight. Note wing-bar.

Yellow Wagtail in flight. Shortest tail of any wagtail. Light plain underwing.

Juvenile Yellow Wagtail. Note buff coloration, dull lemon yellow underparts with blackish 'necklace'

Yellow Wagtail. Some late-autumn Yellow Wagtails are grey and white, lacking green and yellow tones. They are similar to Citrine Wagtail (see text).

1st-winter Citrine Wagtail. Note grey and white plumage, broad supercillium extending around ear-coverts and thick white wing-bars.

Sykes's Wagtail *beema*

Breeds in Kazakhstan and SW Siberia. No longer on the British List; individuals resembling this form are much more likely to be intergrade 'Channel Wagtails' (see above). Genuine male *beema* is superficially similar to *flava*, but is slightly paler on the head with a longer and broader supercilium and whiter and more prominent subocular stripe, sometimes so obvious that most of the central ear-coverts look white. The upper throat and upper malar are also white (usually all-yellow in *flava*). Females resemble female *flava* but show a longer, broader and whiter supercilium, and more prominent subocular stripe.

Eastern Yellow Wagtail *Motacilla tschutschensis*

Genetic research indicates that the e. Asian forms *tschutschensis*, *macronyx* and *taivana* may be best treated as a separate species 'Eastern Yellow Wagtail'. As with other Far Eastern vagrants, birds resembling these forms occur in late autumn, mainly in the second half of October; some have remained to winter, with a penchant for sewage works. It remains to be seen, however, how the records will be treated by the relevant committees; currently, an analysis of their DNA is required. Birds resembling this form have a very consistent appearance, in many ways more reminiscent of first-winter Citrine. They are grey above and white below with a white supercilium, bordered above by a fairly thick dark grey lateral crown-stripe and below by a broad dark grey line across the lores; there is also a small but variable white subocular patch above a fairly discrete dark grey moustachial stripe. The white underparts show grey shading on the breast-sides and very faint streaking on the flanks. Whilst some calls may be similar to Yellow Wagtail, most distinctive is a slightly buzzy, more rasping *spzzeu* or *zeup* suggesting Citrine. A strange sibilant or slightly trilled *spspsp* was heard from one vagrant thought likely to be of this form. It also has a longer hindclaw (although there is overlap) and tends to show a stronger, slightly more dagger-like bill than Yellow, some with horn-coloured cutting edges and lower mandible. Note, however, that some European Yellow Wagtails are colourless grey-and-white birds, thereby clouding the issue. Any such bird seen outside the late autumn/winter occurrence period of Eastern Yellow Wagtail is far more likely to be an aberrant Yellow. See Collinson *et al.* (2013) for further information.

Citrine Wagtail *Motacilla citreola*

Structure and plumage Citrine is similar in shape to Yellow Wagtail, but slightly shorter-tailed; the bill may look longer and sturdier, and the head shape more angular. First-winter is grey above (slightly darker than White Wagtail *M. alba alba*) and whitish below, often with grey breast-sides, and some have a pale peachy tint to the breast. From behind, it resembles a White Wagtail, but from the front lacks White's obvious black necklace across the upper breast. Later-moulting individuals may retain at least some browner juvenile feathering, particularly traces of the juvenile's lateral throat-stripes and necklace, and should be aged as 'juvenile/first-winter'. Conversely, some late autumn first-winters are strongly primrose-yellow on the head. On all birds, the white tips to the median and greater coverts, and the white tertial fringes are much broader than Yellow Wagtail and very striking. Pay particular attention to the head pattern, the following features being characteristic of Citrine. **1** A broad whitish supercilium (sometimes tinged yellow) *extends right around the ear-coverts to form a complete 'ear-covert surround'* (prominence varies individually and may be obscured if, for example, the head is sunk into the shoulders). **2** Usually a pale centre to the ear-coverts, often appearing as a broad pale crescent below the eye ('subocular crescent'). **3** May also show a buff forehead (lacking on Yellow). **4** The supercilium may be emphasised by a narrow dark lateral crown-stripe.

Call Citrine has a very distinctive call: an almost buzzing *dzzeeup*, *tzzeeeep* or *tzzweeeeup*, like a rasping Tree Pipit *Anthus trivialis*. Caution is needed, however, as some calls of Citrine are similar to Yellow and some races of Yellow Wagtail (e.g. Black-headed *feldegg*) and Eastern Yellow Wagtail, give 'buzzing' calls. Note that the latter appears mainly in late October, whereas Citrines appear from early August and peak in September. Eastern Yellows are similar to Citrine in their overall grey-and-white plumage tones, but they *lack Citrine's pale ear-covert surround*. They have more solidly grey ear-coverts (at best only a narrow pale crescent below the eye) and narrower white wing-bars and tertial fringes; their structure is also more typical of Yellow Wagtail.

Spring birds Three races: adult male of the northerly nominate *citreola* has a pale yellow head and grey back, with a black collar; adult male of southerly *werae* is generally paler yellow, rarely has a black collar and little if any grey on the flanks; the even more southerly *calcarata* is darker yellow and has a black back. Spring males in Britain have resembled *citreola*. First-summer males tend to show messy black mottling on the crown and ear-coverts. Spring female is similar in pattern to first-winter (see above) but has bright yellow on the face and duller yellow underparts.

Pied and White Wagtails

Where and when Pied Wagtail *Motacilla alba yarrellii* breeds commonly throughout Britain and Ireland; in winter, it withdraws from high ground and some emigrate. The Continental race *alba* (White Wagtail) is a passage migrant from early March to May and from mid August to October (a few linger into November in milder south-western areas). As most passage birds are en route to and from Iceland, they are commonest in western areas. It occasionally breeds in n. Scotland, sometimes intergrading with Pied.

Spring *Plumage* Adult male Pied is easily identified by its jet-black back and scapulars. Like both sexes of White Wagtail, female Pied's mantle and scapulars are grey but they are darker than White's, varying from blackish-grey, with black feathers admixed, to dark olive-grey; even greyer females look dark and sooty, with little contrast between the back and the head. Both sexes of White Wagtail have a pale, clean looking ash-grey mantle which, unlike Pied, contrasts strongly with the wings and the black-and-white head, producing a smart, clean-cut appearance. Like the back, the rump is pale grey, becoming darker grey towards the upper-tail-coverts, which are blacker. On Pied, the rump is blackish or blackish-grey (but is often cloaked by the folded wings). Female Pied is quite a dark *sooty-grey* on the breast-sides and flanks but White has *pale grey* restricted largely to the breast-sides. In consequence, the flanks appear *largely white*, often with pale grey restricted to a narrow strip along the top of the flanks, immediately below the folded wing (the obviousness of this depends to some extent on how the wing is held). The predominantly white flanks are readily apparent in flight and this enables even migrating White Wagtails to be identified with some degree of confidence. Minor points include White's duller, browner wings, while male White (because of its paler mantle and whiter flanks) may give the *impression* of having a larger bib. **Sexing** Spring adult White Wagtails can often be sexed by reference to the crown. In males, the black is clearly

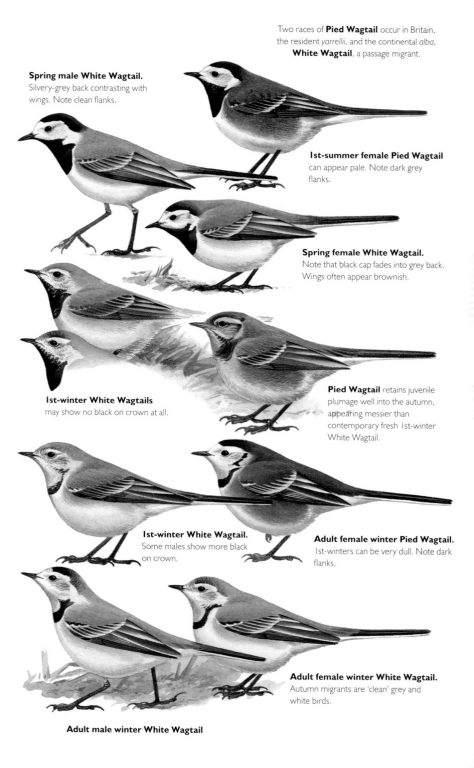

Two races of **Pied Wagtail** occur in Britain, the resident *yarrellii*, and the continental *alba*, **White Wagtail**, a passage migrant.

Spring male White Wagtail. Silvery-grey back contrasting with wings. Note clean flanks.

1st-summer female Pied Wagtail can appear pale. Note dark grey flanks.

Spring female White Wagtail. Note that black cap fades into grey back. Wings often appear brownish.

1st-winter White Wagtails may show no black on crown at all.

Pied Wagtail retains juvenile plumage well into the autumn, appearing messier than contemporary fresh 1st-winter White Wagtail.

1st-winter White Wagtail. Some males show more black on crown.

Adult female winter Pied Wagtail. 1st-winters can be very dull. Note dark flanks.

Adult female winter White Wagtail. Autumn migrants are 'clean' grey and white birds.

Adult male winter White Wagtail

341

demarcated from the mantle, whereas on females it is less clear-cut, sometimes with grey admixed. On first-summer females (and some males) the black is reduced or even lacking, giving them a distinctive pale grey crown and nape.

Autumn *Moult* Although separating Pied and White Wagtails in spring is relatively straightforward, there is often confusion in autumn, when grey-backed juvenile Pied confuses the issue. However, there is a very simple short cut to the separation of the two forms at this time that renders the identification of White Wagtail straightforward. White Wagtails moult before they migrate, but the more resident Pied has a protracted moult throughout much of the autumn. Whereas Pied Wagtails commence their moult in mid July (on average), most do not complete it until the end of September or early October, taking on average 76 days to do so. White Wagtails, however, start their moult in early July but, because they need to moult before they migrate, they take just 46–48 days to complete it (*BWP*). The upshot is that late August and September White Wagtails have already completed their moult and are in 'clean', immaculate and fresh winter plumage at a time when local Pieds are still moulting, appearing scruffy, 'moth-eaten' and dishevelled. As far as young birds are concerned, in late August and September, one is usually comparing immaculate *first-winter* Whites with scruffy, moulting Pieds that retain significant amounts of weak, fluffy *juvenile* body plumage. ***Plumage*** First-winter White has a pale grey crown and mantle, and a neat and contrasting *narrow*, crescent-shaped necklace on the lower throat/upper breast (broader and deeper on adults); this contrasts strongly with the *almost all-white underparts*. Not only are juvenile/first-winter Pieds scruffy at this time, but they still retain their black juvenile central breast-patches. In addition, autumn Pied is of course a darker, sootier grey on the upperparts (adult males are largely black above), but *the* important feature of autumn Pied (even after both species have completed their moult) is its *extensive dark sooty-grey breast-sides and flanks*.

Richard's, Tawny and Blyth's Pipits

Where and when Tawny Pipit is a rare spring and autumn migrant, mainly in May and late August to September, with stragglers into October. Most occur on south and east coasts; it is rare in summer. Richard's Pipit is a late autumn visitor from Siberia, seen mostly at well-watched coastal sites from mid September to November; some overwinter and there are occasional early spring records of wintering birds heading east. Therefore, any large pipit seen before mid September should be a Tawny, whereas, after early October, Richard's is far more likely. Tawny currently averages just 11 records a year, having declined from an average of 36 in the 1980s, with a peak of 56 in 1983. This compares with a current average of 120 Richard's, with a peak of 353 in 1994. Although Blyth's Pipit was first recorded in 1882, the first modern record was in 1988. Since then, it has proved to be an increasingly regular late autumn vagrant between late October and early November, with a few staying into winter. There had been 22 records by 2011, most frequently in Shetland and Scilly.

Richard's *Anthus richardi* and Tawny Pipits *A. campestris*

Structure and behaviour Both are large, slim, rather wagtail-like pipits, with long, orangey legs; they feed with a start-stop action, sometimes wagging the tail (Tawny much more frequently). Whereas both are liable to be seen running on short grass, Richard's tends to prefer longer grass, which Tawny avoids. Like Skylark *Alauda arvensis*, Richard's may hover with a spread tail before alighting (Tawny rarely does this). Richard's can appear almost thrush-like in its feeding behaviour. The flight of both is strong, undulating and rather wagtail-like.

Adult Tawny Pipit

Significantly smaller, slighter, less robust and somewhat slimmer-billed than Richard's (*c.*10% shorter and perhaps 20–30% lighter). Adult Tawny is easily separated by its pale sandy plumage, which lacks significant streaking. It has a whitish supercilium, narrow dark eye-stripe (*including the lores, which are pale on Richard's*), and narrow moustachial and lateral throat-stripes. Most significant is a black 'bar' on the median coverts, which contrasts strongly with the rest of the pale plumage (formed by large black centres to the feathers).

Richard's Pipit

A rather bold, upright pipit, with a sturdy, deep-based, almost thrush-like bill that usually appears uptilted from the face. It has rather an 'open face', with *pale lores*, a dark eye-stripe behind the eye and a thin dark crescent-shaped moustachial stripe below the eye; however, the head is dominated by a broad creamy supercilium that extends around the back of the ear-coverts to form *a complete 'ear-covert surround'* (although note that some Richard's have pale ear-coverts that render the ear-covert surround inconspicuous). The upperparts are well streaked, although in fresh plumage the streaking is subdued as broad brown feather fringes largely conceal the black feather centres. There are two buff wing-bars (wearing whiter) and broad buff tertial fringes. The underparts are creamy-white, often with warm orange-buff flanks. A narrow lateral throat-stripe expands into a large dark blotch at the base of the throat before merging into a gorget of fairly random brown streaks on the upper breast. In flight, it is large and rather long-tailed with a strong, undulating flight. When hovering prior to landing, it often spreads its tail to reveal extensive white in the two outer feathers.

Autumn moult of juvenile Tawny Pipit Identification problems in autumn relate to the fact that Tawny Pipits are often double-brooded; this means that late young may sometimes remain in the nest well into August. Consequently, later fledged juveniles do not have time to complete their post-juvenile body moult prior to autumn migration. They either suspend their moult during migration, or do not start it until arriving in their winter quarters. So, whereas first-brood Tawny Pipits have time to complete their moult and acquire adult-like first-winter plumage by late August, some second-brood birds do not reach this state until January. Because of this, young autumn Tawny Pipits are very variable: they are either in adult-like first-winter plumage, in full juvenile plumage, or various stages between the two. Paradoxically, it is the later occurring Tawny Pipits that tend to look the most juvenile.

Juvenile Tawny Pipit Richard's (all ages) and juvenile Tawny can look surprisingly similar in superficial views. Concentrate on the head pattern, particularly the lores. Tawny has *a distinct dark line between the bill and the eye*, whereas, on Richard's, the lores are plain and rather creamy, producing a rather pleasant, open-faced expression. Juvenile Tawny also has a stronger dark eye-stripe behind the eye, a better-defined dark moustachial stripe and lower border to the ear-coverts, all of which combine to produce a more 'severe' facial expression. Other differences are as follows. (1) On juvenile Tawny, the feathers that form the streaks on the mantle are finely fringed with white, so that the upperparts show lines of *scallops*, rather than solid streaks. However, these fringes wear as autumn progresses so that the scalloped effect is less obvious on later birds. (2) Juvenile Tawny usually lacks Richard's large dark blotch at the base of the lateral throat-stripe, while the breast is *delicately and profusely* streaked, forming a deep gorget of *fine and even streaking*. Richard's breast is more randomly and diffusely streaked. (3) Tawny's slightly smaller size and more delicate structure mean that it can appear somewhat wagtail-like. (4) Tawny has a shorter hindclaw (7–12mm) whereas Richard's is ridiculously long (13.5–19mm; Svensson 1992) best seen when perched on a wire fence.

First-winter Tawny Pipit As first-brood Tawny Pipits may have completed a late summer post-juvenile body moult by August, many first-winters are adult-like (pale, sandy and relatively unstreaked). Consequently, such birds are readily separable from Richard's Pipit.

Ageing in autumn Regardless of variations in their body moult, both species can usually be aged by reference to their median coverts. Young birds usually show a *mix* of old juvenile feathers (black, narrowly and contrastingly fringed white) and new adult feathers (slightly longer, black and *diffusely fringed buff*).

Calls Superficially similar, but readily distinguishable with practice. Richard's classic call is a loud, deep, rather explosive, sparrow-like *chreep* or *chree-up*. When excited, two or three calls may be run together. Conversely, it may sound softer, quieter and sometimes shorter and more clipped (although perception varies according to distance and the wind). These variations may lead to confusion, so do not expect a Richard's to always give classic calls. Tawny's call is also sparrow-like, but is softer, weaker and less rasping: *chee-up*, *chlee-up* or *tree-up*. Like Richard's the call sometimes sounds more musical, a sparrow-like *chlup* or *schlup* while, conversely, it can sound like a harder *trip*.

Other pitfalls Richard's and Tawny should not be confused with other species but mistakes have occurred. Occasional aberrant pale sandy-coloured Rock Pipits *A. petrosus* can suggest Tawny, as can summer-plumaged Water Pipits *A. spinoletta* (but the latter are likely to be seen only in late March/early April). Both species can be eliminated by their blackish legs and *pseep* calls, as well as habitat (rocky coasts and freshwater environments respectively). Tawny has also been confused with dull, buffy juvenile and first-winter Yellow Wagtails *Motacilla flava*.

Blyth's Pipit *Anthus godlewskii*

General Blyth's Pipit shows few diagnostic features. Instead it exhibits a host of minor differences that create the impression of something distinctly 'different'. The confusion species is Richard's Pipit, but Tawny also must be considered, particularly those late autumn birds that retain significant juvenile body plumage (see above). Remember that the one key feature that separates Tawny from Richard's and Blyth's is the eye-stripe: on Tawny it extends from the bill *across the lores* and through the eye, whereas on Richard's and Blyth's it extends from the eye back, leaving the lores plain, pale and buffy.

Size and structure The plumage of Blyth's is similar to Richard's so it is size and structure

Juvenile Tawny Pipit (rare here). Note spotting on flanks and scalloped back.

Adult Tawny Pipit. Pale, poorly marked back and plain breast. Rather wagtail-like. Note short hindclaw. ▶

◀ Adult type

Adult and juvenile feathers.

Richard's Pipit. Lores may appear dark depending on angle of viewing. ▼

Compare Richard's Pipit tail pattern and median coverts with Blyth's Pipit.

Richard's Pipit. Large and robust, rather thrush-like. Note pale lores, wide lateral throat-stripe. Flanks are strong buff rather than creamy. Dark streaked upperparts. Dark tail with white outer edges. Note long hindclaw.

Tawny Pipit moulting from juvenile to 1st-winter. Note dark lores.

1st-winter Blyth's Pipit. Delicate necklace over apricot pectoral band, pale lores. ▼

Blyth's Pipit shows less white in tail than Richard's Pipit. ▼

Adult type ▼

1st-winter Blyth's Pipit. Evenly ▶ proportioned, fine bill more like a smaller pipit. Buffy below fine neat streaking.

Adult and juvenile feathers.

Adult type median coverts squarer than those of Richard's Pipit.

Hindclaw shorter than Richard's Pipit, gently curved.

that are likely to ring 'alarm bells'. Blyth's resembles a rather small, delicate Richard's (in many ways more similar to Tawny in size and structure). Perhaps the most obvious initial structural difference is the bill, which appears distinctly smaller and slimmer than Richard's. Nevertheless, it is quite deep-based and rather short, with a straight culmen and quite a pointed tip. It lacks the heavier, wedge-shaped, more thrush-like bill profile of Richard's. The head too is rather small and it has a slender neck (frequently extended). All of this combines to form a daintier impression than the more robust Richard's. Its relatively small size and slim appearance are also apparent in flight, when it looks slim and compact, lacking the bulky body and pot-bellied appearance often shown by Richard's. In addition, it has a distinctly shorter tail, making it look more evenly proportioned than Richard's, perhaps suggesting a large Tree Pipit *A. trivialis*. Some vagrants feed with flocks of Meadow Pipits *A. pratensis* and, in flight, they can be surprisingly difficult to pick out.

Plumage Basically similar to Richard's but the following are the main differences. **1 PLUMAGE TONE** Distinctly deep pale buff on the underparts (deepest on the flanks) perhaps similar in tone to autumn Whinchat *Saxicola rubetra*. The upperparts are well lined with broad blackish streaks. **2 HEAD PATTERN** Subtly but distinctly different from Richard's. The crown is regularly and evenly lined with black, lacking any hint of the dark lateral crown-stripes of Richard's. The creamy supercilium is quite discrete, being confined to above and immediately behind the eye, and lacking Richard's 'ear-covert surround'. The lateral throat-stripe is much weaker and indistinct compared to most Richard's, being fine and very narrow; most importantly, it ends in a small and rather faint patch (lacking Richard's much more obvious dark neck blotch). **3 BREAST STREAKING** Immediately below the lateral throat-stripes is a deep breast-band of *profuse fine brown 'pencil' streaking*, which reaches quite high on the breast and is deepest at the sides, narrowing in the middle. It is, therefore, distinctly different from the heavier, more random streaking shown by Richard's. The streaking recalls the breast-band of a Skylark.

Calls Very important. Blyth's gives several calls, all of them quiet and soft compared to the more familiar loud, rasping, sparrow-like call of Richard's (*chreep* or *chree-up*). Blyth's utters a soft and squeaky *schleup*, a soft, abrupt *schlup* or *tchlup*, or quite an abrupt, soft, musical *schleeu schleeu*. These may be followed by a quiet but distinctive, subdued *djup djup djup* or a slightly musical *skiew djup djup*. At times, the *djup* calls may sound rather finch-like (recalling Linnet *Linaria cannabina*). An important point to bear in mind is that Richard's sometimes gives softer or atypically abbreviated calls, so it is important to hear a series of calls before drawing firm conclusions.

Behaviour When feeding, Blyth's has an energetic, nervous manner, often raising its neck so that it appears quite upright. Like Richard's it often tilts its head to one side when feeding, looking at the ground in a thrush-like manner. It may feed with flocks of Meadow Pipits.

The 'hard features' Although, with practice, Blyth's is a distinctive bird in its own right, three 'hard features' have traditionally been seen as fundamental to its identification and these should be checked if possible (good photographs will be helpful). **1 MEDIAN COVERTS** In autumn, young Richard's, Blyth's and Tawny Pipits have varying combinations of juvenile and adult median coverts. The juvenile feathers are similar on all three species, being black with a clear-cut white fringe. The black centres are either pointed or rather rounded but with a slight point extending down the shaft. When present, the adult feathers are more diffusely fringed with buff, but the important difference is that, on Richard's, the black centres of the adult feathers are pointed, whereas on Blyth's they are more sharply defined and more squarely cut off (although often with a slight point down the shaft). When the adult median coverts are present, it is advisable to concentrate on the central feathers as these are the most consistently different (Heard 1995). **2** TAIL PATTERN Both species have extensive white on the outer-tail feathers, this being prominent in flight, but the pattern on the penultimate tail feather differs. On Richard's, the white extends up the inner web in a long narrow point, immediately adjacent to the shaft, mirroring the pattern of the outer-tail feather. On Blyth's, the white on the penultimate feather is broader, shorter, more restricted and more roundly or squarely cut off. Some Richard's also show restricted white on the penultimate tail feather but it is still narrow and pointed, not appearing as a distinct broad patch, as it does on Blyth's. **3** HINDCLAW Richard's has a very long and fairly straight hindclaw, showing only a gentle curve. It is usually over 13mm in length and as long as 24.5mm on nominate *richardi*, which occurs in Britain. Blyth's has a shorter and more arched claw that is 9–13.5mm long, and is thus more similar to the hindclaw of Tawny Pipit, which measures 7–12mm (Svensson 1992). The problem is that the hindclaw can be frustratingly difficult to see in the field (best seen if the bird lands on a wire fence).

Plumage variation Like Richard's and Tawny Pipits, Blyth's may vary in the extent of its post-juvenile moult. Some look very fresh, with broadly fringed upperparts feathering, showing broad pale feather fringes that severely reduce the streakiness of their upperparts. Conversely, unmoulted individuals can look darker, worn and streakier, while some rather dishevelled and 'moth-eaten' birds are likely to be in active moult from juvenile to first-winter plumage.

References Bradshaw (1994), Heard (1995), Svensson (1992).

Small pipits: Meadow, Tree, Red-throated, Olive-backed and Pechora Pipits

Where and when Meadow Pipit is an abundant breeding, passage and wintering species throughout much of Britain and Ireland. Tree Pipit is a summer visitor (mid April to October) breeding on heaths, woodland edges, young conifer plantations and rough ground with scattered trees. Most numerous in the north and west (but very rare in Ireland). Red-throated Pipit is a vagrant, currently averaging ten records a year, mainly in May and September/October (with as many as 47 in 1992). Olive-backed Pipit currently averages 13 records a year (with a remarkable influx of 43 in October 1990); it occurs mainly in late September and October, with a few into November; there have also been a few winter and spring records (the latter undoubtedly wintering birds en route back to Siberia). Both species are most frequently recorded at well-watched coastal sites, mainly on the Northern Isles and the British east coast. Formerly a great rarity, Pechora Pipit is now almost annual in late September and early October (with single spring and November records). It averages about three a year, with peaks of ten in 1994 and 2009. It occurs almost exclusively on the Northern Isles, with only the occasional record in the south.

Meadow *Anthus pratensis* and Tree Pipits *A. trivialis*

These two species are similar, their separation being complicated by seasonal and individual variation in plumage tone. However, it becomes less difficult with practice, although many distinguishing features are subtle or inconsistent.

Calls The easiest distinction. Meadow has a familiar *sip sip sip* (number of notes varies, as does the power of delivery: sometimes sounding a lower-pitched and hoarser *ski ski ski ski*). Tree utters a short, incisive *zeep*, *spzeep* or a more scolding *speez* (overhead migrants are easily detected by this call); in flight, it can also give a very soft, barely audible *sip*. On the breeding grounds, both species utter a variety of calls: Meadow gives a dry *si-sip* or a soft, nervous *sidip* anxiety note, particularly when carrying food; Tree Pipit repeatedly utters a soft *sit* alarm call (also when carrying food) or a high-pitched, ringing *stick* (when the young are under threat).

Songs Meadow has a variable delivery, typically a rising sequence of *chi-chi-chi-chi-chi...* notes as it climbs in song flight, often accelerating into a trill before decelerating into a thin and more musical *si-si-si-si-si...* as the bird descends to the ground (Meadow's song is similar to Rock Pipit's *A. petrosus*, but the latter's is typically 'thicker', slightly lower-pitched and simpler). Tree Pipit has a similar sequence but is louder and fuller, vaguely suggesting a Chaffinch *Fringilla coelebs*, and ending with a characteristic loud, far-carrying flourish: *swee seee swee swee tu tu swee swee SEOO SEOO SEOO SEOO*. This is frequently given in a parachuting display flight, often landing in the top of a tree. Particularly when singing from a perch, it may give an abbreviated version, lacking the final *SEOO* flourish: *swee swee swee swee titititititit* (the latter phrase a rapid trill).

Structure Subtle but definite differences. Tree Pipit is slightly larger, longer and sleeker-looking, with a longer, heftier and more wedge-shaped bill angled upwards from the face. Meadow is rounder-headed and less streamlined. Tree's longer wings are readily apparent in flight, producing a slightly stronger, more purposeful flight than Meadow (which is weaker and more hesitant). The length of the hindclaw is diagnostic, but very difficult to see in the field: short and arched on Tree, very long and more gently curved on Meadow.

Behaviour Tree is far more arboreal than Meadow, often singing from the top of a tree, and it may walk along branches, wagging its tail. However, Meadow readily perches on or even in trees and bushes, particularly if flushed. When feeding, Meadow wanders rather aimlessly through vegetation, twitching its tail up and down, rather than gently wagging it; Tree is stealthier and more purposeful, although rather furtive.

Plumage Differences must be evaluated sensibly, bearing in mind that adults of both species show considerable wear by midsummer. In fresh autumn plumage, Meadow has a greenish tint to the upperparts and an olive-buff wash to the underparts. Spring adults are generally browner above and whiter below, showing few green tones; they are colder and more washed-out than autumn birds. Some particularly pale, stripy Meadow Pipits (probably of the race *theresae* from Iceland) pass through western areas in spring and autumn (potentially confusable with Red-throated Pipit; see p. 350). In fresh plumage, Tree is better marked than Meadow, and the following differences are most useful. **1** FACIAL PATTERN Tree has, on average, a more strongly patterned face with a better-marked supercilium from the eye back and more prominent dark eye-stripe behind the eye (and often across the lores); it may also show a pale spot on the rear of the ear-coverts (often referred to as the 'supercilium drop'). On Meadow, the supercilium and eye-stripe are more subdued and the lores are usually plain; this

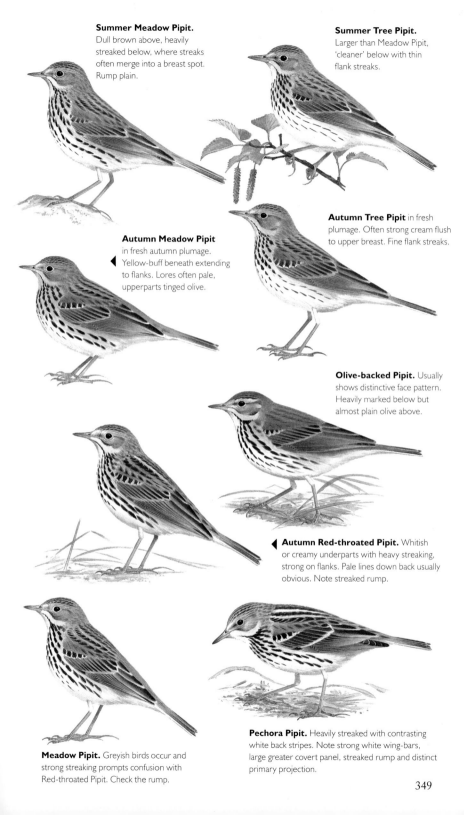

Summer Meadow Pipit. Dull brown above, heavily streaked below, where streaks often merge into a breast spot. Rump plain.

Summer Tree Pipit. Larger than Meadow Pipit, 'cleaner' below with thin flank streaks.

Autumn Meadow Pipit in fresh autumn plumage. Yellow-buff beneath extending to flanks. Lores often pale, upperparts tinged olive.

Autumn Tree Pipit in fresh plumage. Often strong cream flush to upper breast. Fine flank streaks.

Olive-backed Pipit. Usually shows distinctive face pattern. Heavily marked below but almost plain olive above.

Autumn Red-throated Pipit. Whitish or creamy underparts with heavy streaking, strong on flanks. Pale lines down back usually obvious. Note streaked rump.

Meadow Pipit. Greyish birds occur and strong streaking prompts confusion with Red-throated Pipit. Check the rump.

Pechora Pipit. Heavily streaked with contrasting white back stripes. Note strong white wing-bars, large greater covert panel, streaked rump and distinct primary projection.

349

produces a markedly open-faced appearance in which the dark eye and pale eye-ring stand out. **2** THROAT AND BREAST COLOUR In fresh plumage, Tree has the submoustachial stripe, throat and breast strongly tinged orangey-buff, contrasting with the whitish belly. **3** BREAST AND FLANKS STREAKING *The flanks streaking is perhaps the best and most consistent individual plumage difference.* Tree's streaking is confined mainly to the breast, with a gorget of neat, well-defined streaks, giving way on the flanks to *faint pencil streaking* that can be difficult to detect in the field. On Meadow, the breast streaking is more random, and the streaks often coalesce to form a dark spot in the central breast; unlike Tree, the streaking *extends quite strongly onto the flanks.* **4** UPPERPARTS Appear more contrasting on Tree: the wing-bars and tertial fringes are generally more prominent, and the dark-centred median coverts form a blackish bar (equivalent to the dark bar shown by Tawny Pipit *A. campestris*) often highlighted by contrasting white feather fringes. In summer, Meadow is noticeably colder, greyer, plainer and often 'tattier' than Tree Pipit, lacking strong greenish or buffy tones, but Tree Pipit also wears and fades by midsummer, becoming browner and plainer.

Red-throated Pipit *Anthus cervinus*

Call Most likely to be confused with a well-marked Meadow Pipit (see p. 348), but identification most easily confirmed by call, often the first indicator of a Red-throated amongst a flock of Meadow Pipits (records of non-calling individuals are likely to be very critically scrutinised by records committees). The flight call is a very distinctive thin, piercing, metallic *psssst*, trailing off towards the end. This may sound quite hoarse when flushed: *skeez* or *skier*. *Chup* calls are sometimes referred to in the literature but they appear to be given only when breeding.

Plumage *Autumn* Colder and less buff than Meadow, and much more heavily streaked. The lateral throat-stripe ends in a thick blotch on the neck-sides, joining heavy, broad breast striping; note in particular that this extends in two long thick lines down the flanks. The upperparts are strongly striped, with two pale 'tramlines' usually prominent on the sides of the mantle (more so than on many Meadow Pipits). The wings are also more strongly marked, the dark feather centres contrasting with buff or whitish fringes. Unlike Meadow and Tree Pipits, *the rump is heavily streaked* (usually best seen from the side when the tail is depressed while feeding). The bill is also stronger than Meadow, often with a yellow cutting edge and lower mandible. It looks distinctly shorter-tailed than Meadow, particularly in flight. *Summer* Easily identified by its brick-red 'face' and throat (note that autumn and winter adults, particularly males, may show at least a hint of this colouring, sometimes quite strongly). Spring migrants are usually in summer plumage (any retained winter feathers look worn, rather plain and dark, lacking the pale 'tramlines').

Olive-backed Pipit *Anthus hodgsoni*

Plumage Resembles Tree Pipit and, similarly, is often found in or around trees. Easily identified, the following being the most significant characters. **1** FACIAL PATTERN *A prominent broad, creamy supercilium* is highlighted by a thin black lateral crown-stripe and narrow dark eye-stripe, the whole effect vaguely recalling Redwing *Turdus iliacus*. A *whitish spot on the rear ear-coverts* (the 'supercilium drop') and dark lower rear border to the ear-coverts. **2** MANTLE On the race occurring in Britain and Ireland (*yunnanensis*) the mantle has a distinct green tone and is *only faintly streaked*, at a distance appearing uniformly olive-green or, on duller individuals, olive-brown. **3 BREAST** Can be quite buff or even orangey (often including the throat and submoustachial stripe) *heavily streaked with thick black lines*, extending more thinly onto the flanks. **4 TAIL-WAGGING** Wags or 'pumps' its tail more persistently than Tree Pipit.

Call Basically similar to Tree Pipit's but generally slightly shorter, thinner and shriller (with a slight Redwing-like quality), often with more of a consonant at the end and usually given doubly. Transcriptions include a thin *speez*, a shrill *tzzseep…tzzseep*; *psee…psee*; *szip…szip* or a thin, throaty *ski* or *skier*. High-flying individuals may sound more abrupt: *ski…ski*, similar to Tree but 'throatier'.

Pitfalls Not difficult to identify, but note that some Tree Pipits also show a faint 'supercilium drop', while some are atypically plain on the mantle.

Pechora Pipit *Anthus gustavi*

Plumage Like Red-throated, a well-marked pipit, but Pechora is surprisingly distinctive in its own right. An attractive bird, very clean and streaky. Like Red-throated, it has two mantle Vs but these are whiter and even more prominent. Two white wing-bars are broader, more prominent and squared-off to form two parallel 'bands' on the closed wing; it also has prominent white or warm buff tertial fringes. The underparts appear very white with neat black streaking forming a well-defined pectoral band, with two rows of heavy black streaks extending down the flanks. The crown is finely lined and it has a rather bland, open-looking face (often with a distinct ginger tone) in which the black eye and narrow white eye-ring stand out. The bill is mainly dull pinkish and the legs are also noticeably pale pink.

Structure Rather delicate with a proportionately small, somewhat rounded head and quite a long, fine and parallel-edged bill. Concentrate on the primaries: *there is a distinct primary projection* of two or three primary tips beyond the tertials (which completely cloak the primary tips in all other small pipits).

Behaviour and call Tends to be secretive and solitary, often flying a short distance before dropping back into cover. Vagrants invariably silent, but can give an unobtrusive, rather soft *pwit* or a sharp *tswip*.

Reference Mullarney (1987).

Rock, Water and Buff-bellied Pipits

Where and when Rock Pipit is a familiar resident of rocky coastlines, although largely absent as a breeding bird from much of the English east coast between Lincolnshire and Kent. It frequents a variety of coastal environments in winter, when numbers are augmented by visitors of the Scandinavian race *littoralis* (which seem to occur mainly on coasts of e. and s. England). Rock Pipits also occur inland in small numbers, mainly from September to November and again in March, but also occasionally between times. British Water Pipits originate from the mountains of s. Europe and are widespread but local from mid October to mid April, mainly in s. England. They occur in a variety of freshwater habitats, such as reservoirs, sewage farms, cress beds and marshes, but they generally avoid saline environments (although they may resort to coastal marshes in freezing weather). Water Pipits seen in saltwater environments should be identified with caution. Highest numbers occur during mild winters, and hard winters may severely deplete their population. Buff-bellied Pipit is a North American vagrant, increasingly identified mainly in western areas (27 records to 2011).

Rock Pipit *Anthus petrosus*

Habitat and behaviour In many ways rather nondescript, but easily identified in typical habitat: rocky coastlines. Although rarely found far from the inter-tidal zone, wintering or passage individuals may occur in less typical surroundings, such as inland lakes and sewage farms. Here they invariably select an area most akin to their usual habitat, such as a reservoir dam or a stony shoreline. Usually rather solitary and, unlike Water Pipit, often relatively tame.

Size and structure Compared to Meadow Pipit *A. pratensis*, Rock is a larger, bulkier, more upright bird with longer legs and a noticeably longer, more dagger-like bill (all dark or with an extensive orange lower mandible). In flight, longer-winged and longer-tailed, with a more purposeful flight action than the weaker, rather more hesitant Meadow Pipit.

Plumage Heavily colour saturated, producing a dark and oily appearance that blends in with its rocky surroundings. Its smoky-olive upperparts are darker than Meadow Pipit's and relatively unstreaked; the wings show two dull creamy-buff wing-bars. The underparts are rather dull yellowy-cream, with heavy brown streaking covering not only the breast and flanks but also much of the belly. It has a relatively plain face with a narrow creamy eye-ring (usually broken) that is typically more obvious than the subdued creamy supercilium (often indistinct). Two further features eliminate Meadow Pipit: the legs are dark, blackish at long range but dark pinkish-red close-up (bright pinkish-orange on Meadow), and the outer-tail feathers are creamy or pale brown (white on Meadow). In summer, Rock Pipits often become worn, appearing greyer above and whiter below, with a contrastingly black bill and legs; in such plumage, these individuals may suggest Scandinavian *littoralis* (see p. 354). **Juvenile** Recent fledglings may have pale pink legs and extensive pink or orange on the bill, as well as broad, buff wing-bars.

Call A single, loud, shrill *pseep* or *feest*. Meadow Pipits usually give a thinner, weaker *sip sip sip* and, although single calls are not infrequent, the difference in quality is distinctive once learnt.

Song Similar to Meadow Pipit but slightly slower, throatier and simpler.

Water Pipit *Anthus spinoletta*

Habitat and behaviour From damp freshwater habitat a large, timid, streamlined pipit flushes at some distance and rises high into the air, giving a loud, shrill, strident *fsst:* as it gains height, it swings back behind the observer and drops into similar habitat several hundred metres away; its flight is strong and direct, it is longer-winged and longer-tailed than a Meadow Pipit, and its underparts look contrastingly pale as it shoots overhead. Such is a typical encounter with a Water Pipit. Rock Pipit is usually quite tame and, in similar circumstances would have probably flushed at close range, flown low over the water and resettled after a relatively short distance. Although such behavioural differences are not diagnostic,

Rock Pipit. Rather olive-grey, heavily streaked below. Subdued streaking above. Weak supercilium, but pale eye-ring. Note dark legs, buff outer feathers.

Meadow Pipit for comparison. Generally browner and more stripy. Note pale orange legs.

Winter Water Pipit from behind. Note white belly, dark legs.

Winter Water Pipit. Upperparts browner than Rock Pipit and barely streaked. Underparts whiter, with streaking mainly on breast. Usually has strong supercilium. Note dark legs and white outer-tail feathers.

Water Pipits may appear streaky below during spring moult.

Summer Water Pipit. Bluish-grey head, pale supercilium, virtually unstreaked below with pink tinge to breast.

Summer Rock Pipit of race *littoralis*. Slightly bluish tinge to head. Streaked breast and flanks. Vinaceous tint to breast. Outer-tail feathers mainly buff, but may be white at tips. General appearance intermediate between Rock and Water Pipits.

Buff-bellied Pipit. Strong buff underparts with fine streaking, dark legs. Plain back, obvious buff eye-ring. Smaller and slighter than Water Pipit.

353

this is one that invariably holds true. Rock Pipits are also strongly territorial and have a bold demeanour, perching prominently on rocks and boulders, calling loudly and indulging in frequent territorial chases. Water Pipits, in contrast, are much more sociable and sometimes occur in small parties, often with Meadow Pipits. In spring, they may feed in fields and, unlike Rock, readily perch in trees and bushes. In the evening, they may roost communally in reedbeds.

Plumage *Winter* Seen well, Water is smarter and more contrasting, browner above and whiter below, with streaking largely confined to the breast (weaker on the flanks). What usually attracts attention is the whitish supercilium, which typically extends from the bill, over the eye, and tapers towards the nape. Note, however, that this feature varies, some being less well endowed than others, while a small minority shows virtually none at all. The wing-bars, tertial fringes and outer-tail feathers are much whiter than in Rock (not creamy or brownish). In winter, the bill usually has a yellow base to the lower mandible. *Summer* Unlike British Rock Pipits (race *petrosus*), Water Pipit acquires a distinct summer plumage. A body moult occurs from late February to early April and, consequently, birds at this time are often scruffy and dishevelled. However, they leave Britain in full summer plumage, which is quite striking and very attractive, almost wagtail-like, comprising a pale blue-grey head (the white supercilium is retained) and off-white underparts that are virtually plain, with a beautiful pale pink or soft apricot flush to the breast (some streaking may persist on the breast and flanks).

Call Very similar to Rock Pipit's *pseep* but Water Pipit's is perceptibly thinner and weaker: more of a *fssst*. Remember that Water's calls are given singly (although Meadow also give occasional single calls).

Scandinavian Rock Pipit *Anthus petrosus littoralis*

Unlike British birds, which are similar year-round, Scandinavian Rock Pipits (race *littoralis*) acquire a distinct summer plumage that enables them to be distinguished with some certainty before they depart in spring. There appears, however, to be something of a cline between Rock Pipits in n. Britain and those in s. Scandinavia, as well as individual variation, so not all will be certainly identifiable. 'Classic' examples acquire certain plumage characters in spring ordinarily associated with Water Pipit, so a potential Water Pipit in atypical rocky or stony habitat, or outside the normal range, should be carefully checked to eliminate *littoralis* Rock Pipit. The latter has a blue-grey tone to the head and rump, a distinctive creamy-white eye-ring and supercilium (at least from the eye back) and rather whitish wing-bars; the underparts acquire a strong creamy-buff, yellowish or salmon-pink suffusion (off-white, suffused pink, on Water Pipit). However, unlike Water, this is overlain with variable brown streaking on the breast and flanks; they also retain their dark lateral throat-stripe. Some, however, are virtually plain on the breast, so great care is needed. Unlike Water Pipit, *littoralis* Rock has creamy or pale brown outer-tail feathers, although on some the outermost tail feather is whiter towards the tip.

Buff-bellied Pipit *Anthus rubescens*

Most likely to be found with flocks of Meadow Pipits in drier habitats than Rock and Water Pipits (often in fields). Slightly smaller and less sturdy than Rock and Water, with a more rounded head and weaker bill. Entire plumage strongly suffused buff. It has a fairly plain back and scapulars but buff wing-bars and tertial fringes, and brown streaking on its strongly buff underparts. The head is relatively plain with a weak buff supercilium and pale lores; most distinctive is an obvious buff eye-ring. Like Water Pipit, it has dark legs. Summer adults are a soft grey on the head and upperparts with an apricot supercilium and underparts, the latter with delicate upper breast streaking. Its soft calls are reminiscent of Meadow Pipit, but are usually fast, urgent and distinctly two-, three- or four-noted: *si–sip*, *si–si–sip* or *si–si–si–sip*.

References Johnson (1970), Knox (1988), Williamson (1965).

Green finches: Greenfinch, Siskin and Serin

Where and when Greenfinch is a common resident throughout Britain and Ireland, although it shuns very high ground. However, numbers have recently been reduced by disease (Trichomonosis). Siskin breeds in coniferous woodland, mainly in Scotland, Wales and Ireland, and also thinly (but increasingly) in England, as far south as Devon and Hampshire. It is more widespread and often common in winter, although numbers fluctuate annually. It feeds in alder, birch, spruce and larch, and is now regular on feeders in suburban gardens. Serin is a rare visitor, mainly to southern coastal counties, mostly in spring and summer (although recorded throughout the year). It currently averages 55 records a year, with a peak of 99 in 1996, and occasionally breeds. It tends to occur in weedy fields.

Greenfinch *Chloris chloris*

Structure and plumage A familiar garden bird. A large, bulky, sparrow-sized finch with a thick, conical, pale pink bill and pink legs. ***Adults*** Male is green, brightest in summer, with soft grey on the neck-sides and more obviously on the greater coverts and tertials; female duller, with vague and diffuse streaking below, and lacks the male's black lores. Both have yellow flashes in the primaries and tail, generally duller on the female. ***Juveniles*** Often confused by beginners. Juveniles are much browner and diffusely streaked, both above and below, with a brown rump and two broad but diffuse brownish wing-bars. However, the wings and tail also show the adult-like bright yellow patches. This plumage is lost in a partial, or sometimes complete, late summer/autumn moult.

Voice Its call and song are complex but their mellow and rippling quality (especially compared to Linnet *Linaria cannabina*) is distinctive once learnt. The flight call is a rapid, deep, mellow *dji–dji–dji–dji–djit*. The full song is varied and includes rapid mellow trilling (pitch frequently varied, sometimes high-pitched and sibilant) alternated with a higher-pitched *twee twee twee* and frequently interspersed with a very distinctive loud, upslurred, nasal *djuwee* (similar to the song of Brambling *Fringilla montifringilla*). It is often given in a bat-like display flight with slow-motion wingbeats. In the breeding season, males also give a lazy, upslurred, musical *pwoo–ee*. Begging juveniles persistently give a soft, dry *djip djip djip djip….*

Siskin *Spinus spinus*

Behaviour Gregarious, often in large flocks with Lesser *Acanthis cabaret* or Mealy Redpolls *A. flammea*. Markedly arboreal, often hanging tit-like from the ends of branches. Like

Goldfinch *Carduelis carduelis*, it has a light, 'dancing' flight, especially apparent when in flocks.
Structure and plumage A small, compact, delicate green-and-yellow finch with a sharply
pointed bill and deeply forked, rather splayed tail. Easily separated from the larger Greenfinch
by its streaked plumage, prominent yellow wing-bars (with contrasting black feather bases)
and distinctive calls (see below) but, like Greenfinch, also has yellow patches at the base of the
tail. In flight it suggests a green-and-yellow Goldfinch, the thick yellow wing-bars and bases
of the primaries and secondaries combining to form a large yellow mid-wing panel, visible
from above and below. Male less streaked than female and yellower on the face and breast,
with obvious black on the crown and chin (sometimes grey on first-years). Juvenile resembles
the female, but is duller, browner and heavily but finely streaked, with narrower, pale yellow
wing-bars (this plumage is lost in a late summer/early autumn body moult).
Calls Siskins are very vocal, the most common flight call being a very distinctive loud, clear,
musical *slee-u* or *sweeeloo*. Flocks also give a hard chattering and sweet twittering, particularly
when feeding.
Song Discordant but sweet, containing rapid trills, twitters and occasional scratchy notes,
the sweet musical quality of some notes recalling the flight calls. It also includes a thin,
high-pitched, nasal *zweeee*, sometimes heard from wintering flocks in early spring. The song
is often given in a Greenfinch-like display flight.

Serin *Serinus serinus*

This small, yellowish finch needs to be separated from Siskin, and from other small finches
that may escape from captivity.
Behaviour Markedly more terrestrial than Siskin, usually feeding on the ground or in weeds
(although Siskins may feed in similar habitat in the absence of suitable trees).
Structure A small, rather dumpy, round-headed finch, with a rather short tail and *short,
stubby, conical bill* (longer and distinctly pointed on Siskin). Note also that Serin's primaries
are *relatively* short (about equal to tertial length) whereas Siskin's are long and extend well
down the tail (c. 25% longer than the tertials).
Plumage Most important is the bright yellow rump patch which is prominent, clear-cut and
rather rectangular; it positively 'glows' on the adult male but is duller on female and rather buff
on juvenile, although still standing out as distinct pale patch, prominent in flight (on Siskin,
the rump is greenish-yellow, less clear-cut and altogether less eye-catching.). *Male* Bright
yellow forehead and supercilium, broadening on the neck-sides to form a wide ear-covert
surround. The throat and breast are also bright yellow, the latter contrasting with a white belly
and flanks; the breast-sides and flanks are heavily streaked. Narrow yellowish wing-bars are
much less obvious than on Siskin. *Female* Although similar in pattern to the male, it is duller
and less intensely yellow, with rather a plain face in which the beady eye stands out; some,
particularly first-winter females, can be quite buffy, with limited yellow confined mainly to
the face and breast. *Juvenile* Duller again, with the yellow replaced by buff; this plumage
is soon lost in a partial or sometimes complete moult that starts soon after fledging. ***Other
differences* 1** WING-BARS Narrow and can be virtually lost through wear by midsummer
(broad on Siskin, contrasting strongly with the black feather bases). **2** TAIL Lacks Siskin's
yellow bases to the tail-sides. **3** HEAD Male lacks Siskin's black or blackish on the crown and
chin; also, both sexes have a pale crescent immediately below the eye and diffuse pale spot on

Juvenile Greenfinch. Large and bulky with yellow in primaries and tail. Juvenile is browner than adult and lightly streaked on breast.

Male Siskin. Note black on crown and chin, strong wing-bars, black legs.

Male Serin. Yellow on head and breast. Prominent yellow rump in flight. Note small, conical bill.

Female Siskin. Note long, pointed bill, long primaries, thick wing-bars and yellow at base of tail.

Juvenile Serin. Much buffer than female. This plumage is moulted soon after fledging.

Female Serin. Note bill and facial pattern, yellowish rump in flight.

Canary in natural colours. Note short primary projection, and longer tail than Serin.

Male Yellow-fronted Canary or **Green Singing Finch.** Escaped finches need to be eliminated when identifying Serin.

the lower ear-coverts (yellow to yellowish-buff). **4 TERTIALS, SECONDARIES AND PRIMARIES** On Siskin, these feathers are fringed yellow, against which the broad black bases to the secondaries and the plain black primary coverts stand out. **5 MANTLE AND BACK** Serin is browner, Siskin greener. **6 IN FLIGHT** Lacks Siskin's yellow Goldfinch-like wing-panel.

Voice Particularly at migration watchpoints, Serins may be detected by call as they fly overhead in a fast, direct but undulating flight: a very quick, soft, dry, almost trilling *ti-li-li-li-lit* or *psit-it-it-it-it* (number of notes varies). This call is diagnostic and totally different from the musical *slee-u* of Siskin. Serin can give other calls, such as a thin, nasal *tchooo-it*. Juvenile's begging calls include a *dzeep*, vaguely recalling Tree Pipit *Anthus trivialis*, and a high-pitched *seee seee*. Male's song (one of the most familiar sounds of the Mediterranean) may be the first clue to the species' presence during the breeding season: a remarkably rapid, prolonged jumble of high-pitched twittering and jangling notes (vaguely suggesting a high-pitched, fast Corn Bunting *Emberiza calandra*). This is often delivered with great gusto, with the bill fully open, the tail half-cocked and the head stretched up and jinked from side to side. It may also be given in a Greenfinch-like display flight. Some individuals have phrases remarkably similar to the song of a Wren *Troglodytes troglodytes*.

Escapes

Canary

When identifying a Serin, it is essential to consider the possibility of escaped cagebirds, the most likely of which is Canary *S. canaria*. Most escaped Canaries are domesticated forms, many of which are completely yellow and unstreaked. Those resembling their streaky wild ancestors are easily confused with Serin, but the following features should assist recognition. **1 SIZE AND SHAPE** Canary is distinctly larger and more attenuated than Serin. **2 PRIMARY AND TAIL LENGTH** Canary has short primaries, only about two-thirds the length of the overlying tertials (Serin's are about equal). Partly as result, Canary appears much longer-tailed. **3 BILL** Canary's is not as stubby as Serin's. **4 CALL** Canary has a soft *sweet* or *tsooeet*, as well as a pleasant soft, twittering contact call. **5 PLUMAGE** Wild-type Canaries tend to be greyer above, yellower on the belly and duller on the rump.

Yellow-fronted Serin

Another escape that has caused problems is Yellow-fronted Canary or Green Singing Finch *S. mozambicus*, which is yellow below and green above (lightly streaked) and has green wing-bars and fringes to the flight feathers. It also has a bright greenish-yellow rump. The male is easily identified, however, by its grey rear crown, lores and ear-coverts (the grey varies racially), a tapering yellow supercilium and black lateral throat-stripe. The tail is tipped white. The female has a similar pattern, but is duller and greener. It is heavier-billed than Serin and the primary projection is only half the tertial length.

Brown finches: Linnet, Twite and Common Rosefinch

Where and when Linnet is a locally common, widespread but declining breeder, most frequently found on heaths and arable farmland. In winter it avoids upland areas and concentrates in flocks on farmland, rough ground and saltmarshes. Twite replaces Linnet in upland Scotland and the Scottish islands; it also breeds in the Pennines, with a small population in N. Wales (also w. Ireland). In winter, it descends to lower areas, particularly to coastal saltmarshes. Although still numerous in Scotland, it has declined significantly in England, even in its east coast winter strongholds. Elsewhere in England and Wales it is extraordinarily rare and great caution needs to be exercised with sightings away from the east coast. Common Rosefinch is a rare spring and autumn migrant, with most seen at well-watched coastal sites, particularly Shetland. It currently averages 130 records a year, mostly in late May and early June and from late August to mid October. Autumn migrants are most likely to be found in weedy fields, often feeding unobtrusively with flocks of Linnets or other finches. It has bred in Scotland and England on several occasions, with a peak of five pairs in 1992.

Linnet *Linaria cannabina*

Moult Linnets have a single complete post-breeding moult between June and October, so by early winter they are all in a similar fresh winter plumage that is rather nondescript brown, buff and dull grey. This gradually wears and, by late winter, they are duller and colder in tone. The male's pretty summer plumage is acquired not by moult but by the gradual wearing of the dull winter feather fringes.

Structure and plumage A small and rather dumpy finch, the most distinctive features being prominent white fringes to the outer webs of the black primaries and tail feathers. These show as prominent white 'flashes' in flight. In fresh winter plumage, the underparts are rich buff, the upperparts darker brown, both streaked with brown. The head is greyer, looking quite plain at a distance, but at close range there is a pale area above the eye and a noticeable pale 'subocular crescent' below, as well as a buff spot in the centre of the ear-coverts. Buff tips to the greater coverts form only a slight wing-bar (stronger on Twite and the redpolls). The bill is grey and the legs dark reddish. Although the sexes are similar in winter, close scrutiny reveals that adult males are more colour saturated, being slightly chestnut-toned above and somewhat deeper buff below, diffusely streaked dark chestnut. In spring and summer, the male acquires the distinctive grey head, chestnut back and bright pink forehead and breast. Summer female remains similar to its winter appearance, albeit more worn. The loosely plumaged juveniles are well streaked on the breast and upperparts, with two broad rich buff wing-bars.

Calls and song Flocks feeding in low vegetation rise with a distinctive dry twittering. It has a variety of calls, but birds passing overhead give a soft, dry *djit djit dji-dji-djip* etc. Begging juveniles utter a persistent soft whistling or piping *sweet* calls. Males have an elaborate sweet, twittering song, often given from the tops of gorse bushes and similar vegetation.

Juvenile Linnet. Distinctive face pattern, white fringes to primaries.

Summer Twite, worn. Orange-buff tinge and heavily streaked; white primary fringes, yellow bill.

Autumn male Linnet. Grey head, plainish chestnut above, buff below, pink breast obscured by dull fringes.

Winter Twite. Orange-buff face and yellow bill. ▼

Winter female Linnet. Chestnut ▲ above, greyer head. White fringes to primaries.

Note flight patterns: Twite (top), Linnet (middle) and Lesser Redpoll (below).

Spring Common Rosefinch. Sparrow-sized, plain face and large blunt bill. Note two thin wing-bars. Juvenile (below) more olive and buff with slightly stronger streaking below. ▼

◀ **Juvenile Lesser Redpoll.** Pale head. Lacks red 'poll'.

◀ **Summer Lesser Redpoll**, worn. Dark, streaked.

Fresh autumn Lesser Redpoll. Warm tawny-brown, heavily streaked with obvious wing-bars. Note black 'bib'. ◀

Fresh autumn Greenland Redpoll. ▶ Large and dark with contrasting pale rump.

Fresh autumn Mealy Redpoll. Colder brown with white feather fringes, appearing 'frosted'.

Twite *Linaria flavirostris*

Structure and plumage In winter, Twite is superficially similar to Linnet but slightly slimmer and distinctly longer-tailed, with a deeper fork (the tail is *c*.10% longer than Linnet's; *BWP*). It too has obvious white flashes in the primaries and tail, but is readily separated from Linnet by the following features. **1** FACE AND UNDERPARTS A distinctive and rather beautiful orange-buff ground colour to the face and unstreaked throat. Brown streaking extends from the breast-sides rather diffusely onto the flanks, but the belly and undertail-coverts are white. **2** UPPERPARTS Darker brown than Linnet, with heavier and darker streaking; in addition, there is a more prominent wing-bar on the tips of the greater coverts, varying from buff to whitish. Thus, the upperparts are more reminiscent of a redpoll than a Linnet. **3** BILL Obviously yellow in winter, contrasting with the face, but note that it is grey in summer (normally starts to darken in March/April). **4** RUMP The male has a pink rump but, in autumn and winter, it is often partially obscured by pale feather fringes; brighter in summer but even then can be frustratingly difficult to see, even in flight. On female, the rump is brown, streaked black.

Calls and song Like Linnet, Twite has a variety of calls, including twittering notes that hardly differ from Linnet's, but are usually slightly harder and more nasal. Most distinctive is a harsh, disyllabic, rasping, nasal *tchooeek*, this often enabling Twite to be located amongst a flock of Linnets. Like Linnet, it may also give softer calls, such as a soft *tweep tweep tweep*. The song is rather nasal, containing both hard twittering and nasal *tchooeek* notes.

Common Rosefinch *Erythrina erythrina*

Ageing Note that Common Rosefinches do not moult until arriving in their winter quarters, so all young birds in autumn are juveniles, not first-winters. This also means that juveniles can be aged by their very fresh and immaculate plumage. Autumn adults are very rare here, but they are less olive with more heavily worn primaries and tail. Note that adult males (also rare here) do not normally acquire their distinctive red plumage until their second summer; brown first-summer males may sing and hold territory.

Structure and plumage Nondescript, but identification is nevertheless straightforward. A sparrow-sized bird, thickset, blunt-headed and bull-necked, somewhat like Bullfinch *Pyrrhula pyrrhula* in shape. Rather bland and featureless face, with the most distinctive feature being the dark, beady eye; note also the dark, Bullfinch-like bill. The upperparts are cold grey-brown, lightly streaked, the underparts cold greyish-white or buff, more heavily streaked. The most distinctive plumage feature is the two thin but noticeable white wing-bars, and white tertial fringes.

Calls and song A soft, upslurred musical whistling *soo-eeet* almost recalls a loud, slightly strident Willow Warbler *Phylloscopus trochilus*. The song is a quick but simple, soft, penetrating whistling: *seeoo-see-seeoo* (downslurred on the final syllable).

Lesser, Mealy and Arctic Redpolls

Where and when Lesser Redpoll occurs mostly in upland areas of Wales, N. England, Scotland and Ireland. In summer it is associated with young conifer plantations but it favours alders and birches in winter, when it is much more widespread in lowland Britain (although numbers vary annually). Mealy Redpoll (also confusingly referred to as 'Common Redpoll') is a scarce winter visitor, mainly to northern areas and the east coast. Numbers also fluctuate annually, with occasional irruption years (there was a huge invasion in the winter of 1995/96). A few pairs sometimes breed in N. Scotland. Arctic Redpoll is a great rarity, currently averaging seven records a year, with occasional records in the Northern Isles and a few on the east coast. There are occasional influxes, with a remarkable 431 accepted in 1995/96, and a peak in December/January (Votier *et al.* 2000).

General approach

Lesser Redpoll was split from Mealy in 2001, although this decision has remained controversial. The field separation of all of the redpoll forms is fraught with difficulty and presents one of the toughest challenges, some individuals defying identification. It is important to stress that redpolls show considerable individual variation and not all forms will be identifiable, even by experienced observers. It may, however, be easier to reach a conclusion in influx years, when several individuals of the same form are present together (e.g. periodic influxes of Greenland Redpolls into the Northern Isles). The starting point is an understanding of the significant variation in the plumage of Lesser Redpolls, relating to age, sex and plumage wear. It is important to identify individual birds using a *combination* of features.

Ageing A simple way to age redpolls is by the shape of their tail feathers: sharply pointed on first-years, but more rounded on adults (often readily visible on close-feeding birds, but easiest to evaluate when the feathers are fresh). Until their complete late summer body moult (July–September) juveniles lack the red 'poll'.

Sexing Adults completely lacking pink should usually be females, whilst those with pink on the head-sides, breast-sides and flanks should be males (Svensson 1992). Those in between could be either.

Moult Redpolls have only one moult per year (in late summer) so they become very worn by spring, appearing darker above as the brown feather fringes wear away, and dull greyish-white below. They also have narrower and whiter wing-bars, whilst the fringes to the primaries and secondaries may wear off completely. However, feather wear towards spring also reveals more pink. Caution is essential at this season as feather wear means that worn birds will be more difficult to identify than fresh ones. Note in particular that, unlike autumn and winter birds, worn and faded spring Lesser often lack buff tones and appear much more like Mealy.

Lesser Redpoll *Acanthis cabaret* (illustrations on p. 360)

Size, structure and behaviour Compared to Linnet *Linaria cannabina*, a distinctly smaller, dumpier-looking finch, far more arboreal in its behaviour, although it will also feed in low vegetation (and on the ground, particularly in spring). It has a rather finely pointed yellow bill. Typically seen in small flocks, often with Siskins *Spinus spinus*, feeding in a tit-like manner on the outer twigs of birches and alders, or on the top of tall weeds. In flight looks small and dumpy with a rather short, deeply forked tail.

Winter male Arctic Redpoll of race *excilipes*. Pale, especially nape, whitish base colour to mantle, large white rump, sparse think flank streaks.

Winter female Coue's Arctic Redpoll. Less well marked than male. Note short, conical bill, white undertail-coverts.

Hornemann's Arctic Redpoll. Large, strong face pattern.

Winter female Mealy Redpoll. Rump streaked, nape and mantle browner, stronger streaking to flanks and undertail-coverts. Bill less stubby-looking.

Spring male Arctic Redpolls show only a hint of pink. Plumage greyer, rump still clear white.

Arctic Redpolls have undertail-coverts white or with fine streaks (left). Streaks on Mealy Redpoll thicker (right).

Plumage Compared to Linnet, darker and more heavily streaked, above and below. Lacks the white primary and tail flashes of Linnet and Twite *L. flavirostris*, instead having two narrow buff or whitish wing-bars. Adults show a red 'poll' (on forehead) and black bib, but juveniles (before their late summer/autumn body moult) lack these. Like all redpolls, in spring and summer adult males show a strong pinkish-red tinge to the sides of the head, breast and rump (acquired by feather wear, not moult). Adult females can show traces of this, but first-winter males may lack all pinkish-red except the 'poll'. Lesser's separation from other forms of redpoll is dealt with below, but its overall appearance is far more colour saturated than the others, its general plumage tone being very brown (even the wing-bars are often suffused with buff) and it usually lacks obvious pale 'tramlines' down the centre of the mantle. Note, however, that worn and faded spring Lessers may appear much more like Mealies: they often lack strong buff tones, are whiter below, have whiter wing-bars and show white 'tramlines' down the back, with extensive whitish feathering (streaked dark) sometimes intruding onto the rump. Also, they are more likely to feed on the ground at this time.

Call The easiest way to detect Redpolls is by their distinctive flight call: a dry rhythmic, staccato *chi chi chi chi chi*, usually heard as the bird passes high overhead in light, buoyant, undulating flight. Other calls include an upslurred, sweet nasal *chuweet*.

Song Similar to the call but rapid and higher-pitched, interspersed with *twee twee twee twee* notes and punctuated by a high-pitched, sibilant, dry, buzzing *dwzeee* (recalling Greenfinch *C. cholris*) with sweet *chuweet* notes. Often given in a wide-ranging song flight.

Mealy Redpoll *Acanthis flammea*

Mealy Redpoll *A. f. flammea*

Nominate Scandinavian Mealy is typically larger than Lesser with a deeper, more triangular bill. It is paler and 'colder', with a greyish-white ground colour to the head and upperparts, lacking strong buff tones. It tends to be slightly less heavily streaked below (with crisper flank streaking) and often has two whitish 'tramlines' down the back. The wing-bars and fringes to the primaries and secondaries average whiter. The rump is also whiter, usually greyish-white, and well streaked (some adult males show an unstreaked pink or pinkish-red rump). It must be stressed, however, that there is much individual variation in tone, some being browner and buffer than others, so caution is strongly advised. The rhythmic flight call is rather deeper than Lesser, but it would not be advisable to separate redpolls on call alone.

Greenland Redpoll *A. f. rostrata*

An erratic autumn and winter visitor to n. and w. Scotland and Northern Ireland (with a few records as far south as Scilly). Larger, bulkier and longer-winged than nominate Mealy (Linnet-sized) with a rather short, broad tail and heavier, slightly convex bill. Unlike other northern redpolls, its plumage is dark, heavily streaked above and below. Most distinctive is a whitish rump, which can be noticeable in flight. It is this *combination* of dark plumage and whitish rump that is most distinctive. It also has two whitish 'tramlines' on the back, a broad white wing-bar on the greater coverts and rather white underparts. Its calls are deep and 'throaty', some being remarkably similar to those of Twite.

Icelandic Redpoll *A. f. islandica*

The situation in Iceland is confused by the fact that some redpolls there resemble Mealy or even Arctic (with a large, unstreaked white rump), others resemble Greenland (although dark Icelandic birds are smaller; Stoddart 2011). Icelandic Redpolls have been recorded in Britain (mainly in the Northern Isles) but their identification is problematic.

Arctic Redpoll *Acanthis hornemanni*

Two races occur in Britain.

Coues's Arctic Redpoll *C. h. exilipes*

This circumpolar race is the most frequent, our immigrants arriving from Scandinavia. Much of the following is based on Votier *et al.* (2000) and Stoddart (2009). Although some are likely to stand out because of their overall paleness or whiteness, many females and first-winters are browner. If a candidate is found, the following should be evaluated. **1 SHAPE** Largely because of its dense plumage and greater feather mass (to cope with a cold climate), Coues's appears larger and bulkier than Mealy (described as a Mealy 'wrapped in a duvet'). It also tends to appear flat-headed, bull-necked and broad-bodied, with shaggy flanks and belly. When resting in the cold, they often fluff up their body feathering and appear almost spherical, with the extensive and conspicuous white rump standing proud of the closed wings. Dense facial feathering obscures the bill base, producing a characteristically short-billed/flat-faced appearance, almost as if the bird has flown into a wall. On some, the bill can look tiny, with a straight or even concave culmen. **2 RUMP** The extent of the white on the rump varies, but there is usually 10–20mm of unstreaked white. On some (particularly first-winters) the white may be narrower and restricted to the upper rump. Some show fine streaking (sometimes extensive, approaching Mealy), but this may be related to plumage wear (later in winter and spring, the birds may reveal dark feather bases as white feather tips wear off). **3** HEAD Appears plain-faced, the ear-coverts not standing out as a darker area in an otherwise pale face; in consequence, it appears beady-eyed. Females and first-winters often have a warm buff wash to the face. **4** UPPERPARTS Many are pale and rather greyish (especially males) but others are browner, although even these show two obvious white 'tramlines' on the back. **5** UNDERPARTS Very white with fine streaking confined to the upper breast and flanks; inevitably, females and first-winters are more heavily streaked than adult males (but usually less so than Mealy). **5** UNDERTAIL-COVERTS Either completely white and unmarked (especially adult males) or with just a single streak on the longest feather (although they can show 2–3 narrow shaft-streaks). Mealy is usually heavily streaked, the streaks being darker, broader-based and tapering to a point.

Hornemann's Arctic Redpoll *C. h. hornemanni*

Originates in Greenland and NE Canada. A stunningly pale, very large, long-bodied and long-winged redpoll (larger than Twite), with a big head and a short, deep bill. Long flank and belly feathering may completely cloak the legs. Even first-winters (aged by tail feather shape) show a whitish background colour to the upperparts, a broad, crisp white greater covert bar, pure white fringes to the wing and tail feathers, and a huge, unstreaked pure white rump/lower back, which may be fluffed right out over the upperwing at rest. The underparts are also very white (with minimal streaking). The head can show a beautiful, soft orangey-buff tone. Adult males in particular are often described as 'frosty snowballs', which neatly sums them up.

References Stoddart (2009, 2011), Votier *et al.* (2000).

Common, Parrot, Scottish and Two-barred Crossbills

Where and when Common Crossbill occurs in suitable habitat throughout much of Britain, but numbers fluctuate and its distribution varies, being much more numerous during periodic invasion years. Scottish Crossbill occurs only in the Scottish Highlands, with an estimated population of 6,800 pairs in a 2008 survey. In 1991 it was discovered that Parrot Crossbill also breeds in the Scottish Highlands, with *c.* 50 pairs in 2008. Otherwise it is a very rare vagrant from Scandinavia and Russia, recorded almost exclusively during infrequent invasions (e.g. 85 in 1962/63, 100 in 1982/83 and nearly 270 in 1990); most have been in n. Scotland and e. England (it bred in East Anglia following the 1982/83 invasion). Two-barred Crossbill originates in NE Europe and Siberia, and is the rarest crossbill. It is not annual, with rarely more than two or three records a year, but occasional small invasions also occur, with as many as 59 birds in 2008 (58% of them juveniles). Most occur in the Northern Isles in July–September but surprisingly few filter south; those that do occur throughout the year.

Common Crossbill *Loxia curvirostra*

General features Generally found only in coniferous woodland. Gregarious, most likely to be located by its very distinctive calls (see below). Often seen in high-flying rather disparate flocks, the birds appearing large, bulky and short-tailed with a fast, direct and rather sparrow-like flight. Note that Crossbills breed in response to food supplies and so can nest any time from August to June (*BWP*); therefore, song is often heard in winter, with juveniles frequently encountered from late winter onwards.

Voice Very noisy, giving a very distinctive loud, abrupt, rather hollow *twick...twik...* or *djip ... djip...* calls. The song is high-pitched and twittering, interspersed with a variety of harder *tiwee tiwee* notes and more nasal sounds; it is thinner, faster and more ethereal than Parrot Crossbill's. It is often given in a slow-motion song flight, which recalls that of Greenfinch *Chloris chloris* (but slow wingbeats not as exaggerated). Begging juveniles give *jip jip jip...* calls, quieter and more subdued than adults'.

Bill Thick, with tips noticeably crossed, but slimmer than Scottish and Parrot Crossbills (see below). Young juveniles of all species lack crossed tips.

Plumage *Adult* Males are mainly reddish, but some resemble females, showing little red, while others appear distinctly orange in tone (variability apparently unrelated to age); females are mainly grey-green or yellowish-green, yellower on the rump. Note that a very few show prominent whitish wing-bars (see Two-barred Crossbill p. 367). *Juvenile* Well streaked, greyish above and greyish-white below; the rump is slightly paler and it usually has a thin buff double wing-bar. Juvenile plumage can be encountered from midwinter and can be retained until June to October.

Parrot Crossbill *Loxia pytyopsittacus*

Size and structure Noticeably larger and 25–40% heavier than Common but adult males average larger than females and juveniles. Large-billed (see below) and thick-necked (recalling Hawfinch *Coccothraustes coccothraustes*) contributing to a front-heavy appearance. Viewed front-on, it has a distinctly broader head than Common. Large-billed individuals often seem to lack a forehead, with a very flat crown, but Parrots may also show a slight but distinct 'forehead step' between the bill and the head; this is because the bill juts out more horizontally from the forehead, unlike the more curved bill of Common. However, the crown feathers can be raised to produce a dome-headed appearance. Smaller-billed females and juveniles usually show a more pronounced forehead.

Bill Key to the identification is an accurate evaluation of bill structure (a *detailed* description or photograph of the bill would be a prerequisite for acceptance by a records committee). Compared to Common's, noticeably heavier, deeper and *proportionately* shorter – indeed more 'parrot-like'. The upper mandible is prominently but evenly arched, whilst the lower bulges at the gonys before angling up to the tip. Note especially that, compared to Common, the tips *do not look obviously crossed in profile*, the bill therefore looking blunter and stubbier; the tip of the upper mandible is more obvious than that of the lower. Seen front-on, the bill is broader-based than Crossbill's. Bill size apparently increases with age, so not only do males have larger bills than females, but older adult males are those most likely to have classically huge 'parrot-like' bills; nevertheless, even females usually look larger-billed than male Commons. Juveniles in particular may be altogether less impressive, so a careful assessment of bill *shape* is essential. Parrot Crossbill also has noticeable pale ivory cutting edges to the mandibles, which expand to form two 'half moons', deepest and broadest on the lower mandible, but not reaching the tip. These are not always present on Common but, when they are, they tend to reach the bill tip (Bowey & Westerberg 1994).

Adult male Common Crossbill. All crossbills share the same plumage characteristics. ▶

Large head and bill of Parrot Crossbill obvious from some distance.

1st-year male Common Crossbill is orangey, but so are some older males.

Adult female Common Crossbill. Greyish-green upperparts, yellowish underparts and strong yellow rump. ▶

Juvenile Common Crossbill. Greyish-green and heavily streaked. Bill not yet fully formed.

Bill size and shape critical to identification.

Juvenile Common Crossbill. The bill is still growing.

Two-barred Crossbill

Common Crossbill

Scottish Crossbill

Some crossbills (1 in 1000) show whitish wing-bars. They are narrow, often diffuse in males. ▼

Compare pattern of tip of central tertial; diffuse on Common, sharply defined on Two-barred Crossbill.

Adult male Two-barred Crossbill. Pinkish rather than orange-red. Note shape of white feather tips in wing, especially the tertials.

Parrot Crossbill. Bull-necked and large-headed. Note bulging lower mandible and tips of bill not obviously crossed in profile.

Voice The call is slightly lower-pitched, less 'clipped' and 'mellower' than Common Crossbill's, but interpretation depends on familiarity with the latter's call. Parrot's metallic and echoing *chyioop* has been described as diagnostic. Juvenile calls may be higher-pitched than the adult's. The song is weak and hesitant: *chit, chit, chit-chit-chit tcho-ee tcho-ee* (with calls interspersed), deeper, slower and better enunciated than Common's.

Plumage No reliable differences from Crossbill, but males are usually a duller, deeper crimson with a greyer nape and mantle. Females tend to look duller and greyer, especially on the head and mantle (Catley & Hursthouse 1985).

Scottish Crossbill *Loxia scotica*

This form has at various times and by various authorities been considered a subspecies of either Common Crossbill or Parrot Crossbill, but it is now treated as a full species. Identification of Scottish Crossbill remains one of the most challenging of any species on the British List. Formerly considered easiest to identify by locality, any large-billed crossbill seen in the Scottish Highlands almost certainly being this species. However, the recent discovery of Parrot Crossbills also breeding in the Highlands has confused the situation. It is, however, important to maintain a sense of perspective. A survey in N. Scotland in 2008 suggested the presence of 27,100 individual Common, 13,600 Scottish, but only 100 Parrot (Holling *et al.* 2010). Statistically, therefore, any large-billed crossbill seen in N. Scotland is far more likely to be Scottish (by a ratio of 136 : 1). That being the case, unless the bird(s) show(s) a bill structure indicative of Parrot, then the balance of probability favours Scottish (although, of course, the more individuals examined the better). The bill of Scottish is intermediate in size and structure between Common and Parrot Crossbills (see illustrations).

Two-barred Crossbill *Loxia leucoptera*

Owing to the occurrence of Common Crossbills with wing-bars, the identification of this species requires care and a combination of all of the relevant features should be used. It is essential to note the *exact size and shape* of the wing-bars and tertial tips, and differences in structure and plumage tone.

Structure Generally looks smaller, slimmer and more compact than Common, with a smaller, flatter head; also less bull-necked, while the wings and, particularly, the tail usually appear proportionately longer. Note, however, that there is size overlap with Common.

Bill Usually looks slimmer than Common's, but again there is overlap.

Call An abrupt, dry, slightly metallic *tyip tyip*, distinctly softer, weaker and higher-pitched than Common's. It also gives a distinctive soft, nasal trumpet call, vaguely reminiscent of a Bullfinch *Pyrrhula pyrrhula*.

Plumage Double white wing-bar and white tertial tips, but it must be stressed that Commons can also show wing-bars (indeed, in 1850, Common Crossbills with wing-bars were described as a separate species: '*L. rubrifasciata*'). Adult Two-barred's wing-bars are *huge, broad* and pure white; they *bulge prominently on the inner coverts* but taper to a point on the outer feathers (the upper median covert bar may be obscured by overlying feathers). The wing-bars are 5–12mm deep at their broadest (Svensson 1992) and consist of white feather tips neatly squared off, creating a solid-looking bar; white tips to the tertials are also prominent. Juveniles show narrower wing-bars (2.6–6mm) but are nevertheless still very obvious;

they too have white tertial tips. On Common Crossbill, any wing-bars and tertial tips are usually narrower (1–2.5mm, exceptionally up to *c.* 5mm) and normally consist of relatively narrow crescent-shaped fringes to the feather tips that do not form a solid bar (appearing diffuse and blurred in the field). However, there appears to be overlap in the exact *pattern* of white on the wing-coverts and tertials: on both species, it may slightly extend up the outer web of the feather. In the hand at least, Common's wing-bars tend to be slightly buffer than those of Two-barred. The body plumage of male Two-barred tends to be slightly paler and more *pinkish-red* than Common (but some are similar in tone) and males also tend to show pale-edged brownish or even blackish scapular feathers (although they can lack these).

References van den Berg & Blankert (1980), Bowey & Westerberg (1994), Catley & Hursthouse (1985), Holling *et al.* (2010), Hudson *et al.* (2009), Svensson (1992), Summers & Buckland (2011).

Cirl Bunting and Yellowhammer

Where and when Yellowhammer *Emberiza citrinella* is widespread and locally common in all kinds of open country, particularly arable farmland. In winter, it withdraws from high ground and concentrates in flocks, again mainly on arable farmland. Numbers have, however, declined enormously in recent decades and it is now uncommon in many areas. Cirl Bunting *E. cirlus* is rare and unlikely to be encountered away from a narrow coastal belt in S. Devon, between Exeter and Plymouth. However, intensive conservation efforts have seen a recent population increase, with 862 pairs estimated in 2009. They have also been reintroduced into S. Cornwall. Outside this region, identification, particularly of females, should be attempted with the utmost caution.

Structure and general hints Compared to Yellowhammer, Cirl is a slightly slimmer, more compact, bunting; however, relaxed birds can look quite plump and thickset. Females of the two are superficially similar: pay particular attention to call, plumage tone, facial pattern and, especially, the rump.

Plumage *Male* Cirl Bunting is easily identified by its black eye-stripe, large black throat, green breast-band, chestnut upperparts and breast-sides, and the yellow background to the head and underparts (fresh early winter plumage is duller). *Female* **1** PLUMAGE TONE Whereas female Yellowhammer has its plumage strongly saturated with yellow, female Cirl is duller, browner and nondescript, being buffy-brown and grey-brown, lacking *strong* yellow or chestnut tones. However, some females are yellower than others, many showing a pale yellow or dull yellow ground colour to the head and underparts. **2** HEAD PATTERN Note that its rather stripy face strongly reflects the male's distinctive pattern: a thick brown stripe from the eye back and a broad brown moustachial stripe along the lower border of the ear-coverts, parallel to the eye-stripe, contrast strongly with the buff or dull yellow supercilium, lores/cheek line, submoustachial stripe and throat (the ear-coverts may show a small yellow spot in the rear corner). In comparison, female Yellowhammer's yellow ground colour to the face considerably reduces the stripy effect of the brown facial lines. Female Cirl's eye

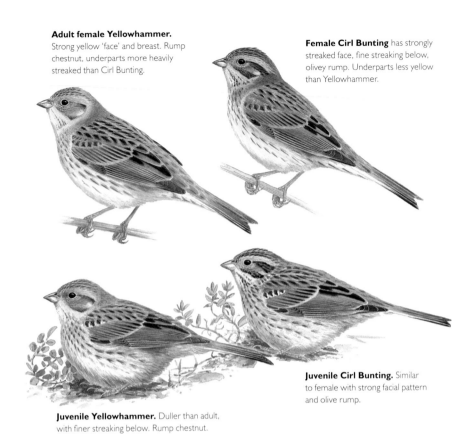

Adult female Yellowhammer. Strong yellow 'face' and breast. Rump chestnut, underparts more heavily streaked than Cirl Bunting.

Female Cirl Bunting has strongly streaked face, fine streaking below, olivey rump. Underparts less yellow than Yellowhammer.

Juvenile Cirl Bunting. Similar to female with strong facial pattern and olive rump.

Juvenile Yellowhammer. Duller than adult, with finer streaking below. Rump chestnut.

is dark and prominent, and accentuated by a noticeable buff or dull yellow eye-ring. Also, it lacks the obvious crown-stripe shown by many (but not all) female Yellowhammers, the crown appearing relatively uniform, evenly and finely streaked darker. **3 UNDERPARTS** Dull buff to pale yellow or dark, deep yellow (never attaining the bright, intense yellow of Yellowhammer). There is a band of fairly well-defined streaking on the breast and flanks. On female Yellowhammer, the ground colour to the head and underparts is obviously yellow, with diffuse chestnut-brown streaking on the breast (forming something of a band) extending onto the belly and flanks; on juveniles and some first-winter females the amount of yellow may be severely reduced but is still readily apparent about the face and neck. **4 UPPERPARTS** Cirl's upperparts are duller and browner, with the only chestnut usually confined to the scapulars. **5 RUMP** In particular, note that Cirl's rump *is a dull olive-brown or greyish-brown, concolorous with the back and quite unlike the bright chestnut rump of Yellowhammer*. *Juvenile* Prior to their post-juvenile autumn body moult, juvenile Cirl Buntings are particularly brown and nondescript, with heavy streaking on the upperparts, breast, belly and flanks (although they still show the female's basic head pattern). Juvenile Yellowhammers are also much streakier on the head and breast but nevertheless show distinct yellow tones to these areas and, like adults, have a chestnut rump.

Call A most useful character: Yellowhammer has a familiar loud, hard, metallic *tchik, tzik* or *tillip* (sometimes reminiscent of Grey Wagtail *Motacilla cinerea*). Cirl has a totally different, soft, barely audible *st* or *sip*, perhaps suggesting a very soft Song Thrush *Turdus philomelos*, occasionally strung out to *sit–sit* or *sissi–sissi–sip*.

Song If a lone female Cirl is encountered, a nearby male's song may assist in the identification. Two song types that may be alternated: a fast, high-pitched, dry, insect-like trill, or a slower, lower-pitched, more 'dribbling' Yellowhammer-like trill that can be strangely unobtrusive. The latter may recall Lesser Whitethroat *Sylvia curruca*.

Behaviour Cirl is more elusive than Yellowhammer, often feeding unobtrusively in a secretive, mouse-like manner, hidden even in short vegetation. Singing males are often associated with tall evergreen trees.

Reference Davis (1982).

Reed, Little, Rustic and Lapland Buntings

Where and when Despite recent declines, Reed Bunting is a common and widespread breeder in all kinds of marshy habitats and even in drier areas (such as young conifer plantations); in winter, it retreats from high ground and concentrates in freshwater and coastal marshes, and on farmland (sometimes visiting garden feeders). Little Bunting is a rare late spring and autumn migrant, currently averaging 32 records a year (peak 59 in 2000), mostly in Shetland and at east and south coast migration watchpoints, west to Scilly. There are also winter and early spring records, often inland, usually amongst finch and bunting flocks. Rustic Bunting is an annual vagrant, increasing during the 1970s and 1980s to an annual average of 20 a year in the 1990s (with a remarkable peak of 50 in 1993); however, since 2002 it has become much rarer, currently averaging only three a year. Its geographical and temporal occurrence patterns mirror those of Little Bunting, but it is much less frequent in winter. Lapland Bunting is a scarce coastal passage migrant mainly from late September to November, but small numbers winter, mainly on the east coast from s. Scotland to Kent. There are occasional large influxes, with several thousand in 2010–11 thought to have originated in Greenland (Pennington *et al.* 2012). It also bred in Scotland in 1977–80.

Reed *Emberiza schoeniclus* and Little Buntings *E. pusilla*

When seen well, Little is not difficult to identify. Pay particular attention to the call and fine details of the head pattern. They are quite variable in plumage tone, with adult males *tending* to be brightest and most chestnut.

Size and structure On first view, Little appears a small, delicate, compact bunting, often creeping around rather like a Dunnock *Prunella modularis*, a similarity enhanced by nervous wing-flicking. The bill has a straight or slightly concave culmen, combining to produce a pointed head and bill profile.

Head pattern Little's facial pattern and colour should initially attract attention, and the following differences from female Reed Bunting should be carefully noted (in rough order of significance). **1 LORES AND EAR-COVERTS** Distinctly chestnut in tone, but intensity variable, some being very dull, but spring males are very bright; note also that the lores are plain, lacking the dark loral line of Reed Bunting (although Reed can also occasionally lack it). **2 EYE-RING** Narrow but obvious buff eye-ring, the presence of which should be clearly established. **3 EYE-STRIPE** A black eye-stripe encircles the rear of the ear-coverts (broadening at the rear) but peters out on the lower border of the ear-coverts roughly below the eye. **4 MOUSTACHIAL STRIPE** Because the ear-covert border peters out below the eye, this means that there is no dark moustachial stripe from the base of the bill backwards along the lower border of the ear-coverts. (Do not confuse the moustachial and lateral throat-stripes: see 'Topography' at front of book). **5 EAR-COVERT SPOT** There is usually a distinct pale spot at the rear of the ear-coverts. **6 SUPERCILIUM** Prominent pale buff supercilium (often tinged orangey or chestnut), most prominent behind the eye and squarer-ended than on Reed. **7 CROWN** Black or brown lateral crown-stripes with a grey, buff, orangey-buff or chestnut central crown-stripe; thus the crown is more contrasting than on Reed Bunting. The combination of pale eye-ring, pale lores and lack of a dark moustachial stripe produces a distinctly more 'open-faced' impression than the rather streaky-faced Reed Bunting.

Other plumage features 1 WING-BARS Differences are useful: on Reed, the wing-bars are buff and do not stand out from the buff tones of the rest of the plumage; on Little, whitish tips to the median and, sometimes, the greater coverts produce obvious wing-bar(s). **2** LESSER COVERTS On Little these are dull olive-brown (chestnut on Reed), although they are usually hidden in the field. **3** UNDERPARTS The upperpart tone is variable, but the underparts are often very white compared with Reed, with finer, neater and more delicate breast streaking.

Call An important difference. Little typically gives a weak *tic*, reminiscent of Robin *Erithacus rubecula*, or a slightly softer *twsik*, more suggestive of Song Thrush *Turdus philomelos*. This is totally different from the familiar but variable calls of Reed Bunting: a short *schwe*, mournful upslurred *schwee* or clearer, more cheerful, downslurred *sweeoo*. The Robin-like call of Little will enable confident identification even if the fine detail is not observed (but note that Rustic Bunting *E. rustica* and rarer Siberian buntings have a similar call).

Tail-flicking A behavioural difference is Reed's habit of flicking open its tail sideways, revealing the white outer-tail feathers. Although Little occasionally does this, it is not so deliberate or habitual. Reed also frequently twitches its tail, even in flight over short distances.

Ageing and sexing The ageing and sexing of Little Buntings is not recommended (see Svensson 1992).

Rustic Bunting *Emberiza rustica*

Size and structure Similar to Reed Bunting but with a distinctive crest, producing a pronounced peak to the rear crown, with a rather ragged nape.

Plumage *Adult male* Very distinctive in spring, with a black crown and ear-coverts, narrow and ill-defined white crown-stripe and striking white supercilium from the eye back (plus a small white spot at the rear of the ear-coverts). The underparts (including chin and throat) are strikingly white, with a broad, rich chestnut breast-band extending onto the flanks as a series of thick streaks. In autumn (when fresh) the chestnut breast feathers may be fringed buff. The nape, mantle, scapulars, rump and uppertail-coverts are also rich chestnut. It has two narrow white wing-bars (very buff on Reed). Note that some summer males resemble adult females, whilst some summer females resemble males (Svensson 1992). *Adult female summer* Rather like a dull version of the adult male, sharing the rich chestnut breast-band,

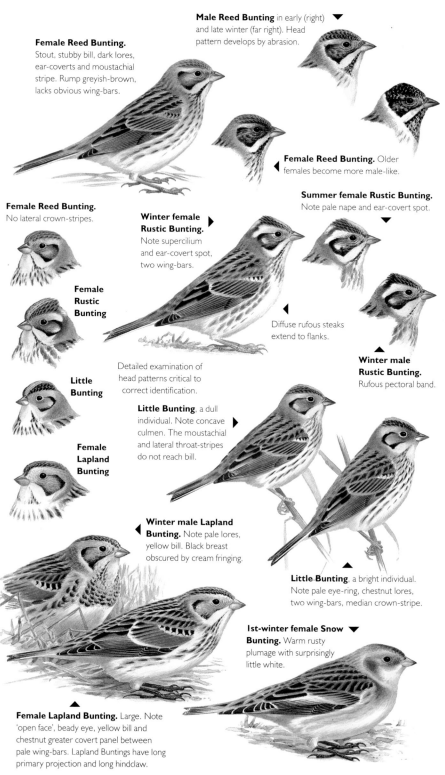

Female Reed Bunting. Stout, stubby bill, dark lores, ear-coverts and moustachial stripe. Rump greyish-brown, lacks obvious wing-bars.

Male Reed Bunting in early (right) and late winter (far right). Head pattern develops by abrasion.

Female Reed Bunting. Older females become more male-like.

Female Reed Bunting. No lateral crown-stripes.

Winter female Rustic Bunting. Note supercilium and ear-covert spot, two wing-bars.

Summer female Rustic Bunting. Note pale nape and ear-covert spot.

Female Rustic Bunting

Diffuse rufous steaks extend to flanks.

Winter male Rustic Bunting. Rufous pectoral band.

Detailed examination of head patterns critical to correct identification.

Little Bunting

Little Bunting, a dull individual. Note concave culmen. The moustachial and lateral throat-stripes do not reach bill.

Female Lapland Bunting

Winter male Lapland Bunting. Note pale lores, yellow bill. Black breast obscured by cream fringing.

Little Bunting, a bright individual. Note pale eye-ring, chestnut lores, two wing-bars, median crown-stripe.

1st-winter female Snow Bunting. Warm rusty plumage with surprisingly little white.

Female Lapland Bunting. Large. Note 'open face', beady eye, yellow bill and chestnut greater covert panel between pale wing-bars. Lapland Buntings have long primary projection and long hindclaw.

373

flank streaking, nape, rump and uppertail-coverts, but the head is stripy, with a white rear crown-stripe, white supercilium from the eye back and dark border to the brown ear-coverts (interrupted by a small white spot at the rear). *First-winter* Autumn vagrants retain the basic pattern of summer adults but are much browner and more similar to Reed Bunting. However, even at a distance or in flight they are colder and far more contrasting than Reed Bunting, the darker upperparts contrasting with white underparts (autumn Reed is very buff overall). At closer range, Rustic has a prominent creamy-buff supercilium from the eye back, pale centre to the crown (becoming whiter on the rear crown) and plain ear-coverts (usually bordered with dark brown), with a small pale spot at the rear; also, a narrow white or buff eye-ring. Blackish lateral throat-stripes spread into a distinctive band of brown streaking on the breast and flanks (often strongly tinged chestnut), mirroring the male's bright chestnut spring feathering. The ground colour of the underparts is white (much whiter than Reed Bunting). There are two noticeable pale buff or white wing-bars (much more obvious than Reed Bunting's). The rump is chestnut but, when fresh, partly obscured by buff feather fringes. The call is similar to Little Bunting: a soft *stic*, recalling Song Thrush. Ageing and sexing is often difficult in autumn (Svensson 1992) but first-winter has pointed tail feather tips, adult more rounded. Inevitably, brighter, more chestnut birds are likely to be males.

Lapland Bunting *Calcarius lapponicus*

Once known, Lapland is simply not confusable with Reed Bunting, but observers unfamiliar with Lapland may be thrown by out-of-context Reeds.

Size, structure and habitat Compared to Reed, Lapland is *a large, bulky, broad-beamed, almost lark-like bunting,* similar in size, shape and structure to Snow Bunting *Plectrophenax nivalis.* It feeds on the ground (not in trees or bushes) where it may run mouse-like through the vegetation. Often very approachable, it usually occurs on coastal saltmarshes and fields, or on windswept headlands on migration, when it often feeds in short heather. Long wings are clearly apparent at rest: *the primary projection is very long, approximately equal in length to the overlying tertials* (on Reed, the projection is short: only about half the tertial length).

Plumage *Autumn/winter* Head pattern distinctive. A black eye-stripe extends from the eye back, surrounding the ear-coverts (thickest at the corners) and extending forward along the lower edge of the ear-coverts to the bill; thus, the lores are plain buff, as is the central ear-coverts and supercilium. This produces a very open-faced expression in which the dark eye is prominent. It has a buffish or chestnut 'shawl' over the nape. White tips to the median and greater coverts form a double wing-bar, and sandwiched between the bars is a distinctive chestnut panel on the greater coverts. Adult male usually has extensive areas of black on the throat, but these feathers have pale fringes in winter so the obviousness of the black depends on the degree of abrasion. The bill is pale, usually pink (yellow in summer). *Summer* Male very striking, with a black 'face', throat and breast, and creamy supercilium extending down the neck-sides to form a broad white border to the black breast; also a chestnut shawl on the lower nape/upper mantle, and a yellow bill. The female also has heavy black blotching on the breast.

Ageing Adult has broad, rounded tail feather tips, first-year narrow and pointed.

Calls When flushed, often gives a dry, rattling call, variously transcribed as *ti-li-li-lit*, *t-t-tip*, or *t-t-t-t-tik*, and often followed by soft, clear *peu* or *teu*. Note that Snow Bunting has a similar *teu* call, but this is loud and ringing, and often preceded by a soft, liquid rippling.

Flight identification A bulky, long-winged, blunt-headed, lark-like bunting, with a strong, purposeful flight. In comparison, the smaller Reed Bunting has a very distinctive action in which the short, rounded wings are erratically closed into the body to produce a hesitant flight, giving the impression that the bird is not quite sure where it is going.

References Fraser *et al.* (2007), Pennington *et al.* (2012), Svensson (1992), Wallace (1976).

General bibliography

Cramp, S. & Simmons, K. E. L. (eds) (1977–88) *The Birds of the Western Palearctic*. Vols 1–5. Oxford University Press, Oxford. (*BWP*)

Grant, P. J. (1986) *Gulls: A Guide to Identification*. Second edn. T. & A. D. Poyser, Calton.

Harrison, P. (1983) *Seabirds: An Identification Guide*. Croom Helm, Beckenham.

Hayman, P., Marchant, J. & Prater, T. (1986) *Shorebirds: An Identification Guide to the Waders of the World*. Croom Helm, London & Sydney.

Madge, S. C. & Burn, H. (1987) *Wildfowl: An Identification Guide to the Ducks, Geese and Swans of the World*. Christopher Helm, London.

Porter, R. F., Willis, I., Christensen, S. & Nielsen, B. P. (1976) *Flight Identification of European Raptors*. Second edn. T. & A. D. Poyser, Calton.

Svensson, L. (1992) *Identification Guide to European Passerines*. Fourth edn. Privately published, Stockholm.

Svensson, L., Mullarney, K. & Zetterström, D. (2009) *Collins Bird Guide*. Second edn. HarperCollins, London.

Statistics

Rare bird statistics are taken from the 'Report on rare birds in Great Britain', scarce migrant totals are taken from the 'Report on scarce migrant birds in Britain', and data on rare breeding birds are taken from the annual report 'Rare breeding birds in the UK'. All of these reports are published annually in *British Birds*.

Specific bibliography

British Birds is *the* monthly journal for the keen birdwatcher and, over the years, it has made a unique contribution to bird identification. Those papers marked with an asterisk were also published in *Frontiers of Bird Identification* edited by J. T. R. Sharrock (Macmillan, 1980).

Andrews, R. M., Higgins, R. J. & Martin, J. P. (2006) American Black Tern at Weston-super-Mare: new to Britain. *Brit. Birds* 99: 450–459.

Appleby, R. H., Madge, S. C. & Mullarney, K. (1986) Identification of divers in immature and winter plumages. *Brit. Birds* 79: 365–391.

Baker, K. (1988) Identification of Siberian and other forms of Lesser Whitethroat. *Brit. Birds* 81: 382–390.

Batty, C. & Lowe, T. (2001) Vagrant Canada Geese in Britain and Ireland. *Birding World* 14: 57–61.

Beaman, M. & Madge, S. (1998) *The Handbook of Bird Identification for Europe and the Western Palearctic.* Christopher Helm, London.

van den Berg, A. (2004) Population growth and vagrancy potential of Ross's Goose. *Dutch Birding* 26: 107–111.

van den Berg, A. & Blankert, J. J. (1980) Crossbills with prominent double wing-bar. *Dutch Birding* 2: 33–36.

Bowey, K. & Westerberg, S. S. (1994) Identification of Parrot Crossbill. *Brit. Birds* 87: 398–401.

Bradshaw, C. (1994) Blyth's Pipit identification. *Brit. Birds* 87: 136–142.

Bradshaw, C., Kehoe, C. & Pitches, A. (2004) The Carl Zeiss Award 2004. *Brit. Birds* 97: 542–544.

Broome, A. (1987) Identification of juvenile Pomarine Skua. *Brit. Birds* 80: 426–427.

Broughton, R. K. (2009) Separation of Willow Tit and Marsh Tit in Britain: a review. *Brit. Birds* 102: 604–616.

Burn, D. M. & Mather, J. R. (1974) The White-billed Diver in Britain. *Brit. Birds* 67: 257–296.

Burton, J. F. & Johnson, E. D. H. (1984) Insect, amphibian or bird? *Brit. Birds* 77: 87–104.

Callahan, D. (2012) Back-crossed butcherbirds. *Birdwatch* 238: 31.

Campbell, O. (2012) The status of Isabelline Shrike taxa in the United Arab Emirates. *Brit. Birds* 105: 417–420.

Catley, G. P. & Hursthouse, D. (1985) Parrot Crossbills in Britain. *Brit. Birds* 78: 482–505.

Chapman, M. S. (1984) Identification of Short-toed Treecreeper. *Brit. Birds* 77: 262–263.

Charlwood, R. H. (1973) Hybrid Swallow × House Martin. *Brit. Birds* 66: 398–400.

Collinson, J. M., Smith, A., Waite, S. & McGowan, R. Y. (2013) British records of 'Eastern Yellow Wagtail'. *Brit. Birds* 106: 36–41.

Davenport, D. L. (1987) Behaviour of Arctic and Pomarine Skuas and identification of immatures. *Brit. Birds* 80: 167–168.

Davis, A. H. (1982) Mystery photographs 66: Cirl Bunting. *Brit. Birds* 75: 283–285.

Davis, A. H. & Prytherch, R. J. (1976) Field identification of Long-eared and Short-eared Owls. *Brit. Birds* 69: 281–287.

Dawson, J. (1994) Ageing and sexing of King Eiders. *Brit. Birds* 87: 37–40.

Dean, A. R. (1985) Review of British status and identification of Greenish Warbler. *Brit. Birds* 78: 437–451.

Dean, A., Bradshaw, C., Martin, J., Stoddart, A. & Walbridge, G. (2010) The status in Britain of 'Siberian Chiffchaff'. *Brit. Birds* 103: 320–338.

De Knijff, P., van der Spek, V. & Fischer, J. (2012) Genetic identity of grey chiffchaffs trapped in the Netherlands in autumns of 2009–11. *Dutch Birding* 34: 386–392.

Dennis, R. H. & Wallace, D. I. M. (1975) Field identification of Short-toed and Lesser Short-toed Larks. *Brit. Birds* 68: 238–241.*

Dickinson. E. C. (2008) The name of the easternmost population of Common Nightingale *Luscinia megarhynchos*. *Bull. Brit. Orn. Club* 128: 141–142.

Dubois, P. J. & Yésou, P. (1984) Identification of juvenile Yellow-legged Herring Gulls. *Brit. Birds* 77: 344–348.

van Duivendijk, N. (2011) *Advanced Bird ID Handbook. The Western Palearctic*. New Holland, London.

Ekins, G. (unpubl.) The separation of *carbo* and *sinensis* races of Cormorant in the field.

Ellis, P. (1994) Ageing and sexing of King Eiders. *Brit. Birds* 87: 36–37.

Fisher, E. A. & Flood, R. L. (2010) Scopoli's Shearwater off Scilly: new to Britain. *Brit. Birds* 103: 712–717.

Flood, R. L. (2010) Storm-petrels. *Birdwatch* 217: 29–32.

Flood, R. L., Hudson, N. & Thomas, B. (2007) *Essential Guide to Birds of the Isles of Scilly*. Privately published.

Forsman, D. (1999) *The Raptors of Europe and the Middle East. A Handbook of Field Identification*. T. & A. D. Poyser, London.

Fraser, P. A. & Ryan, J. F. (1994) Scarce migrants in Britain and Ireland. Part 2. Numbers during 1986–92: gulls to passerines. *Brit. Birds* 87: 605–612.

Fraser, P. A., Rogers, M. J. & the Rarities Committee (2007) Report on rare birds in Great Britain in 2005. Part 1: non-passerines. *Brit. Birds* 100: 16–61.

Garner, M. (2008) *Frontiers in Birding*. BirdGuides Ltd., Sheffield.

Gibbins, C., Small, B. J. & Sweeney, J. (2010) From the Rarities Committee's files. Identification of Caspian Gull. Part 1: typical birds. *Brit. Birds* 103: 142–183.

Gibbins, C., Neubauer, G. & Small, B. J. (2011) From the Rarities Committee's files. Identification of Caspian Gull. Part 2: phenotypic variability and the field characteristics of hybrids. *Brit. Birds* 104: 702–742.

Glass, T., Lauder, A. W., Oksien, M. & Shaw, K. D. (2006) Masked Shrike: new to Britain. *Brit. Birds* 99: 67–70.

Grant, P. J. (1972) Field identification of Richard's and Tawny Pipits. *Brit. Birds* 65: 287–290.

Grant, P. J. (1981) Mystery photographs 51: Brünnich's Guillemot. *Brit. Birds* 74: 144–145.

Grant, P. J. (1983) The 'Marsh Hawk' problem. *Brit. Birds* 76: 373–376.

Grant, P. J. (1987) Wing shape of Chough and Alpine Chough. *Brit. Birds* 80: 116–117.

Grant, P. J. & Jonsson, L. (1984) The identification of stints and peeps. *Brit. Birds* 77: 293–315.

Grant, P. J. & Scott, R. E. (1969) Field identification of juvenile Common, Arctic and Roseate Terns. *Brit. Birds* 62: 297–299.*

Grant, P. J., Scott, R. E. & Wallace, D. I. M. (1971) Further notes on the 'portlandica' phase of terns. *Brit. Birds* 64: 19–22.*

van Grouw, H. (2013) What colour is that bird? The causes and recognition of common colour aberrations in birds. *Brit. Birds* 106: 17–29.

Hanson, H. C. (2006) *The White-cheeked Geese: Taxonomy, Ecophysiographic Relationships, Biogeography, and Evolutionary Considerations*. Vol. 1. Avvar Books, Blythe, California.

Hanson, H. C. (2007) *The White-cheeked Geese: Taxonomy, Ecophysiographic Relationships, Biogeography, and Evolutionary Considerations*. Vol. 2. Avvar Books, Blythe, California.

Harrison, J. M. & Harrison, J. G. (1966) Hybrid Grey Lag × Canada Goose suggesting influence of Giant Canada Goose in Britain. *Brit. Birds* 59: 547–550.

Harrison, J. M. & Harrison, J. G. (1968) Wigeon × Chilöe Wigeon hybrid resembling American Wigeon. *Brit. Birds* 61: 169–171.

Harrop, H. R., Mavor, R. & Ellis, P. M. (2008) Olive-tree Warbler in Shetland: new to Britain. *Brit. Birds* 101: 82–88.

Harvey, P. V. & Heubeck, M. (2012) Changes in the wintering population and distribution of Slavonian Grebes in Shetland. *Brit. Birds* 105: 704–715.

Harvey, W. G. (1981) Pallid Swift in Kent. *Brit. Birds* 74: 170–178.

Heard, C. D. R. (1995) Unravelling the mystery. *Birdwatch* 41: 20–24.

Hellström, M. & Wærn, M. (2011) Field identification and ageing of Siberian Stonechats in spring and summer. *Brit. Birds* 104: 236–254.

Hirschfeld, E. (1985) Further comments on treecreeper identification. *Brit. Birds* 78: 300–302.

Holling, M. & the Rare Breeding Birds Panel (2010) Rare breeding birds in the United Kingdom in 2008. *Brit. Birds* 103: 482–538.

Holling, M. & the Rare Breeding Birds Panel (2012) Rare breeding birds in the United Kingdom in 2010. *Brit. Birds* 105: 352–416.

Holt, C., Austin, G., Calbrade, N., Hearn, R., Mellan, H., Stroud, D., Wotton, S. & Musgrove, A. (2012) *Waterbirds in the UK 2010/11. The Wetland Bird Survey*. British Trust for Ornithology, Royal Society for the Protection of Birds, Joint Nature Conservation Committee and Wildfowl and Wetlands Trust, Thetford.

Howell, S. N. G. & Dunn, J. (2007) *A Reference Guide to Gulls of the Americas*. Houghton Mifflin, New York.

Hudson, N. & the Rarities Committee (2008) Report on rare birds in Great Britain in 2007. *Brit. Birds* 101: 516–577.

Hudson, N. & the Rarities Committee (2009) Report on rare birds in Great Britain in 2008. *Brit. Birds* 102: 528–601.

Hudson, N. & the Rarities Committee (2012) Report on rare birds in Great Britain in 2011. *Brit. Birds* 105: 556–625.

Hume, R. A. & Grant, P. J. (1974) The upperwing pattern of adult Common and Arctic Terns. *Brit. Birds* 67: 133–136.*

Jännes, H. (2003) *Calls of Eastern Vagrants*. Hannu Jännes/Earlybird Tours, Helsinki.

Johnson, I. G. (1970) The Water Pipit as a winter visitor to the British Isles. *Bird Study* 17: 297–319.

Jonsson, L. (1984) Identification of juvenile Pomarine and Arctic Skuas. *Brit. Birds* 77: 443–446.

Kemp, J. B. (1982) Field identification of Long-eared and Short-eared Owls. *Brit. Birds* 75: 227.

Knox, A. (1988) Taxonomy of the Rock/Water Pipit superspecies. *Brit. Birds* 81: 206–211.

Lidster, J. A. (2009) From the Rarities Committee's files. The Green Farm Booted Warbler. *Brit. Birds* 102: 617–621.

Madge, S. & Quinn, D. (1997) Identification of Hume's Warbler. *Brit. Birds* 90: 571–575.

Martin, J. (2008) Northern Harrier on Scilly: new to Britain. *Brit. Birds* 101: 394–407.

Mather, J. R. (1981) Mystery photographs 54: Long-tailed Skua. *Brit. Birds* 74: 257–259.

Mather, J. R. (1991) Guillemots with dark neck bands. *Brit. Birds* 84: 439–441.

Mather, J. R. (2010) Pacific Diver: new to Britain and the Western Palearctic. *Brit. Birds* 103: 539–545.

Mitchell, D., & Vinicombe, K. (2011) *Birds of Britain: The Complete Checklist*. 4th ed. Warners Group Publications, Lincolnshire.

Morova, I. M., Federov, V. V., Shipilina, D. A. & Alekseev, V. N. (2009) Genetic and vocal differentiation in hybrid zones of passerine birds: Siberian and European Chiffchaffs (*Phylloscopus* [*collybita*] *tristis* and *Ph.* [*c.*] *abietinus*) in the Southern Urals. *Doklady Biol. Sci.* 427: 384–386.

Mullarney, K. (1987) Mystery photographs 124: Tree Pipit. *Brit. Birds* 80: 158–160.

Musgrove, A. J., Austin, G. E., Hearn, R. D., Holt, C. A., Stroud, D. A. & Wotton, S. R. (2011) Overwinter population estimates of British waterbirds. *Brit. Birds* 104: 364–397.

O'Brien, M., Crossley, R. & Karlson, K. (2006) *The Shorebird Guide*. Houghton Mifflin, New York.

Oddie, W. E. (1980) Leg colour and calls of Spotted Sandpiper. *Brit. Birds* 73: 185–186.

Ogilvie, M. A. & Wallace, D. I. M. (1975) Identification of grey geese. *Brit. Birds* 68: 57–67.

Olsen, K. M. & Christensen, S. (1984) Field identification of juvenile skuas. *Brit. Birds* 77: 448–450.

Olsen, K. M. & Larsson, H. (2003) *Gulls of North America, Europe and, Asia*. Christopher Helm, London.

Panov, E. N. (2009) On the nomenclature of the so-called Isabelline Shrike. *Sandgrouse* 31: 163–170.

Pearson, D. J., Svensson, L. & Frahnert, S. (2012) Further on the type series and nomenclature of the Isabelline Shrike *Lanius isabellinus*. *Bull. Brit. Orn. Club* 132: 270–276.

Pennington, M. G., Riddington, R. & Miles, W. T. S. (2012) The Lapland Bunting influx in Britain and Ireland in 2010/11. *Brit. Birds* 105: 654–673.

Pym, A. (1982) Identification of Lesser Golden Plover in Britain and Ireland. *Brit. Birds* 75: 112–124.

Pym, A. (1985) Bill coloration of treecreepers. *Brit. Birds* 78: 303.

Robertson, I. S. (1977) Identification and European status of eastern Stonechats. *Brit. Birds* 70: 237–245.*

Robertson, I. S. (1982) Field identification of Long-eared and Short-eared Owls. *Brit. Birds* 75: 227–229.

Rodriguez de los Santos, M. (1985) Notes on Short-toed Treecreepers from southern Spain. *Brit. Birds* 78: 298–299.

Rogers, M. J. & the Rarities Committee (2004) Report on rare birds in Great Britain in 2003. *Brit. Birds* 97: 558–625.

Rowlands, A. (2010) From the Rarities Committee's files. Identification of eastern Woodchat Shrike. *Brit. Birds* 103: 385–395.

Rowlands, A. (2012) The Carl Zeiss Award 2012. *Brit. Birds* 105: 474–478.

Rumsey, S. J. R. (1984) Identification pitfalls: Aquatic Warbler. *Brit. Birds* 77: 377.

Scott. R. E. & Grant, P. J. (1969) Uncompleted moult in *Sterna* terns and the problems of identification. *Brit. Birds* 62: 93–97.*

Sharrock, J. T. R. (ed.) (1980) *Frontiers of Bird Identification*. Macmillan, London.

Sharrock, J. T. R. & Nightingale, B. (2010) Identification of Willow Tit and Marsh Tit. *Brit. Birds* 103: 121–122.

Sibley, D. (2003) *Field Guide to the Birds of Eastern North America*. Christopher Helm, London.

Sibley, D. (2004) Distinguishing Cackling and Canada Geese. www.sibleyguides.com/2007/07/identification-of-cackling-and-canada-goose/.

Slack, R. (2009) *Rare Birds Where and When*. Vol. 1. Rare Birds Books, York.

Stoddart, A. (2008) Dark-bellied and Pale-bellied Brent Geese. *Birdwatch* 197: 26–27.

Stoddart, A. (2009) Mealy and Arctic Redpolls. *Birdwatch* 201: 31–33.

Stoddart, A. (2009) Arctic and Greenish Warblers. *Birdwatch* 206: 29–31.

Stoddart, A. (2011) Redpolls photo guide. *Birdwatch* 234: 41–46.

Stoddart, A. (2012) Juvenile Pomarine, Arctic and Long-tailed Skuas. *Birdwatch* 242: 45–50.

Stoddart, A. (2012) Juvenile harriers photo guide. *Birdwatch* 244: 37–42.

Summers, R. W. & Buckland, S. T. (2011) A first survey of the global population size and distribution of the Scottish Crossbill *Loxia scotica*. *Bird Conserv. Intern.* 21: 186–198.

Svensson, L., Shirihai, H., Frahnert, S. & Dickinson, E. C. (2012) Taxonomy and nomenclature of the Stonechat complex *Saxicola torquatus sensu lato* in the Caspian region. *Bull. Brit. Orn. Club* 132: 260–269.

Thorpe, J. P. (1988) Juvenile Hen Harriers showing 'Marsh Hawk' characters. *Brit. Birds* 81: 377–382.

Tucker, L. A. (1984) Possible use of bill colour in separating Short-toed Treecreeper and Treecreeper. *Brit. Birds* 77: 263–264.

Ullman, M. (1984) Field identification of juvenile Pomarine Skua. *Brit. Birds* 77: 446–448.

Vinicombe, K. E. (1980) Tern showing mixed characters of Black Tern and White-winged Black Tern. *Brit. Birds* 73: 223–225.

Vinicombe, K. E. (1988) Unspecific Golden Plover in Avon. *Birding World* 1: 54–56.

Vinicombe, K. (2008) The Scilly shrike. *Birdwatch* 200: 29–32.

Votier, S. C., Steele, J., Shaw, K. D. & Stoddart, A. M. (2000) Arctic Redpoll *Carduelis hornemanni exilipes*: an identification review based on the 1995/96 influx. *Brit. Birds* 93: 68–84.

Wallace, D. I. M. (1970) Identification of Spotted Sandpipers out of breeding plumage. *Brit. Birds* 63: 168–173.*

Wallace, D. I. M. (1976) Distinguishing Little and Reed Buntings. *Brit. Birds* 69: 465–473.*

Williamson, K. (1963) The identification of the larger pipits. *Brit. Birds* 56: 285–292.*

Williamson, K. (1965) Moult and its relation to taxonomy in Rock and Water Pipits. *Brit. Birds* 58: 493–504.

Yésou, P., Paterson, A. M., Mackrill, E. J. & Bourne, W. R. P. (1990) Plumage variation and identification of the 'Yelkouan Shearwater'. *Brit. Birds* 83: 299–319.

Index

Figures in *italic* refer to the illustrations; figures in **bold** to main text references. Significant subspecies have been included

D

H

I

J

K

L

R

T

X

Y

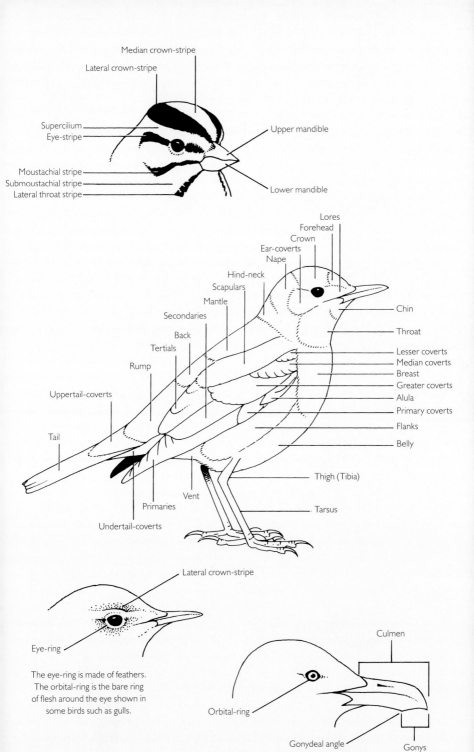

Median crown-stripe

Lateral crown-stripe

Supercilium
Eye-stripe

Upper mandible

Moustachial stripe
Submoustachial stripe
Lateral throat stripe

Lower mandible

Lores
Forehead
Crown
Ear-coverts
Nape
Hind-neck
Scapulars
Mantle
Secondaries
Back
Tertials
Rump
Uppertail-coverts
Tail

Chin
Throat
Lesser coverts
Median coverts
Breast
Greater coverts
Alula
Primary coverts
Flanks
Belly

Thigh (Tibia)

Vent
Primaries
Undertail-coverts

Tarsus

Lateral crown-stripe

Eye-ring

The eye-ring is made of feathers.
The orbital-ring is the bare ring
of flesh around the eye shown in
some birds such as gulls.

Culmen

Orbital-ring

Gonydeal angle

Gonys

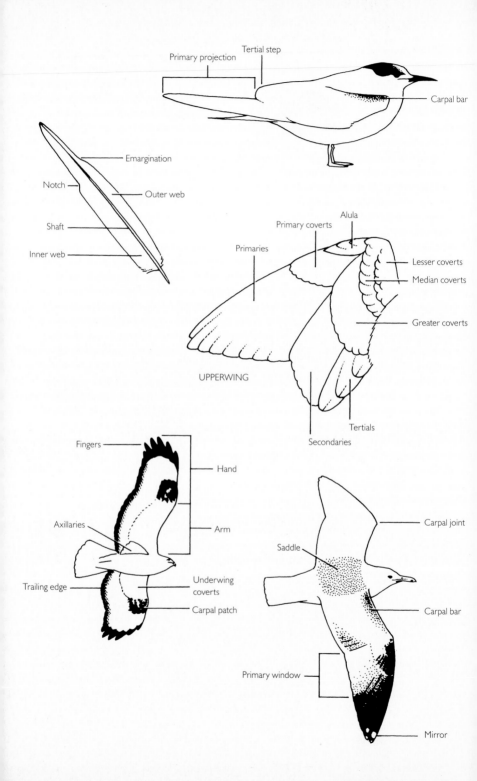

Primary projection

Tertial step

Carpal bar

Emargination

Notch

Outer web

Shaft

Inner web

Primary coverts

Alula

Primaries

Lesser coverts

Median coverts

Greater coverts

UPPERWING

Secondaries

Tertials

Fingers

Hand

Arm

Axillaries

Trailing edge

Underwing coverts

Carpal patch

Carpal joint

Saddle

Carpal bar

Primary window

Mirror